RNAi for Plant Improvement and Protection

_____

Funded by the Horizon 2020 Framework Programme
of the European Union

cost
EUROPEAN COOPERATION
IN SCIENCE & TECHNOLOGY

iPlanta

RNAi for Plant Improvement and Protection

Editors:

Bruno Mezzetti

Professor Plant Breeding and Biotechnology, Department of Agricultural, Food and Environmental Sciences – Università Politecnica delle Marche, Italy

Jeremy Sweet

Director, Sweet Environmental Consultants, Willingham, Cambridge, UK

Lorenzo Burgos

Profesor de Investigación at CEBAS-CSIC. Head of Fruit Biotechnology Group, Department of Fruit Breeding, Campus Universitario de Espinardo, Murcia, Spain

CABI

CABI is a trading name of CAB International

CABI
Nosworthy Way
Wallingford
Oxfordshire OX10 8DE
UK

Tel: +44 (0)1491 832111
Fax: +44 (0)1491 833508
E-mail: info@cabi.org
Website: www.cabi.org

CABI
WeWork
One Lincoln St
24th Floor
Boston, MA 02111
USA

T: +1 (617)682-9015
E-mail: cabi-nao@cabi.org

A catalogue record for this book is available from the British Library, London, UK.

References to Internet websites (URLs) were accurate at the time of writing.

ISBN-13:   978 1 78924 889 0 (hardback)
           978 1 78924 880 6 (epdf)
           978 1 78924 891 3 (epub)

DOI: 10.1079/9781789248890.0000

Commissioning Editor: David Hemming
Editorial Assistant: Lauren Davies
Production Editor: James Bishop

Typeset by Exeter Premedia Services Pvt Ltd, Chennai, India

Contents

Contributors

Salvatore Arpaia, ENEA, TERIN-BBC, Research Centre Trisaia, Rotondella (MT), Italy. Email: salvatore.arpaia@enea.it

Atanas Atanassov, Joint Genomic Center Ltd, Sofia, Bulgaria. Email: atanas_atanassov@jgc-bg. org

Maria Luisa Badenes, Centre of Citriculture and Plant Production, Valencian Institute for Agricultural Research (IVIA), Valencia, Spain. Email: badenes_mlu@gva.es

Pascal Briard, Unite Mixte de Recherches 1332, INRAE-Bordeaux; Villenave d'Ornon, 33882, CS 20032, France. Email : pascal.briard@inrae.fr

Lorenzo Burgos, Grupo de Biotecnología de Frutales. Departamento de Mejora. CEBAS-CSIC. Campus Universitario de Espinardo, Edificio n° 25, 30100 Murcia, Spain. Email: burgos@cebas. csic.es

Olivier Christiaens, Department of Plants and Crops, Ghent University, Ghent, Belgium. olchrist. Email: olchrist.christiaens@UGent.be

Kristof De Schutter, Department of Plants and Crops, Ghent University, Ghent, Belgium. Email: kristof.deschutter@UGent.be

Antje Dietz-Pfeilstetter, Julius Kühn-Institut (JKI), Institute for Biosafety in Plant Biotechnology, Braunschweig, Germany. Email: antje.dietz@julius-kuehn.de

Christina Dixelius, Swedish University of Agricultural Sciences, Department of Plant Biology, Uppsala BioCenter, Linnean Center for Plant Biology, PO Box 7080, S-75007 Uppsala, Sweden. Email: Christina.Dixelius@slu.se

Matthias Fladung, Thuenen-Institute of Forest Genetics, 22927 Grosshansdorf, Germany. Email: matthias.fladung@thuenen.de

Dario G. Frisio, Department of Environmental Science and Policy, Università degli Studi di Milano, Italy. Email: dario.frisio@unimi.it

Achim Gathmann, Department of Plant Protection Products, Federal Office of Consumer Protection and Food Safety, Braunschweig, Germany. Email: achim.gathmann@bvl.bund.de

Hely Häggman, University of Oulu, Oulu, Finland. Email: hely.haggman@oulu.fi

Kristian Persson Hodén, Swedish University of Agricultural Sciences, Department of Plant Biology, Uppsala BioCenter, Linnean Center for Plant Biology, PO Box 7080, S-75007 Uppsala, Sweden. Email: kristian.persson.hoden@slu.se

Huw D. Jones, IBERS Aberystwyth University, UK. Email: hdj2@aber.ac.uk

Gijs Kleter, RIKILT Wageningen University & Research, Wageningen, The Netherlands. Email: gijs.kleter@wur.nl

Anna Kolliopoulou, Institute of Biosciences & Applications, National Centre for Scientific Research 'Demokritos', Aghia Paraskevi, Greece. Email: a.kolliopoulou@bio.demokritos.gr

Dimitrios Kontogiannatos, Institute of Biosciences & Applications, National Centre for Scientific Research 'Demokritos', Aghia Paraskevi, Greece. Email: dim_kontogiannatos@yahoo.gr

Paul Henning Krogh, Department of Bioscience, Aarhus University, Denmark. Email: phk@bios.au.dk

Zhen Liao, Swedish University of Agricultural Sciences, Department of Plant Biology, Uppsala BioCenter, Linnean Center for Plant Biology, PO Box 7080, S-75007 Uppsala, Sweden. Email: zhen.liao@slu.se

Bruno Mezzetti, Department of Agricultural, Food and Environmental Sciences, Università Politecnica delle Marche, Via Brecce Bianche, 60100 Ancona, Italy. Email: b.mezzetti@univpm.it

Laura Miozzi, Institute for Sustainable Plant Protection, National Research Council of Italy, Torino, Italy. Email: laura.miozzi@ipsp.cnr.it

Barbara Molesini, Department of Biotechnology, University of Verona, Strada Le Grazie 15, 37134 Verona, Italy. Email: barbara.molesini@univr.it

Hanspeter Naegeli, University of Zürich, Institute of Veterinary Pharmacology and Toxicology, Zürich, Switzerland. Email: hanspeter.naegeli@vetpharm.uzh.ch

Emanuela Noris, Institute for Sustainable Plant Protection, National Research Council of Italy, Torino, Italy. Email: emanuela.noris@ipsp.cnr.it

Hilde-Gunn Opsahl-Sorteberg, BIOVIT, NMBU, N 1432 – Ås, Norway. Email: hildop@nmbu.no

Tiziana Pandolfini, Department of Biotechnology, University of Verona, Strada Le Grazie 15, 37134 Verona, Italy. Email: tiziana.pandolfini@univr.it

Ivelin Pantchev, Department of Biochemistry, Sofia University, Sofia, Bulgaria. Email: ipantchev@abv.bg

Kimberly Parker, Department of Energy, Environmental, and Chemical Engineering, Washington University in St Louis, St Louis, Missouri, USA. Email: kmparker@wustl.edu

Ángela Polo-Oltra, Centre of Citriculture and Plant Production, Valencian Institute for Agricultural Research (IVIA), Valencia, Spain. Email: a.polo@btc.upv.es

Goritsa Rakleova, Joint Genomic Center Ltd, Sofia, Bulgaria. Email: grakleova@gmail.com

Michel Ravelonandro, Unite Mixte de Recherches 1332, INRAE-Bordeaux; Villenave d'Ornon, 33882, CS 20032, France. Email: michel.ravelonandro@wanadoo.fr

Angela Ricci, Department of Agricultural, Food and Environmental Sciences, Università Politecnica delle Marche, Ancona, Italy. Email: angela.ricci@pm.univpm.it

Silvia Sabbadini, Department of Agricultural, Food and Environmental Sciences, Università Politecnica delle Marche, Ancona, Italy. Email: s.sabbadini@staff.univpm.it

Werner Schenkel, Department of Genetic Engineering, Federal Office of Consumer Protection and Food Safety, Braunschweig, Germany. Email: werner.schenkel@bvl.bund.de

Guy Smagghe, Department of Plants and Crops, Ghent University, Ghent, Belgium. Email: Guy. Smagghe@UGent.be

Suvi Sutela, Natural Resources Institute Finland, Helsinki, Finland. suvi.sutela@luke.fi

Jeremy Sweet, Sweet Environmental Consultants, 6 Green St, Cambridge CB24 5JA, UK. Email: jeremysweet303@aol.com,

Luc Swevers, Institute of Biosciences & Applications, National Centre for Scientific Research 'Demokritos', Aghia Paraskevi, Greece. Email: swevers@bio.demokritos.gr

Clauvis Nji Tizi Taning, Department of Plants and Crops, Ghent University, Ghent, Belgium. Email: tiziclauvis.taningnji@ugent.be

Vera Ventura, Department of Civil, Environmental, Architectural Engineering and Mathematics, Università degli Studi di Brescia, Italy. Email: vera.ventura@unibs.it

Elena Zuriaga, Centre of Citriculture and Plant Production, Valencian Institute for Agricultural Research (IVIA), Valencia, Spain. Email: garcia_zur@gva.es

Acknowledgements

This book is based upon work from COST Action iPLANTA (CA15223), supported by COST (European Cooperation in Science and Technology). The iPlanta COST Action CA15223 'Modifying plants to produce interfering RNA' (https://iplanta.univpm.it/) was established with the objective of bringing together experts from a wide range of fields to develop a deeper understanding of the science of RNA, the applications of RNAi, the biosafety of these applications and the socio-economic aspects of these potential applications. Most importantly this COST Action was designed to communicate its findings to the wider community, both scientific and those with a general interest in this relatively new area of science. The Editors therefore thank COST for providing the finance which has enabled this book to be produced as an open access e-Book, with a wider distribution.

We also acknowledge the contributions of the authors to their chapters in this book. The authors are from a wide range of countries, organizations and disciplines and present a range of perspectives on RNAi. We thank them for imparting their experience and expertise.

Funded by the Horizon 2020 Framework Programme of the European Union

cost
EUROPEAN COOPERATION
IN SCIENCE & TECHNOLOGY

iPlanta

1 Introduction to RNAi in Plant Production and Protection

Bruno Mezzetti[1]*, Jeremy Sweet[2] and Lorenzo Burgos[3]

[1]*Università Politecnica delle Marche, Ancona, Italy;* [2]*Sweet Environmental Consultants, Cambridge; UK;* [3]*CEBAS-CSIC, Murcia, Spain*

RNA interference (RNAi) has the potential to have a major impact on agriculture, horticulture and forestry with many different applications for plant improvement in terms of both quality of products and productivity. In addition, crop protection applications are being developed which provide 'green' alternatives to conventional pest control methods. RNAi is a naturally occurring process present in plants and animals, in which double-stranded RNA (dsRNA) molecules interfere with homologous RNA. It allows genes to be targeted to remove unwanted products in plants and improve plant productivity and quality of plant products. These RNAi mechanisms were only discovered and described 20 years ago and their discovery led to a Nobel prize in 2006. RNAi is now being developed within plants to silence genes often described as host-induced gene silencing (HIGS). Also, external and topical applications, such as sprays and seed treatments, are being developed to substitute for other types of pesticides or growth regulator treatments. An example is the spray-induced gene silencing (SIGS) approach for targeting pest and pathogen genes and for manipulating endogenous gene expression in plants. Examples of plant improvement applications include: improving fatty acid profiles of soybeans; delayed ripening and improved shelf life of fruits such as apples and tomatoes; or removing unwanted compounds, toxins and allergens from crop products such as decaffeinated coffee, gossypol in cotton seeds and hypoallergenic fruits and cereals.

For pest and disease control applications, dsRNA can be selected for silencing essential genes in pests, pathogens and viruses, expressed either in transformed plants or in exogenous applications. dsRNA can be very specifically targeted at genetic sequences in these targets so that off-target effects are avoided or minimized. Recent advances in genomics and transcriptomics have provided sequence data that enable the design of highly targeted dsRNAs, providing efficient silencing while minimizing the risk of effects on off-target genes or the silencing of gene expression in non-target organisms. Due to the involvement of RNA in virus replication, several virus-resistant plants have been developed (e.g. papaya, plum, squash and tomato) and many more virus control applications are in the pipeline. More recently, plant resistance

*Corresponding author: b.mezzetti@univpm.it

© CAB International 2021. *RNAi for Plant Improvement and Protection*
(eds B. Mezzetti *et al.*)
DOI: 10.1079/9781789248890.0001

to a range of other pests and fungal diseases is being developed, including insect pests such as Colorado potato beetle (*Leptinotarsa decemlineata*) and insect vectors of viruses. The fungal disease targets include a range of diseases such as cereal rusts and *Botrytis* grey mould on fruit. In the USA, maize transformed to express a dsRNA targeting a gene in corn rootworm (*Diabrotica* spp.) has been developed and commercialized.

RNAi provides additional options for plant breeders to improve plant varieties compared with other new breeding techniques (NBTs) such as clustered regularly interspaced short palindromic repeats/CRISPR-association protein (CRISPR/Cas) or transcription activator-like effector nucleases (TALENs). For example, RNAi provides a method for reducing gene expression (knockdown) rather than complete blocking of the expression (knockout). This is important for providing reduced levels of gene expression, or when a specific stage in a physiological process is to be targeted. Another important feature of RNAi is that dsRNA molecules can be highly mobile in plants. Therefore, dsRNA produced in part of the plant (e.g. rootstock) can have the potential to spread into the grafted parts of the plant to confer resistance to disease to the whole plant, including fruit. This results in fruits that are not genetically modified but protected by the presence of target-specific degradable small RNA molecules (Limera *et al.*, 2017). In addition, dsRNA molecules can be formulated and applied as a topical treatment to plants to change their physiology or combat pests and pathogens. This approach will avoid genetically modified organism (GMO) regulations if no GMOs are present in the products.

Research on RNAi is being conducted mainly in Europe, the USA and China. However, in Europe and some regions of the world the technology and its applications are being held back by policies and legislation on biotechnologies, by failures in the implementation of GMO regulations and by failure to develop appropriate methods for the regulation and assessment of novel plant protection products. This is inhibiting investment in research and development (R&D) on novel 'green' applications of RNAi, as can be seen by the reduction in patent applications in Europe. It has been shown that RNAi has the potential to make major contributions towards sustainable crop production and protection with minimal environmental impacts compared with other technologies. In regions where legislation prevents the use of RNAi technology, farmers will not have access to the technology and important options for improving productivity and economic competitiveness (Taning *et al.*, 2019; Mezzetti *et al.*, 2020). Ironically this will be at a time when governments are trying to introduce more sustainable 'green' agricultural practices and when food demand is increasing and food supplies are at risk from climate change, new invasive species and urbanization.

In 1971 a European Cooperation in Science and Technology (COST) programme had been created. In 2016 the iPlanta COST Action CA15223 'Modifying plants to produce interfering RNA' (available at https://iplanta.univpm.it, accessed 1 November 2020) was established with the objective of bringing together experts from a wide range of fields to develop a deeper understanding of the science of RNA, the applications of RNAi, the biosafety of these applications and the socio-economic aspects of these potential applications. This book contains a series of chapters by experts from many countries, who are participating in iPlanta, to review the current scientific knowledge on RNAi, methods for developing RNAi systems in GM plants and a range of applications for crop improvement, crop production and crop protection. Chapters examine both endogenous systems in GM plants and exogenous systems where interfering RNAs are applied to target plants, pests and pathogens. The biosafety of these different systems is examined and methods for risk assessment for food, feed and environmental safety are discussed. Finally, aspects of the regulation of technologies exploiting RNAi and the socio-economic impacts of RNAi technologies are discussed.

References

Limera, C., Sabbadini, S., Sweet, J.B. and Mezzetti, B. (2017) New biotechnological tools for the genetic improvement of major woody fruit species. *Frontiers in Plant Science* 8, 1418. DOI: 10.3389/fpls.2017.01418.

Mezzetti, B., Smagghe, G., Arpaia, S., Christiaens, O., Dietz-Pfeilstetter, A. *et al.* (2020) RNAi: what is its position in agriculture? *Journal of Pest Science* 93(4), 1125–1130. DOI: 10.1007/s10340-020-01238-2.

Taning, C.N., Arpaia, S., Christiaens, O., Dietz-Pfeilstetter, A., Jones, H. *et al.* (2019) RNA-based biocontrol compounds: current status and perspectives to reach the market. *Pest Management Science* 76(3), 841–845. DOI: 10.1002/ps.5686.

2 Gene Silencing to Induce Pathogen-derived Resistance in Plants

Elena Zuriaga, Ángela Polo-Oltra and Maria Luisa Badenes*
Centre of Citriculture and Plant Production, Valencian Institute for Agricultural Research (IVIA), Valencia, Spain

---

2.1 Introduction: Concept and Historical Overview of the Use of Pathogen-derived Resistance in Plants

The discovery and use of RNA interference (RNAi) and pathogen-derived resistance (PDR) in plants has a large history that has been previously reviewed by Gottula and Fuchs (2009), Lindbo (2012) and Rosa *et al.* (2018), among others. The concept of PDR was introduced by Sanford and Johnston (1985) describing the use of a pathogen's own genome to confer resistance via genetic engineering as an alternative strategy to avoid problems in identifying and isolating host resistance genes, the polygenic control of the resistance or, directly, the lack of available resistance genes. This approach is based upon the disruption of parasite-encoded cellular functions that are essential to the parasite but not to the host. As a model, Sanford and Johnston (1985) used genes of the bacteriophage Qß to confer resistance in *Escherichia coli* against this bacteriophage. Before the discovery and description of RNAi, transgenic tobacco plants expressing the coat protein (CP) gene of the tobacco mosaic virus (TMV) were the first demonstration of PDR against a plant virus (Abel *et al.*, 1986). As a result of these experiments, some

transgenic lines showed no symptoms, or a delay in the development of the disease. Afterwards, numerous studies were conducted using CP genes and also other viral sequences (reviewed by Gottula and Fuchs, 2009), but the mechanism of the engineered resistance was not well understood at the time. It was suggested that the expression of the viral CP in a transgenic plant interfered with the virus replication, translation or virion assembly. Later, during an experiment to obtain plants resistant to the tobacco etch virus (TEV), transgenic lines expressing the TEV CP were obtained, and also other lines that expressed a non-translatable, sense-stranded mRNA for the TEV CP that were called RNA control (RC) lines (Lindbo and Dougherty, 1992). Surprisingly, during the TEV challenge, several of the RC lines were immune to the infection. In these plants, the accumulation of antisense RNA was responsible for this protection and not the ectopic expression of a viral protein, but, once again, at this time the cellular mechanism was not fully understood.

RNAi was first recognized in plants in the late 1980s and early 1990s. During experiments to increase the pigment content in purple petunia flowers using genetic engineering, some transgenic plant lines had flowers that were totally white or variegated (Napoli *et al.*, 1990; van

---

*Corresponding author: badenes_mlu@gva.es

© CAB International 2021. *RNAi for Plant Improvement and Protection*
(eds B. Mezzetti *et al.*)
DOI: 10.1079/9781789248890.0002

der Krol *et al.*, 1990). These authors called this phenomenon 'cosuppression' or 'gene silencing' of both the transgene and the homologous endogenous genes. However, the mechanisms involved were still unknown. Lindbo and collaborators, following their experiments with TEV-resistant transgenic plants, proposed that cytoplasmic activity targeting specific RNA sequences was responsible for the virus resistance in these plants, as transgene mRNA levels were 12- or 22-fold higher in unchallenged transgenic tissues compared with recovered transgenic plants of the same developmental stage (Lindbo *et al.*, 1993). In this publication, the authors proposed a mechanism for post-transcriptional gene silencing (PTGS)/RNA silencing, where the RNA-dependent RNA polymerase (RdRP, also known as RDR) used the overexpressed viral transgene as a template to produce small RNAs that could rebind to new target RNA (viral and transgene) sequences. This model was further expanded by Dougherty and Parks (1995) suggesting that 10–20 nucleotide (nt) RNAs, generated from aberrant or overexpressed transgenes, were part of a cellular sequence-specific RNA targeting and degradation system. In fact, Hamilton and Baulcombe (1999) detected ~25 nt antisense RNAs, complementary to targeted mRNAs, in four types of transgene- or virus-induced PTGS in plants, that were likely synthesized from an RNA template. These authors suggested that these 25 nt antisense RNAs were components of the systemic signal and specificity determinants of PTGS.

Studies with other biological systems contributed to a deeper understanding of the mechanism of PTGS. The discovery of double-stranded RNA (dsRNA) as a potent inducer of PTGS in plants (Waterhouse *et al.*, 1998) and nematodes (Montgomery and Fire, 1998) was a key contribution. Waterhouse *et al.* (1998) transformed tobacco and rice with gene constructs that produce RNAs capable of duplex formation to confer virus immunity or gene silencing to plants. In parallel, Fire *et al.* (1998) demonstrated that the direct injection into adult animals of dsRNA molecules was substantially more effective in producing interference effects than either strand was individually, and just a few molecules were required per affected cell. These authors described dsRNA as a potent trigger for RNAi. The use of direct dsRNA injection

was suggested as a new tool for gene function studies in *Caenorhabditis elegans*, but also for other nematodes, other invertebrates and, potentially, in vertebrates and plants. As the genetic screens got easier, the identification of the genes required for RNAi in *C. elegans*, and their comparison with the ones required for gene silencing in *Drosophila*, plants and fungi, showed the existence of a common underlying mechanism (Mello and Conte, 2004). In 2006, Andrew Z. Fire and Craig C. Mello were awarded the Nobel Prize in Physiology or Medicine for their discovery of 'RNA interference – gene silencing by double-stranded RNA'.

RNA silencing, or RNAi, is a conserved regulatory mechanism of gene expression in eukaryotic organisms that involves both transcriptional and post-transcriptional regulation. Different classes of small non-coding RNA molecules (sRNAs) are generated from dsRNAs by an RNase III-like nuclease called Dicer or Dicer-like (DCL). The guide strand of the dsRNA binds an Argonaute (AGO) protein to form the mature RNA-induced silencing complex (RISC), while the passenger strand of the duplex is selectively degraded (Fang and Qi, 2016). The sRNAs function as a guide to direct RISCs to RNA or DNA targets through base-pairing. Moreover, in some eukaryotes, including plants, RDRs can convert the targeted mRNAs into dsRNAs, generating secondary sRNAs (de Felippes, 2019). This could produce an amplification of the silencing signal, both against the initial target and/or by silencing new ones. Additionally, although a basic structure of RNAi pathways is maintained throughout eukaryotes, the evolution of the DCL, AGO and RDR gene families, including gene duplication or loss, has increased the diversity of these pathways (Molnar *et al.*, 2011). Plants seem to share a core set of primarily four DCL proteins (DCL1–4) (Rosa *et al.*, 2018), while the number of AGO family members varies greatly in different species (10 in *Arabidopsis*, 15 in poplar, 17 in maize and 19 in rice) (Fang and Qi, 2016). At a broader level, Pyott and Molnar (2015) classified the RNAi mechanisms based on the source of the dsRNA initiator as endogenous (within the host genome) or exogenous (outside the host genome). By contrast, Fang and Qi (2016) explained the mechanisms according to the role of the different AGO family members involved. As a result, several sRNA species that differ in biogenesis and functions have been characterized in plants and are discussed in the next sections.

2.2 Use of PDR for Basic Research

RNAi is widely used for functional analysis of plant genes. This approach can be achieved via generating stable transformants but also transient assays, avoiding difficult drawbacks that typically affect the stable transformation protocols. Also, this RNAi approach can be utilized for gene functional analysis in protoplasts (Zhai *et al.*, 2009). Different types of constructs have been employed to achieve gene silencing purposes that have become more complex over time as knowledge on RNAi mechanisms has been advanced (Quintero *et al.*, 2013; Baulcombe, 2015; Khalid *et al.*, 2017). Inspired by the PDR concept, the sense gene-induced PTGS strategy was the first employed in trying to confer resistance to viruses by overexpression of a viral protein. For these types of constructs, a fragment of the viral sequence was directly cloned in sense. Although resistance was successfully achieved in many cases, an RNA-mediated PTGS was actually the mechanism responsible (Lindbo *et al.*, 1993). Also, before the RNAi mechanism was well understood, the expression of antisense viral sequences was also tested for conferring resistance. Prins *et al.* (1996) investigated the RNA-mediated resistance to tomato spotted wilt virus (TSWV) using randomly selected sequences (sense and antisense) of the viral genome to confer resistance.

A second generation of constructs led to the hairpin RNA-induced PTGS strategy. For this, a sense and an antisense viral fragment are cloned (separated by a fragment, usually an intron) to produce transcripts that can fold into dsRNA due to the complementarity of both fragments. This approach has been widely used for gene silencing in plants. As reviewed by Singh *et al.* (2019), intron-spliced hairpin RNA (ihpRNAs) constructs derived from viral proteins have been used, for instance, to confer resistance to plum pox virus (PPV) in plum, prunus necrotic ringspot virus (PNRSV) in cherry, or banana bunchy top virus (BBTV) in banana. Gaffar and Koch (2019) also provided a broad list of examples of

the use of this method to control viral pathogens in different plant families, such as Solanaceae (tobacco, tomato, or potato), Cucurbitaceae (melon, cucumber), Fabaceae (soybean, common bean, cowpea and white clover), Poaceae (rice, wheat, maize and barley), Euphorbiaceae (cassava and poinsettia), or Rutaceae (*Citrus macrophylla*, Mexican lime and sweet orange).

Moreover, virus-induced gene silencing (VIGS) can be used as an alternative to exploit the innate plant defence system of PTGS against viral infections. The development and use of VIGS vectors have been recently reviewed by Dhir *et al.* (2019), and also previously by Lange *et al.* (2013), or Robertson (2004). Nowadays, the validation of gene functions is the major bottleneck in functional genomics. For this purpose, VIGS can be used as a fast method for screening candidate genes. To obtain a VIGS vector, a fragment of a target gene is inserted into a plant virus that upon infection of a plant host induces PTGS of the target gene. For instance, Gunupuru *et al.* (2019) used a barley stripe mosaic virus (BSMV) VIGS vector for functional characterization of disease resistance genes in barley seedlings. Moreover, VIGS can be used to silence genes from the host plant but also from other plant pathogens during co-infections. As an example, Lee *et al.* (2015) adapted the latest generation of binary BSMV VIGS vectors for functional analysis of wheat genes involved in susceptibility and resistance to *Zymoseptoria tritici*, a filamentous ascomycete fungus. Different methods to deliver the viral vectors to the plant have been employed, like agro-inoculation, or mechanical or biolistic inoculation. The utility of a virus as a VIGS vector will be determined by its ability to infect more or fewer species. For instance, Kawai *et al.* (2016) used the apple latent spherical virus (ALSV) vector in seven *Prunus* species, including apricot, sweet cherry, almond, peach, Japanese apricot, Japanese plum and European plum, with different efficiency depending on the species and/or cultivar used.

After microRNAS (miRNAs) became known, a new revolution began and new tools appeared. Transcription of *MIR* genes produces long non-coding transcripts with internal self-complementary regions that allow them to fold back and form an imperfect dsRNA stem-loop structure (primary miRNA, or pri-miRNA). They are recognized and cleaved by DCL1 to produce a 21 nt dsRNA heteroduplex in the canonical pathway (Pyott and Molnar, 2015). From them, the guide strand is loaded into the AGO protein to produce a mature miRNA that can silence the target gene. According to Khalid *et al.* (2017), the artificial miRNAs (amiRNAs) are the third generation of constructs. For this purpose, the mature miRNA sequences in a natural pri-miRNA transcript are replaced with specific RNA sequences that are complementary to target viruses/genes. The first attempts to confer viral resistance using this strategy were reported in *Arabidopsis* and tobacco. Niu *et al.* (2006) modified an *Arabidopsis thaliana* miR159 precursor to express amiRNAs targeting viral mRNA sequences encoding the P69 of turnip yellow mosaic virus (TYMV) and the helper-component proteinase (HC-Pro) of turnip mosaic virus (TuMV). Qu *et al.* (2007) used an amiRNA targeting sequences encoding the silencing suppressor 2b of cucumber mosaic virus (CMV) in transient expression assays.

Later, the discovery of secondary small interfering RNAs (siRNAs) allowed the development of new tools, as recently reviewed by Carbonell (2019) and de Felippes (2019). In some cases, an miRNA-loaded RISC activity on a target transcript results in the production of dsRNA via RDR6 activity and can produce secondary siRNAs by successive DCL processing. They are called miRNA-triggered secondary siRNAs and can act by reinforcing the initial silencing signal (acting in *cis*) or affecting new targets (in *trans*) (de Felippes, 2019). The latter are known as trans-acting siRNAs (tasiRNAs). To date, four families of tasiRNA-producing loci (TAS1–4) have been described in *Arabidopsis thaliana* (TAS1 and TAS2 targeted by miR173, TAS3 by miR390, and TAS4 by miR828), and another six TAS genes (TAS5–10) have been described or predicted in other species (de Felippes, 2019). In order to use this process as a tool, artificial tasiRNAs (atasiRNAs), also known as synthetic tasiRNAs (syn-tasiRNAs) and miRNA-induced gene silencing (MIGS) constructs were developed (see Figure 3 in de Felippes, 2019). As described by this author, to obtain atasiRNAs, one or more of the tasiRNAs in the TAS gene was replaced by a fragment of the target gene. In the case of MIGS, constructs can be generated by placing the sequence recognized by an miRNA that can start transitivity in front of a fragment of the target gene (de Felippes *et al.*, 2012). According to Carbonell (2019) the use of silencing tools based on secondary siRNAs in plants will continue despite the emergence of clustered regularly interspaced short palindromic repeats (CRISPR) technologies, due to their advantages such as high specificity, possibility of multi-targeting, spatio-temporal control of silencing or the ability to target genes whose complete knockout induces lethality.

2.3　Use of PDR for Commercial Purposes

According to FAO (2017), the threats posed by climate change and the upsurge in transboundary pests and diseases are part of the ten key challenges to eradicate hunger and poverty while making agriculture and food systems sustainable. Climate change is modifying the dynamics of pest populations and creating new ecological niches for the emergence or re-emergence and spread of pests and diseases. The impacts of transboundary plant pests and diseases vary from region to region and year to year. In some cases, they result in total crop failure. Recently, Savary *et al.* (2019) estimated that the yield losses worldwide caused by 137 individual crop pests and pathogens on five major crops (wheat, rice, maize, potato and soybean) ranged between 17% and 23% for all five crops, except rice, for which the estimate is 30%.

Crop pests and pathogens include a wide diversity of organisms, such as viruses and viroids, bacteria, fungi and oomycetes, nematodes, arthropods, molluscs, vertebrates and parasitic plants. The development and use of resistant crops are the most efficient strategies to mitigate the impact of these pests and diseases and to improve yield stability. Traditional breeding has been the way to obtain resistant varieties by classically identifying new resistance sources and introgressing them into economically important crops (Piquerez *et al.*, 2014). However, for some cases this is not possible, as no resistant sources are available, or is too difficult, as in the case of species with a long reproductive cycle. Transgenic approaches can solve these situations and one of the strategies for that is the use of PDR RNAi, in which transgenic plants produce a dsRNA that silences a critical pathogen gene. Moreover, non-transformative strategies, such as the use of topical applications of dsRNA (e.g. spray-induced gene silencing) could be applied in some cases (Wang and Jin, 2017; Taning *et al.*, 2020). The study of the molecular mechanisms underlying plant–pathogen interactions provides new opportunities to identify putative target genes (Dong and Ronald, 2019). Singh *et al.* (2019) reviewed the use of RNAi against viruses in perennial fruit plants, and Gaffar and Koch (2019) reviewed the potential of RNA silencing strategies to protect plants of various major plant families. Additionally, Mamta and Rajam (2017) and Liu *et al.* (2020) discussed the use of RNAi for crop pest control. In this section, some of the commercially approved varieties will be presented; further information is detailed in the suggested reviews. Although this approach has been widely used for basic research, the number of varieties approved for commercial production is not very high, probably due to the impressive regulation and public opinion against GMOs. Dong and Ronald (2019) reviewed the genetically engineered food crops with resistance to microbial pathogens approved by at least one international regulatory agency. Varieties approved for commercial production include squash (*Cucurbita* sp.), papaya, potato, sweet pepper, tomato, plum and bean, in chronological order.

In 1994, a squash variety expressing the CP genes of watermelon mosaic virus 2 (WMV2) and zucchini yellow mosaic virus (ZYMV) and, as a result, resistant to both of these potyviruses, received exemption status from the US Department of Agriculture's Animal and Plant Health Inspection Service (APHIS) (Tricoll *et al.*, 1995). Named as Freedom II, it was the first commercially available virus-resistant and disease-resistant transgenic crop released in the USA. Another interesting example was the development of papaya varieties resistant to the potyvirus papaya ringspot virus (PRSV) (Ferreira *et al.*, 2002). In 1992, PRSV appeared in a district of Hawaii where 95% of the production was located. During a field trial, the transgenic papaya line 55-1, with a single insert of the *CP* gene of a mild strain of PRSV, showed resistance to a Hawaiian isolate of PRSV. From this line, the cultivars 'Rainbow' and 'SunUp' were developed. Their release to growers in May 1998 saved the papaya industry in Hawaii.

Another perennial fruit, the 'HoneySweet' plum cultivar, is resistant to sharka disease caused by plum pox virus (PPV), a very limiting factor for stone fruit production worldwide. This cultivar was originated by a transformation experiment aimed at the insertion of *PPV-CP* gene in the plum cultivar 'Bluebyrd', by using hypocotyl slices as starting explants (Scorza *et al.*, 1994). As a result, the transgenic clone C5 (later named as 'HoneySweet') was highly resistant to this potyvirus. Further analysis explained that the insertion event produced a hairpin of the PPV-CP transgene, resulting in a PTGS event (Scorza *et al.*, 2001; Hily *et al.*, 2005). This cultivar was made freely available for fruit production and as a source of PPV resistance for plum breeding in the USA. Currently, 'HoneySweet' plum has not received approval for cultivation in the European Union (EU) or other locations outside the USA. However, field tests have been developed in Europe, where PPV is endemic, and the effectiveness and safety of 'HoneySweet' have been demonstrated (Polák *et al.*, 2017).

Regarding DNA viruses, to date, bean golden mosaic virus (BGMV), a single-stranded DNA virus of the genus *Begomovirus* (family *Geminiviridae*), is the only example of a deregulated genetically engineered crop showing resistance to a virus with this kind of genetic material (Dong and Ronald, 2019). BGMV is the largest constraint to bean production in Latin America and causes significant yield losses (40–100%) in South and Central America, Mexico and the USA (Bonfim *et al.*, 2007). After initial research efforts using traditional breeding techniques and different transgenic approaches (Aragão and Faria, 2009), Bonfim *et al.* (2007) decided to silence the *rep* viral gene to interfere with viral DNA replication using an intron-hairpin construct. As a result, a transgenic common bean line with superior agronomic performance in field trials was selected and registered as cultivar 'BRS FC401 RMD', becoming the first transgenic common bean cultivar in the world. In Brazil, the EMBRAPA researchers pointed out that this work is an example of a public sector effort to develop useful traits resistance to a devastating disease in an 'orphan crop' cultivated by poor farmers throughout Latin America (Aragão and Faria, 2009).

2.4 Limitations and Tools

Although RNAi technology has been widely used for functional analysis and development of crop varieties due to its many advantages, non-specific effects, often referred to as off-target gene silencing, should be considered. Senthil-Kumar and Mysore (2011) discussed the potential problems of off-target gene silencing in plants and considered possibilities that favour this effect. Du *et al.* (2011) also reviewed the off-target effects in mammals, classifying them into sequence-dependent effects and sequence-independent effects. The first type refers to the possibility that partial sequence homology can lead to the degradation of non-target mRNAs, while the second one refers to any unwanted effect at different steps during the PTGS pathway.

RNAi has been widely used as a reverse genetic tool for gene function characterization in plants. However, these off-target effects introduce uncertainty in gene function studies. According to Xu *et al.* (2006), 50–70% of gene transcripts in *Arabidopsis* plants have potential off-targets when used as a silencing trigger for PTGS and this can obscure experimental results. In fact, 50% of the potential off-targets identified using an siRNA Scan computational tool were actually silenced when tested experimentally. Their results suggest that a high risk of off-target gene silencing exists during PTGS in plants. This problem was also observed in 2003 by using microarray analysis in mammalian cells (Du *et al.*, 2011). Another important aspect to consider is the possibility that the off-targeting could also affect exposed non-target organisms, causing environmental and biosafety issues.

Off-target effects can occur in different steps of the silencing process. Senthil-Kumar and Mysore (2011) indicated the steps that can or cannot be manipulated to increase specificity according to knowledge of the respective mechanism. According to these authors, the most troublesome points are the Dicer cleavage and siRNA production, the siRNA amplification and transitive silencing, and the target gene mRNA recognition and degradation. In order to prevent them, the gene fragment used for producing dsRNA (the trigger) should be chosen to be as specific as possible, taking into account that sequence complementarity of only 14 nt or less can lead to inhibition of gene expression. The use of vectors with tissue-specific and inducible promoters is another solution suggested by these authors. Moreover, excessive siRNA production could also lead to off-target effects, so that the use of appropriate promoters could be really important as well as the number of transgene copies introgressed into the host genome.

Computational prediction tools can be used to design RNAi constructs and to screen potential off-target effects. Fakhr *et al.* (2016) reviewed various algorithms for efficient siRNA design and listed the pros and cons of different online software, but mainly focused on mammalian gene silencing. Interestingly, some of the steps of the scoring system suggested by these authors could also be applied in plants, such as the simultaneous use of various online designing tools to identify the more favourable siRNAs, or the use of specific parameters of Basic Local Alignment Search Tool (BLAST) algorithms (Altschul *et al.*, 1990) to take into account the alignment of small sequences. Moreover, these authors suggested the design of at least three siRNAs for any experiment to achieve the best silencing results. Regarding gene silencing specifically in plants, Ahmed *et al.* (2015) reviewed the most popular computational and experimental approaches. As a first step, a specific region of the target gene should be selected, and the BLAST algorithms could be used to find regions of local similarity against the whole genome of the species. The reduction in sequencing costs has made it possible to have an increasing number of complete genomes of different species that can be accessed using different public databases, like Phytozome (Goodstein *et al.*, 2012), Plaza 4.0 (Van Bel *et al.*, 2018), or PlantGDB (Duvick *et al.*, 2008), among others. Regarding specific RNAi-related databases, PVsiRNAdb holds detailed information related to plant virus-derived small interfering RNAs (vsiRNAs) from 20 different viral strains infecting 12 different plant species (Gupta *et al.*, 2018). Additionally, as sequencing costs have fallen, high-throughput sequencing has become an important tool for sRNA discovery and profiling. The UEA small RNA Workbench is a suite of tools for analysing miRNA and other small RNA data from high-throughput sequencing devices (Stocks *et al.*, 2018).

There are a lot of online tools for siRNA design (Fakhr *et al.*, 2016), but their main use is in mammals. Regarding the tools designed for plants, as an example we could cite some of them, like P-SAMS (Fahlgren *et al.*, 2016) and si-Fi21 (Lück *et al.*, 2019). The Plant Small RNA Maker Site (P-SAMS) is a web tool for efficient and specific targeted gene silencing in plants using two applications: P-SAMS amiRNA and P-SAMS syn-tasiRNA Designers, for the simple and automated design of artificial miRNAs and synthetic trans-acting small interfering RNAs, respectively (Fahlgren *et al.*, 2016). According to Lück *et al.* (2019), si-Fi21 offers efficient prediction of RNAi sequences and off-target search and it is specifically intended for long double-stranded RNAi constructs including virus-, microRNA-, and host-induced gene silencing (HIGS).

Given the interest generated by this topic, numerous reviews are available. We encourage the reading of those that have been cited here as well as other chapters of this book in order to have a deeper knowledge.

Acknowledgment

This work has been supported by the Instituto Nacional de Investigación y Tecnología Agraria y Alimentaria (INIA)-FEDER (grant no. RTA2017-00011-C03-01). Ángela Polo was funded by a fellowship co-financed by the Generalitat Valenciana and European Social Fund (2019–2022) (DOGV 8524, 08.04.2019).

References

Abel, P.P., Nelson, R.S., De, B., Hoffmann, N., Rogers, S.G. *et al.* (1986) Delay of disease development in transgenic plants that express the tobacco mosaic virus coat protein gene. *Science* 232(4751), 738–743. DOI: 10.1126/science.3457472.

Ahmed, F., Dai, X. and Zhao, P.X. (2015) Bioinformatics tools for achieving better gene silencing in plants. In: Mysore, K.S. and Senthil-Kumar, M. (eds) *Plant Gene Silencing. Methods and Protocols in Methods in Molecular Biology series.* Vol. 1287. Springer Science + Business Media, New York, pp. 43–60.

Altschul, S.F., Gish, W., Miller, W., Myers, E.W. and Lipman, D.J. (1990) Basic local alignment search tool. *Journal of Molecular Biology* 215(3), 403–410. DOI: 10.1016/S0022-2836(05)80360-2.

Aragão, F.J.L. and Faria, J.C. (2009) First transgenic geminivirus-resistant plant in the field. *Nature Biotechnology* 27(12), 1086–1088. DOI: 10.1038/nbt1209-1086.

Baulcombe, D.C. (2015) VIGS, HIGS and FIGS: small RNA silencing in the interactions of viruses or filamentous organisms with their plant hosts. *Current Opinion in Plant Biology* 26, 141–146. DOI: 10.1016/j.pbi.2015.06.007.

Bonfim, K., Faria, J.C., Nogueira, E.O.P.L., Mendes, E.A. and Aragão, F.J.L. (2007) RNAi-mediated resistance to bean in genetically engineered common bean (*Phaseolus vulgaris*). *Molecular Plant–Microbe Interactions* 20(6), 717–726. DOI: 10.1094/MPMI-20-6-0717.

Carbonell, A. (2019) Secondary small interfering RNA-based silencing tools in plants: an update. *Frontiers in Plant Science* 10, 687. DOI: 10.3389/fpls.2019.00687.

de Felippes, F.F. (2019) Gene regulation mediated by microRNA-triggered secondary small RNAs in plants. *Plants (Basel)* 8(5), 112. DOI: 10.3390/plants8050112.

de Felippes, F.F., Wang, J.-W. and Weigel, D. (2012) MIGS: miRNA-induced gene silencing. *The Plant Journal* 70(3), 541–547. DOI: 10.1111/j.1365-313X.2011.04896.x.

Dhir, S., Srivastava, A., Yoshikawa, N. and Khurana, S.M.P. (2019) Plant viruses as virus induced gene silencing (VIGS) vectors. In: Khurana, S. and Gaur, R. (eds) *Plant Biotechnology: Progress in Genomic Era.* Springer, Singapore, pp. 517–526.

Dong, O.X. and Ronald, P.C. (2019) Genetic engineering for disease resistance in plants: recent progress and future perspectives. *Plant Physiology* 180(1), 26–38. DOI: 10.1104/pp.18.01224.

Dougherty, W.G. and Parks, T.D. (1995) Transgenes and gene suppression: telling us something new? *Current Opinion in Cell Biology* 7(3), 399–405. DOI: 10.1016/0955-0674(95)80096-4.

Du, Q., Huang, H. and Liang, Z. (2011) siRNA: the specificity and off-target effects. In: Wang, B., Zhang, L.H., Xi, Z. and Chattopadhyaya, J. (eds) *Medicinal Chemistry of Nucleic Acids*. John Wiley & Sons, Inc., Hoboken, New Jersey, pp. 405–422.

Duvick, J., Fu, A., Muppirala, U., Sabharwal, M., Wilkerson, M.D. *et al.* (2008) PlantGDB: a resource for comparative plant genomics. *Nucleic Acids Research* 36(Database issue), D959–965. DOI: 10.1093/nar/gkm1041.

Fahlgren, N., Hill, S.T., Carrington, J.C. and Carbonell, A. (2016) P-SAMS: a web site for plant artificial microRNA and synthetic trans-acting small interfering RNA design. *Bioinformatics* 32(1), 157–158. DOI: 10.1093/bioinformatics/btv534.

Fakhr, E., Zare, F. and Teimoori-Toolabi, L. (2016) Precise and efficient siRNA design: a key point in competent gene silencing. *Cancer Gene Therapy* 23(4), 73–82. DOI: 10.1038/cgt.2016.4.

Fang, X. and Qi, Y. (2016) RNAi in plants: an Argonaute-centered view. *The Plant Cell* 28(2), 272–285. DOI: 10.1105/tpc.15.00920.

FAO (2017) The future of food and agriculture – trends and challenges. Food and Agriculture Organization, Rome. Available at: http://www.fao.org/3/a-i6583e.pdf (accessed 11 November 2020).

Ferreira, S.A., Pitz, K.Y., Manshardt, R., Zee, F., Fitch, M. *et al.* (2002) Virus coat protein transgenic papaya provides practical control of papaya ringspot virus in Hawaii. *Plant Disease* 86(2), 101–105. DOI: 10.1094/PDIS.2002.86.2.101.

Fire, A., Xu, S., Montgomery, M.K., Kostas, S.A., Driver, S.E. *et al.* (1998) Potent and specific genetic interference by double-stranded RNA in *Caenorhabditis elegans*. *Nature* 391(6669), 806–811. DOI: 10.1038/35888.

Gaffar, F.Y. and Koch, A. (2019) Catch me if you can! RNA silencing-based improvement of antiviral plant immunity. *Viruses* 11(7), 673. DOI: 10.3390/v11070673.

Goodstein, D.M., Shu, S., Howson, R., Neupane, R., Hayes, R.D. *et al.* (2012) Phytozome: a comparative platform for green plant genomics. *Nucleic Acids Research* 40(Database issue), D1178–D1186. DOI: 10.1093/nar/gkr944.

Gottula, J. and Fuchs, M. (2009) Toward a quarter century of pathogen-derived resistance and practical approaches to plant virus disease control. *Advances in Virus Research* 75, 161–183. DOI: 10.1016/S0065-3527(09)07505-8.

Gunupuru, L.R., Perochon, A., Ali, S.S., Scofield, S.R. and Doohan, F.M. (2019) Virus-Induced gene silencing (VIGS) for functional characterization of disease resistance genes in barley seedlings. In: Harwood, W. (ed.) *Barley. Vol. 1900 in Methods in Molecular Biology series*. Humana Press, New York, pp. 95–114.

Gupta, N., Zahra, S., Singh, A. and Kumar, S. (2018) PVsiRNAdb: a database for plant exclusive virus-derived small interfering RNAs. *Database 2018* Article bay105, 1–8. DOI: 10.1093/database/bay105.

Hamilton, A.J. and Baulcombe, D.C. (1999) A species of small antisense RNA in posttranscriptional gene silencing in plants. *Science* 286(5441), 950–952. DOI: 10.1126/science.286.5441.950.

Hily, J.-M., Scorza, R., Webb, K. and Ravelonandro, M. (2005) Accumulation of the long Clas of siRNA is associated with resistance to plum in a transgenic woody perennial plum tree. *Molecular Plant-Microbe Interactions* 18(8), 794–799. DOI: 10.1094/MPMI-18-0794.

Kawai, T., Gonoi, A., Nitta, M., Yamagishi, N., Yoshikawa, N. *et al.* (2016) Virus-induced gene silencing in various *Prunus* species with the Apple latent spherical virus vector. *Scientia Horticulturae* 199, 103–113. DOI: 10.1016/j.scienta.2015.12.031.

Khalid, A., Zhang, Q., Yasir, M. and Li, F. (2017) Small RNA based genetic engineering for plant viral resistance: application in crop protection. *Frontiers in Microbiology* 8, 43. DOI: 10.3389/fmicb.2017.00043.

Lange, M., Yellina, A.L., Orashakova, S. and Becker, A. (2013) Virus-induced gene silencing (VIGS) in plants: an overview of target species and the virus-derived vector systems. In: Becker, A. (ed.) *Virus-Induced Gene Silencing: Methods and Protocols. Vol. 975 in Methods in Molecular Biology series*. Springer Science+Business Media, New York, pp. 1–14.

Lee, W.-S., Rudd, J.J. and Kanyuka, K. (2015) Virus induced gene silencing (VIGS) for functional analysis of wheat genes involved in *Zymoseptoria tritici* susceptibility and resistance. *Fungal Genetics and Biology* 79, 84–88. DOI: 10.1016/j.fgb.2015.04.006.

Lindbo, J.A. and Dougherty, W.G. (1992) Untranslatable transcripts of the tobacco etch virus coat protein gene sequence can interfere with tobacco etch virus replication in transgenic plants and protoplasts. *Virology* 189(2), 725–733. DOI: 10.1016/0042-6822(92)90595-G.

Lindbo, J.A., Silva-Rosales, L., Proebsting, W.M. and Dougherty, W.G. (1993) Induction of a highly specific antiviral state in transgenic plants: implications for regulation of gene expression and virus resistance. *The Plant Cell* 5(12), 1749–1759. DOI: 10.2307/3869691.

Lindbo, J.A. (2012) A historical overview of RNAi in plants. In: Watson, J.M., Wang, M. and Totowa, N.J. (eds) *Antiviral Resistance in Plants: Methods and Protocols*. Humana Press, New York, pp. 1–16.

Liu, S., Jaouannet, M., Dempsey, D'Maris Amick., Imani, J., Coustau, C. *et al.* (2020) RNA-based technologies for insect control in plant production. *Biotechnology Advances* 39, 107463. DOI: 10.1016/j.biotechadv.2019.107463.

Lück, S., Kreszies, T., Strickert, M., Schweizer, P., Kuhlmann, M. *et al.* (2019) siRNA-Finder (si-Fi) software for RNAi-target design and off-target prediction. *Frontiers in Plant Science* 10, 1023. DOI: 10.3389/fpls.2019.01023.

Mamta, B. and Rajam, M.V. (2017) RNAi technology: a new platform for crop pest control. *Physiology and Molecular Biology of Plants* 23(3), 487–501. DOI: 10.1007/s12298-017-0443-x.

Mello, C.C. and Conte, D. (2004) Revealing the world of RNA interference. *Nature* 431(7006), 338–342. DOI: 10.1038/nature02872.

Molnar, A., Melnyk, C. and Baulcombe, D.C. (2011) Silencing signals in plants: a long journey for small RNAs. *Genome Biology* 12(1), 215. DOI: 10.1186/gb-2010-11-12-219.

Montgomery, M.K. and Fire, A. (1998) Double-stranded RNA as a mediator in sequence-specific genetic silencing and co-suppression. *Trends in Genetics* 14(7), 255–258. DOI: 10.1016/S0168-9525(98)01510-8.

Napoli, C., Lemieux, C. and Jorgensen, R. (1990) Introduction of a chimeric chalcone synthase gene into petunia results in reversible co-suppression of homologous genes in trans. *The Plant Cell* 2(4), 279–289. DOI: 10.2307/3869076.

Niu, Q.-W., Lin, S.-S., Reyes, J.L., Chen, K.-C., Wu, H.-W. *et al.* (2006) Expression of artificial microRNAs in transgenic *Arabidopsis thaliana* confers virus resistance. *Nature Biotechnology* 24(11), 1420–1428. DOI: 10.1038/nbt1255.

Piquerez, S.J.M., Harvey, S.E., Beynon, J.L. and Ntoukakis, V. (2014) Improving crop disease resistance: lessons from research on Arabidopsis and tomato. *Frontiers in Plant Science* 5, 671–671. DOI: 10.3389/fpls.2014.00671.

Polák, J., Kundu, J.K., Krška, B., Beoni, E., Komínek, P. *et al.* (2017) Transgenic plum *Prunus domestica* L., clone C5 (cv. HoneySweet) for protection against sharka disease. *Journal of Integrative Agriculture* 16(3), 516–522. DOI: 10.1016/S2095-3119(16)61491-0.

Prins, M., Resende, RdeO., Anker, C., van Schepen, A., de Haan, P. *et al.* (1996) Engineered RNA-mediated resistance to tomato spotted wilt virus is sequence specific. *Molecular Plant-Microbe Interactions* 9(5), 416–418. DOI: 10.1094/MPMI-9-0416.

Pyott, D.E. and Molnar, A. (2015) Going mobile: non-cell-autonomous small RNAs shape the genetic landscape of plants. *Plant Biotechnology Journal* 13(3), 306–318. DOI: 10.1111/pbi.12353.

Qu, J., Ye, J. and Fang, R. (2007) Artificial MicroRNA-mediated virus resistance in plants. *Journal of Virology* 81(12), 6690–6699. DOI: 10.1128/JVI.02457-06.

Quintero, A., Pérez-Quintero, A.L. and López, C. (2013) Identification of ta-siRNAs and cis-nat-siRNAs in cassava and their roles in response to cassava bacterial blight. *Genomics Proteomics & Bioinformatics* 11(3), 172–181. DOI: 10.1016/j.gpb.2013.03.001.

Robertson, D. (2004) VIGS vectors for gene silencing: many targets, many tools. *Annual Reviews of Plant Biology* 55, 495–519. DOI: 10.1146/annurev.arplant.55.031903.141803.

Rosa, C., Kuo, Y.-W., Wuriyanghan, H. and Falk, B.W. (2018) RNA interference mechanisms and applications in plant pathology. *Annual Reviews of Phytopathology* 56, 581–610. DOI: 10.1146/annurev-phyto-080417-050044.

Sanford, J.C. and Johnston, S.A. (1985) The concept of parasite-derived resistance-deriving resistance genes from the parasite's own genome. *Journal of Theoretical Biology* 113(2), 395–405. DOI: 10.1016/S0022-5193(85)80234-4.

Savary, S., Willocquet, L., Pethybridge, S.J., Esker, P., McRoberts, N. *et al.* (2019) The global burden of pathogens and pests on major food crops. *Nature Ecology & Evolution* 3, 430–439. DOI: 10.1038/s41559-018-0793-y.

Scorza, R., Ravelonandro, M., Callahan, A.M., Cordts, J.M., Fuchs, M. *et al.* (1994) Transgenic plums (*Prunus domestica* L.) express the plum pox virus coat protein gene. *Plant Cell Reports* 14(1), 18–22. DOI: 10.1007/BF00233291.

Scorza, R., Callahan, A., Levy, L., Damsteegt, V., Webb, K. *et al.* (2001) Post-transcriptional gene silencing in plum pox virus resistant transgenic European plum containing the plum pox potyvirus coat protein gene. *Transgenic Research* 10(3), 201–209. DOI: 10.1023/A:1016644823203.

Senthil-Kumar, M. and Mysore, K.S. (2011) Caveat of RNAi in plants: the off-target effect. In: Kodama, H. and Komamine, A. (eds) *RNAi and Plant Gene Function Analysis. Vol. 744 in Methods in Molecular Biology series*. Humana Press/Springer Science+Business Media, New York, pp. 13–25.

Singh, K., Dardick, C. and Kumar Kundu, J. (2019) RNAi-mediated resistance against viruses in perennial fruit plants. *Plants* 8(10), 359. DOI: 10.3390/plants8100359.

Stocks, M.B., Mohorianu, I., Beckers, M., Paicu, C., Moxon, S. *et al.* (2018) The UEA sRNA Workbench (version 4.4): a comprehensive suite of tools for analyzing miRNAs and sRNAs. *Bioinformatics* 34(19), 3382–3384. DOI: 10.1093/bioinformatics/bty338.

Taning, C.N., Arpaia, S., Christiaens, O., Dietz-Pfeilstetter, A., Jones, H. *et al.* (2020) RNA-based biocontrol compounds: current status and perspectives to reach the market. *Pest Management Science* 76(3), 841–845. DOI: 10.1002/ps.5686.

Tricoll, D.M., Carney, K.J., Russell, P.F., McMaster, J.R., Groff, D.W. *et al.* (1995) Field evaluation of transgenic squash containing single or multiple virus coat protein gene constructs for resistance to cucumber mosaic virus, watermelon mosaic virus 2, and zucchini yellow mosaic virus. *Bio/Technology* 13, 1458–1465.

Van Bel, M., Diels, T., Vancaester, E., Kreft, L., Botzki, A. *et al.* (2018) PLAZA 4.0: an integrative resource for functional, evolutionary and comparative plant genomics. *Nucleic Acids Research* 46(D1), D1190–D1196. DOI: 10.1093/nar/gkx1002.

Van der Krol, A.R., Mur, L.A., Beld, M., Mol, J.N. and Stuitje, A.R. (1990) Flavonoid genes in petunia: addition of a limited number of gene copies may lead to a suppression of gene expression. *The Plant Cell* 2(4), 291–299. DOI: 10.1105/tpc.2.4.291.

Wang, M. and Jin, H. (2017) Spray-induced gene silencing: a powerful innovative strategy for crop protection. *Trends in Microbiology* 25(1), 4–6. DOI: 10.1016/j.tim.2016.11.011.

Waterhouse, P.M., Graham, M.W. and Wang, M.B. (1998) Virus resistance and gene silencing in plants can be induced by simultaneous expression of sense and antisense RNA. *Proceedings of the National Academy of Sciences* 95(23), 13959–13964. DOI: 10.1073/pnas.95.23.13959.

Xu, P., Zhang, Y., Kang, L., Roossinck, M.J. and Mysore, K.S. (2006) Computational estimation and experimental verification of off-target silencing during posttranscriptional gene silencing in plants. *Plant Physiology* 142(2), 429–440. DOI: 10.1104/pp.106.083295.

Zhai, Z., Sooksa-nguan, T. and Vatamaniuk, O.K. (2009) Establishing RNA interference as a reverse-genetic approach for gene functional analysis in protoplasts. *Plant Physiology* 149(2), 642–652. DOI: 10.1104/pp.108.130260.

3 Exogenous Application of RNAs as a Silencing Tool for Discovering Gene Function

Barbara Molesini* and Tiziana Pandolfini

Department of Biotechnology, University of Verona, Verona, Italy

3.1 Introduction

RNA silencing is a powerful technique to unravel the function of genes by inhibiting gene expression at the post-transcriptional level. This technique is particularly appropriate for studying developmental processes such as fruit setting and growth that require a tight organ/tissue and time-specific regulation of target genes expression. Gene silencing in plants is usually achieved by the stable or transient expression of genetic constructs producing hairpin (hp) RNA or microRNA (miRNA). The use of exogenously applied small RNAs (sRNAs) and long double-stranded RNAs (dsRNAs) for transient gene silencing in whole plant and/or detached organs would allow a much higher number of genes to be analysed in a shorter time. The successful application of this technique requires efficient systems for sRNA delivery as well as methods to enhance RNA stability in plant cells.

3.2 Methods Used for Establishing the Function of a Specific Gene Altering Gene Expression at either the Genomic or Post-transcriptional Level

In the past decades, over 300 plant species have been sequenced, improving considerably our understanding of the overall structure and dynamics of plant genomes. However, despite the large number of genes identified, the functional role for the vast majority remains to be uncovered. The most widely used strategy to study gene function exploits reverse genetics, a gene-driven approach that links the alteration in the expression of a target gene with the full range of phenotypes controlled by the gene itself. Information on the role of a gene could be obtained by increasing its expression beyond the norm (i.e. overexpression), or by expressing the gene in a cell type and/or developmental stage or condition in which it is normally not

*Corresponding author: barbara.molesini@univr.it

DOI: 10.1079/9781789248890.0003

expressed (i.e. misexpression), or by diminishing (i.e. knockdown) or completely abolishing (i.e. knockout) its expression. The complete suppression is generally obtained by introducing mutations at the genomic DNA level. In this regard, mutant plants have been obtained by X-rays and γ-ray irradiation, by chemical mutagens such as ethyl methane sulfonate (EMS), by targeting induced local lesion in genomes (TILLING) which couples random chemical mutagenesis with PCR-based screening, by transposon-mediated gene disruption, or by T-DNA insertion. The most recent approach for generating precise modifications of genome sequences is targeted genome editing, carried out by using either engineered nucleases such as zinc-finger nuclease (ZFN) and transcription activator-like effector nuclease (TALEN), or RNA-guided nucleases based on the naturally occurring type II *Clustered Regularly Interspaced Short Palindromic Repeats*/Cas9 (CRISPR/Cas9) system (Chen *et al.*, 2019). Besides the approaches aimed at modifying genomic DNA, tools directed to the mRNA/protein level of the gene of interest, such as post-transcriptional gene silencing (PTGS), have been extensively used for functional studies (McGinnis, 2010). The methods acting post-transcriptionally rarely cause complete loss of function of the target gene, but generally result in various degrees of downregulation with residual levels of the targeted mRNA/protein still detectable. PTGS, a natural mechanism used by the plant for protection against viruses and other invading nucleic acids and as a system to regulate gene expression, is activated by dsRNA molecules which are processed by Dicer-like (DCL) enzymes into 21–24 sRNAs (Martínez de Alba *et al.*, 2013). The two principal classes of sRNAs are the small interfering RNAs (siRNAs) and miRNAs (Axtell, 2013). These sRNAs are loaded into Argonaute-containing silencing effector complexes and guide the sequence-specific cleavage of their mRNA targets. Thus, the expression of a dsRNA homologous to the gene of interest is sufficient to elicit the RNA silencing pathway against the target gene. Using hp or artificial miRNA-based constructs, dsRNA can be expressed in plants and used to silence both specific endogenous genes and genes of invading pathogens. The siRNA population can be increased, and the silencing signal amplified, in a process known as transitivity (Himber *et al.*, 2003). An RNA-dependent RNA polymerase (RdRP) uses the cleavage products of the target mRNA, which include sequences outside the initial homology region present in the silencing construct, as substrate to generate a new population of siRNAs, called secondary siRNAs. The silencing signal that originates from primary and secondary siRNAs is not cell autonomous and can move to adjacent cells and systemically via the phloem (Melnyk *et al.*, 2011). The systemic spread of the silencing seems to compromise the possibility of obtaining an siRNA silencing restricted to specific cells/tissues. However, silencing spread and amplification have been documented with viral sequences and overexpressed transgenes (Luo and Chen, 2007; Melnyk *et al.*, 2011), whereas several studies support a tissue-specific siRNA silencing for endogenous genes when the hp construct is under the control of tissue/developmental specific-promoters (Davuluri *et al.*, 2005; Okabe *et al.*, 2019). For pleiotropic genes, the downregulation obtained with RNA silencing could partially unveil the activity of the target gene, whereas a complete genetic ablation would reveal the full functional role. However, the loss of a vital gene can often lead to embryo lethality as well as severe developmental abnormalities, which preclude the assessment of its role in adult vegetative and reproductive phases. To circumvent these problems, a fine-tuned downregulation of the target gene expression via PTGS could be preferable to a complete knockout.

In this regard, biological processes related to plant growth and development are under the control of complex networks of transcription factors which downstream regulate multiple signalling pathways. This implies a strict modulation of the gene expression brought about by tissue- and time-specific promoters. The use of strategies based on genomic silencing (e.g. T-DNA insertion, genome editing) to identify the function of genes implicated in developmental processes could result in complex phenotypic alterations or embryo lethality. On the other hand, PTGS constructs associated to appropriate promoters may offer a clearer phenotypical output. In this chapter, we discuss different RNAi strategies for studying the genes implicated in fruit set and growth, focusing on the use of ectopically applied sRNAs.

3.3 RNA Silencing as a Tool for Studying Genes Implicated in the Early Phases of Tomato Fruit Development

Tomato (*Solanum lycopersicum*) represents the model species for the study of fleshy fruit development and a great deal of biochemical and genetic information on the different phases of development from flowering to fruit maturation is available (Gapper *et al.*, 2014). The tomato fruit originates from the ovary, the enlarged basal portion of the pistil. Fruit set is the earliest phase of fruit growth and represents the transition from the static condition of the ovary before fertilization to that of the rapidly growing fruit after fertilization (Fig. 3.1). The presence of fertilized ovules generally sustains the development of the ovary into a fruit, and the number of fertilized ovules usually determines the fruit growth rate (Gillaspy *et al.*, 1993). Following fertilization, cell division is activated in the ovary and continues for about 7–10 days. After the period of cell division, fruit growth is mainly due to an increase in cell volume. During the period of rapid cell expansion, the embryos/seeds mature, showing well-developed cotyledons and established

root–shoot axis. Rapid growth continues in the mature green fruit stage. The terminal stage of development is the ripening and initiates after seed maturation has been completed.

In parthenocarpic plants, fruit develops without fertilization, indicating that the ovary growth inhibitory factors have been released before fertilization. Indeed, several genes proved to be repressors of fruit set display a sharp downregulation during the transition from pre-anthesis to fertilized flowers. The genetic factors that repress ovary growth before fertilization can be identified by RNA silencing. The downregulation of the expression of components of auxin (e.g. IAA9, ARF7) and gibberellin signalling pathways (DELLA) and auxin transport (PIN4) induced parthenocarpy (Wang *et al.*, 2005; Goetz *et al.*, 2006, 2007; Martí *et al.*, 2007; Chaabouni *et al.*, 2009; de Jong *et al.*, 2009; Mounet *et al.*, 2012). Other positive regulators of fruit set have been identified by RNA silencing, since their downregulation determines reduced fruit set and increased fruit abortion (e.g. AtNAOD) (Molesini *et al.*, 2015).

Transcription regulators such as members of the Aux/IAA protein family display distinct functions in plant growth and development and some play a role both in vegetative and

Fig. 3.1. Principal phases of tomato fruit development. Fruit set represents the onset of ovary growth after successful fertilization of the ovules. The subsequent fruit growth occurs by cell division and cell expansion. Fruit ripening represents the terminal stage of development.

reproductive development as for *S. lycopersicum* Aux/IAA transcription factor IAA9 (Wang *et al.*, 2005). The use of a strong and constitutive promoter (CaMV35S) to drive the expression of an IAA9 silencing construct, as well as the knockout of IAA9 obtained by CRISPR, induced the parthenocarpic development of the tomato fruit accompanied by alterations in leaf morphology (Wang *et al.*, 2005; Ueta *et al.*, 2017). Thus, a constitutive silencing can produce the same effects as a knockout mutation and can be employed as a strategy to unravel the role of a target gene when the pleiotropic effects are present in different organs and easily distinguishable. However, when multiple phenotypic alterations are manifested in the same organ, for instance when the parthenocarpic trait is associated with modifications of flower morphology (Ampomah-Dwamena *et al.*, 2002; da Silva *et al.*, 2017; Takei *et al.*, 2019), the approaches based on genome modifications or constitutive silencing of the target gene can be cumbersome. Besides, from an applied perspective, plants harbouring a desired trait but also showing unintended pleiotropic effects are not marketable.

3.3.1 RNA silencing obtained by either stable or transient transformation

Stable transformation of plants with siRNA-generating constructs is an efficient method for activating the RNA silencing pathway, but this procedure is time consuming and labour intensive. It includes: genetic transformation via *Agrobacterium* or through biolistic methods; regeneration and selection of stable transgenic plants; molecular analysis of the transgenic state of several independent lines; and phenotypic analysis of subsequent plant generations (T1, T2). To obtain the first hints about the function of a target gene, thus reducing the time of functional analysis, methods for transient RNA silencing have been developed. Transient gene expression is also useful, since it is not influenced by position effects of the transgene in the genome, does not require selection and can be utilized in differentiated plant tissues. It is largely applied for production of high amounts of foreign proteins and for gene silencing by PTGS. The transient

transformation can be obtained by infiltration into the plant cells of *Agrobacterium tumefaciens* harbouring viral vectors or binary vectors. Concerning the study of genes implicated in the development of fruit, which is the last organ produced by a plant, strategies based on agro-injection of virus-induced gene silencing (VIGS) vectors have been developed (Liu *et al.*, 2002). A few examples are reported on the use of agro-injection of VIGS vector in studying fruit development and ripening in Solanaceae species (Fu *et al.*, 2005; Orzaez *et al.*, 2006; Wang and Fu, 2018). In these studies, fruit infiltration was performed *in planta* using a syringe to inject the bacteria into the carpopodium of tomato fruits at 10 days after pollination (Fu *et al.*, 2005), into the stalks of eggplant fruits 5 cm long (Wang and Fu, 2018) and into the stylar apex of tomato fruits at the beginning of mature green stage (20–25 days post-anthesis), respectively (Orzaez *et al.*, 2006). The phenotypes of the fruits were scored about 10–20 days post-inoculation, a temporal window that allows the evaluation of the transient silencing effect. The previous examples refer to genes whose silencing produces a visible phenotype (e.g. impaired synthesis of pigments). To apply transient silencing to genes with no expected visible phenotype and to overcome the problem of irregular distribution of VIGS, a strategy based on a visual reporter of VIGS in tomato fruit was developed (Orzaez *et al.*, 2009; España *et al.*, 2014). These methods, although effective, have some drawbacks; for example, the massive injection of bacteria or virus-derived sequences might induce unintended and not specific effects. Most importantly, the available methods seem feasible to study genes involved in later stages of fruit development and ripening but do not appear ideal to study very early phases of ovary/fruit growth. Orzaez *et al.* (2006) noted deleterious side effects after agro-injection in young ovaries/fruits (from 7 to 20 days post-anthesis) consisting of growth arrest, premature ripening and abscission. Topical application of dsRNA/siRNAs can be an appealing alternative to genetically modify crops for the functional characterization of genes involved in early stages of fruit set and growth.

3.3.2 Applications of exogenously supplied RNA silencing effector molecules to plant tissues

The possibility of exploiting exogenous RNAs for gene functional analyses in plants and for crop improvement is supported by many studies on sRNA metabolism carried out over the past few years in plants. sRNAs can move cell to cell, presumably via plasmodesmata, and over long distances through the vasculature (Brosnan and Voinnet, 2011; Melnyk et al., 2011; Brunkard and Zambryski, 2017). In addition, plant cells can take up exogenously supplied dsRNA and sRNAs (Koch et al., 2016; Wang et al., 2017). sRNAs and dsRNAs can enter the plant through stomata and through wounded or abraded surfaces (Wang et al., 2016) and move away from the initial point of application (Faustinelli et al., 2018). The binding of dsRNAs to nanoparticles, besides increasing their stability (Mitter et al., 2017), could facilitate their penetration (Sanzari et al., 2019). Movement of sRNAs and dsRNAs can take place also between interacting organisms (e.g. plants and fungi) in a process called 'cross-kingdom RNAi' (Wang et al., 2016; Cai et al., 2018).

The variables to be considered for ectopic applications of sRNA are numerous and mainly concern: the type of RNA molecules; the origin of RNA molecules (in vitro, chemically, or bacterially synthesized); the choice of the delivery method; and the use of sRNA carriers. The optimization of these parameters will vary depending on the aim of the study and the plant organ to be treated.

Many studies have proved that the topical application of RNA molecules on plant tissues represents an efficient system for inducing resistance against viruses, fungi and insects (for a comprehensive overview see Dubrovina and Kiselev, 2019; Dalakouras et al., 2020). Fewer examples have been reported on the efficacy of exogenously applied sRNA for silencing of transgenes and endogenous genes (Dubrovina and Kiselev, 2019; Dalakouras et al., 2020). In accordance with the observations made on plants stably expressing silencing constructs, the silencing capacity of ectopic RNAs seems more effective with transgenes rather than endogenous genes (Dubrovina and Kiselev, 2019). This phenomenon can be due to

the higher expression of the transgenes that is usually driven by strong promoters, the high frequency of aberrant transcripts, and the absence of introns and 5' and 3' UTR sequences which contribute to the mRNA stabilization (Luo et al., 2007; Dadami et al., 2014).

In the studies describing the use of external RNAs for the silencing of endogenous genes, the sRNAs or dsRNAs were topically applied on leaves and roots and only in a single case on reproductive organs (Sammons et al., 2011; Numata et al., 2014; Lau et al., 2015; Li et al., 2015). The study by Lau et al. (2015) described the silencing of the MYb1 gene in flower buds of Dendrobium hybrida. The DhMyb1 gene encodes a transcription factor, expressed during flower development, which is putatively involved in flower morphogenesis. To obtain MYb1 silencing, they applied a crude lysate of RNaseIII-deficient Escherichia coli cells expressing dsRNA corresponding to 430 bp DhMyb1 cDNA, on very young flower buds (≤ 0.5 mm in length), by gently rubbing. The treatment was repeated every 5 days and the phenotype recorded 25–29 days after the first treatment. The transcript level of DhMyb1 was reduced approximately two- to fourfold in the treated flower buds as compared with that in the untreated ones. At the phenotypic level, RNA-treated and untreated flower buds appeared indistinguishable, but microscopic analyses revealed that the dsRNA treatment caused changes in the epidermal cells, which had a flattened instead of conical shape.

Interestingly, the suppression of genes involved in reproductive development can be obtained also by systemic silencing after ectopic dsRNA application, as demonstrated in the paper by Li et al. (2015). In this study, 2-week-old Arabidopsis thaliana roots were soaked with a solution containing in vitro synthesized dsRNA of 554 bp in length for the silencing of MOB kinase activator-like 1A (Mob1A). Arabidopsis Mob1A is required for organ growth and reproduction, since its suppression resulted in reduced growth of vegetative organs and defects in seed set (Pinosa et al., 2013). Two weeks after root soaking, a reduction in Mob1A expression was observed as well as impaired bolting and flowering. These results indicate that the silencing in the reproductive organs can occur because of the movement of the silencing signal from the root to the shoot via the vascular tissues.

3.3.3 Perspective on the use of exogenous sRNAs and long dsRNAs for the silencing of genes involved in fruit growth and development

In this section, we will discuss the possibility of utilizing ectopic RNAs as a fast system for the functional analysis of the genes involved in tomato fruit growth and development.

Fruit development is characterized by two important transition phases: the first, from pre-anthesis to the fertilized flower with the consequent activation of ovary growth (fruit set); and the second, from the end of the growth phase to the start of fruit ripening (Fig. 3.1). These transitions involve marked hormonal and biochemical modifications resulting from changes in the expression of many genes. The exogenous sRNAs or dsRNAs could be applied *in planta* on flowers and fruits or *in vitro* under sterile conditions on detached reproductive organs.

In this regard, Nitsch (1950) showed that auxin exogenously supplied to culture medium of pre-anthesis tomato flower buds is sufficient to guarantee the growth of the ovary/fruit up to the ripening phase. More recently, we observed that genetically engineered flower buds with increased auxin synthesis can be grown *in vitro* after emasculation (i.e. stamen detachment) up to fruit ripening without phytohormones in the culture medium (Pandolfini *et al.*, 2010). Therefore, this *in vitro* system can be used to evaluate the genes involved both in fruit setting and in growth and ripening phases.

To assay the efficiency of the exogenous treatment, it is essential in the setting up of the experiments to include some positive controls, consisting of genes whose silencing obtained via stable transformation produces the expected phenotype (e.g. changes in pigment production or fruit set efficiency) (Molesini *et al.*, 2009; Osorio *et al.*, 2012).

The choice of RNA type is the first variable to be considered. Both long dsRNAs (generally 200–800 bp) and short siRNAs (22–24 nt) have been proved to induce the silencing of endogenous genes efficiently (Dubrovina and Kiselev, 2019; Dalakouras *et al.*, 2020). dsRNA, when processed by Dicer within the cell, produces a pool of effector molecules (siRNAs) homologous to different portions of the target mRNA, thus increasing the

probability that the RNAi silencing complex recognizes and cleaves the target mRNA. However, the complexity of the siRNA pool generated from a dsRNA substrate may increase the possibility of having partial and/or perfect matching with off-target mRNAs. When using a single siRNA molecule, an accurate *in silico* design can diminish the probability of unintended matching, but the high dosage of siRNAs needed for silencing may lead to off-target effects.

In our case, since exogenous RNAs will be applied to specific organs, pleiotropic and off-target effects are presumably limited. In addition, considering that the target mRNA structure can affect the accessibility and consequently the gene silencing ability of siRNAs (Gredell *et al.*, 2008), the use of long dsRNAs appears to be an appropriate choice since, once processed, it generates a heterogeneous pool of siRNAs targeting different portions of the transcript.

For dsRNA production, it is preferable to use an *in vitro* transcription system because it guarantees high yield and purity of dsRNAs, rather than raw bacterial lysates, which contain RNA of bacterial origin besides other contaminants.

The choice of the RNA delivery system mainly depends on the anatomy of the tissue/organ being treated. In our case, the effector molecules must penetrate the ovary or the growing/mature fruit (Fig. 3.2). Regarding the ovary, one of the possible routes can be the stylus, therefore RNAs could be applied on the stigma. An alternative method could be the injection of the RNAs in the pedicel of the flower through a syringe, or by deposition after abrasion of the tissue. Direct injection into the ovary must be avoided, as it has been observed that this practice causes damage to the growing fruit. It is also possible to spray the sRNAs directly on the entire surface of the flower bud, in which case the entry of sRNAs might also occur through other natural openings (e.g. stomata) of the flower organs (sepals, petals, stamens) (Fig. 3.2).

It has been observed that, after the delivery of siRNAs to the petiole of *Nicotiana benthamiana* as well as after the injection of dsRNAs in the trunk of *Vitis vinifera* (Dalakouras *et al.*, 2018), the RNAs are transported in the xylem and restricted to the apoplast, while high-pressure spraying is effective in the delivery of siRNAs to the symplast (Dalakouras *et al.*, 2018). This last technique could allow sRNAs to be efficiently conveyed within the

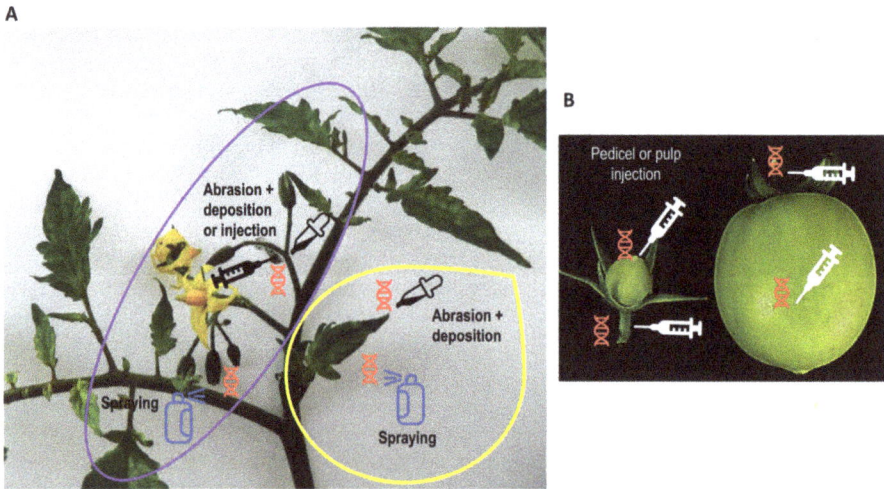

Fig. 3.2. Exogenous sRNA application for the *in planta* silencing of genes involved in fruit growth and development. (A) sRNAs can be applied to young flower buds either by spraying or by abrasion of the pedicel followed by deposition or by injection into the flower pedicel (circled in purple). Using the same application methods (circled in yellow) sRNAs can also be delivered to the young leaves just below the flower trusses. The movement of RNA silencing to the flowers located upstream occurs systemically. (B) sRNAs can be injected using a syringe into the pulp or pedicel of growing fruits. To avoid damage, the treated fruits should have a diameter greater than ~1 cm.

ovary. High-pressure spraying could also be used on the leaf immediately below the flower bud, in which case sRNAs could be systemically transported to the ovary, avoiding any mechanical damage to the floral tissues (Fig. 3.2).

If *in vitro* cultivated flower buds are used, sRNAs can also be applied to the cutting surface of the pedicel (Fig. 3.3). In this case, to facilitate the entry of the effector molecules, the tissue could be subjected to air stress in a laminar airflow hood (Faustinelli *et al.*, 2018).

The application of sRNAs to young ovaries allows the functional study of the genes involved in fruit setting and in fruit development (Fig. 3.3). If the specific target of the investigation is the parthenocarpy, the methods described can be used on flower buds after stamen excision (emasculation). However, in the absence of pollination and fertilization, the tomato ovary ceases cell division and abscises in a few days, therefore sRNAs should be loaded on emasculated very young buds and the treatment should be repeated several times.

Regarding the treatment of growing or mature fruits, it should be considered that the presence of the cuticle can be an obstacle to the entry of sRNAs. Therefore, the application of sRNAs by injection appears a more suitable method than spraying. Injection in the peduncle or in the pulp of the fruit has been used several times for transient expression of transgenes without producing damage to the fruit (Fig. 3.2).

One of the major problems linked to the use of ectopically delivered sRNAs is due to the instability of naked RNA molecules (e.g. action of nucleases and/or environmental conditions such as excessive sunlight). A recent paper demonstrated that the use of dsRNAs loaded on layered double hydroxide (LDH) clay nanosheets improved the stability of the ectopically delivered dsRNA molecules, resulting in a prolonged silencing effect (Mitter *et al.*, 2017).

As previously mentioned, the problem of sRNA penetration is critical when the target organ is the female gametophyte or the ovule, since the sRNAs loaded on the surface of the flower

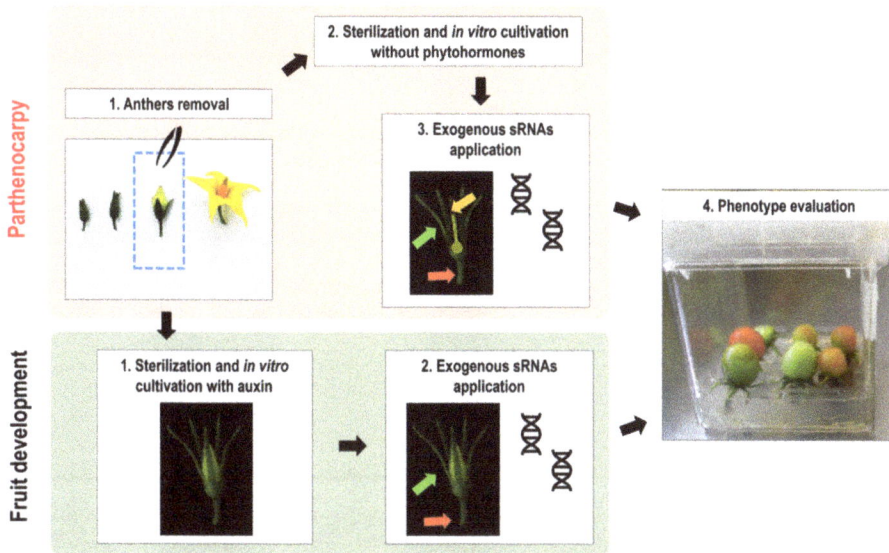

Fig. 3.3. sRNA application on *in vitro* cultivated flower buds. For the evaluation of genes putatively involved in parthenocarpy (upper image), flower buds collected before anthesis are emasculated, sterilized and cultured *in vitro* in a medium not supplemented with phytohormones. The sRNAs could be applied to the stigma (yellow arrow), to the cut surface of the pedicel (red arrow) and to the whole flower buds (green arrow). The growth of the ovary confirms the role of the silenced target gene as repressor of fruit setting. For the evaluation of genes putatively involved in fruit development (lower image), flower buds collected before anthesis are sterilized and cultured *in vitro* in a medium supplemented with auxin. The sRNAs could be applied to the cut surface of the pedicel (red arrow) and to the whole flower buds (green arrow). After approximately 30 days of cultivation, fruits start to ripen.

bud should cross several cell layers before being effective. In this case, the use of nanoparticles as sRNA carriers can be advantageous in favouring the distribution of the effector molecules, as well as sRNA stability. In fact, there is evidence that nanoparticles can passively enter natural plant openings (stomata, stigma, etc.) and those of reduced length (3–50 nm) can also pass through the cell wall (Sanzari *et al.*, 2019), The cuticle is normally a strong barrier to the nanoparticles' diffusion, although TiO_2 particles are reported to produce holes in the cuticle, thus favouring sRNA penetration (Larue *et al.*, 2014).

3.4 Conclusions

The utilization of ectopic sRNAs as a tool for the discovery of gene function is in its infancy and we need future research efforts to test the efficacy of this system. However, it is an attractive perspective for the study of genes involved in developmental processes such as flowering and fruit growth. From a biotechnological point of view, the use of sRNAs not only to improve the crop's defence against pathogens and pests but also to modulate productivity is an exciting challenge.

References

Ampomah-Dwamena, C., Morris, B.A., Sutherland, P., Veit, B. and Yao, J.L. (2002) Down-regulation of TM29, a tomato SEPALLATA homolog, causes parthenocarpic fruit development and floral reversion. *Plant Physiology* 130(2), 605–617. DOI: 10.1104/pp.005223.

Axtell, M.J. (2013) Classification and comparison of small RNAs from plants. *Annual Review of Plant Biology* 64(1), 137–159. DOI: 10.1146/annurev-arplant-050312-120043.

Brosnan, C.A. and Voinnet, O. (2011) Cell-to-cell and long-distance siRNA movement in plants: mechanisms and biological implications. *Current Opinion in Plant Biology* 14(5), 580–587. DOI: 10.1016/j.pbi.2011.07.011.

Brunkard, J.O. and Zambryski, P.C. (2017) Plasmodesmata enable multicellularity: new insights into their evolution, biogenesis, and functions in development and immunity. *Current Opinion in Plant Biology* 35, 76–83. DOI: 10.1016/j.pbi.2016.11.007.

Cai, Q., He, B., Kogel, K.H. and Jin, H. (2018) Cross-kingdom RNA trafficking and environmental RNAi — nature's blueprint for modern crop protection strategies. *Current Opinion in Microbiology* 46, 58–64. DOI: 10.1016/j.mib.2018.02.003.

Chaabouni, S., Jones, B., Delalande, C., Wang, H., Li, Z. *et al.* (2009) Sl-IAA3, a tomato Aux/IAA at the crossroads of auxin and ethylene signalling involved in differential growth. *Journal of Experimental Botany* 60(4), 1349–1362. DOI: 10.1093/jxb/erp009.

Chen, K., Wang, Y., Zhang, R., Zhang, H. and Gao, C. (2019) CRISPR/Cas genome editing and precision plant breeding in agriculture. *Annual Review of Plant Biology* 70(1), 667–697. DOI: 10.1146/annurev-arplant-050718-100049.

da Silva, E.M., Silva, G.F.F.E., Bidoia, D.B., da Silva Azevedo, M. and de Jesus, F.A. (2017) microRNA159-targeted SlGAMYB transcription factors are required for fruit set in tomato. *Plant Journal* 92, 95–109. DOI: 10.1111/tpj.13637.

Dadami, E., Dalakouras, A., Zwiebel, M., Krczal, G. and Wassenegger, M. (2014) An endogene-resembling transgene is resistant to DNA methylation and systemic silencing. *RNA Biology* 11(7), 934–941. DOI: 10.4161/rna.29623.

Dalakouras, A., Jarausch, W., Buchholz, G., Bassler, A., Braun, M. *et al.* (2018) Delivery of hairpin RNAs and small RNAs into woody and herbaceous plants by trunk injection and petiole absorption. *Frontiers in Plant Science* 9, 1253. DOI: 10.3389/fpls.2018.01253.

Dalakouras, A., Wassenegger, M., Dadami, E., Ganopoulos, I., Pappas, M.L. *et al.* (2020) Genetically modified organism-free RNA interference: exogenous application of RNA molecules in plants. *Plant Physiology* 182(1), 38–50. DOI: 10.1104/pp.19.00570.

Davuluri, G.R., van Tuinen, A., Fraser, P.D., Manfredonia, A., Newman, R. *et al.* (2005) Fruit-specific RNAi-mediated suppression of DET1 enhances carotenoid and flavonoid content in tomatoes. *Nature Biotechnology* 23(7), 890–895. DOI: 10.1038/nbt1108.

de Jong, M., Wolters-Arts, M., Feron, R., Mariani, C. and Vriezen, W.H. (2009) The *Solanum lycopersicum* auxin response factor 7 (SlARF7) regulates auxin signaling during tomato fruit set and development. *Plant Journal* 57(1), 160–170. DOI: 10.1111/j.1365-313X.2008.03671.x.

Dubrovina, A.S. and Kiselev, K.V. (2019) Exogenous RNAs for gene regulation and plant resistance. *International Journal of Molecular Sciences* 20(9), 2282. DOI: 10.3390/ijms20092282.

España, L., Heredia-Guerrero, J.A., Reina-Pinto, J.J., Fernández-Muñoz, R., Heredia, A. *et al.* (2014) Transient silencing of CHALCONE SYNTHASE during fruit ripening modifies tomato epidermal cells and cuticle properties. *Plant Physiology* 166(3), 1371–1386. DOI: 10.1104/pp.114.246405.

Faustinelli, P.C., Power, I.L. and Arias, R.S. (2018) Detection of exogenous double-stranded RNA movement in *in vitro* peanut plants. *Plant Biology (Stuttg)* 20(3), 444–449. DOI: 10.1111/plb.12703.

Fu, D.Q., Zhu, B.Z., Zhu, H.L., Jiang, W.B. and Luo, Y.B. (2005) Virus-induced gene silencing in tomato fruit. *Plant Journal* 43(2), 299–308. DOI: 10.1111/j.1365-313X.2005.02441.x.

Gapper, N.E., Giovannoni, J.J. and Watkins, C.B. (2014) Understanding development and ripening of fruit crops in an 'omics' era. *Horticulture Research* 1, 14034. DOI: 10.1038/hortres.2014.34.

Gillaspy, G., Ben-David, H. and Gruissem, W. (1993) Fruits: a developmental perspective. *The Plant Cell* 5(10), 1439–1451. DOI: 10.2307/3869794.

Goetz, M., Vivian-Smith, A., Johnson, S.D. and Koltunow, A.M. (2006) AUXIN RESPONSE FACTOR8 is a negative regulator of fruit initiation in *Arabidopsis*. *Plant Cell* 18(8), 1873–1886. DOI: 10.1105/tpc.105.037192.

Goetz, M., Hooper, L.C., Johnson, S.D., Rodrigues, J.C.M., Vivian-Smith, A. *et al.* (2007) Expression of aberrant forms of AUXIN RESPONSE FACTOR8 stimulates parthenocarpy in *Arabidopsis* and tomato. *Plant Physiology* 145(2), 351–366. DOI: 10.1104/pp.107.104174.

Gredell, J.A., Berger, A.K. and Walton, S.P. (2008) Impact of target mRNA structure on siRNA silencing efficiency: a large-scale study. *Biotechnology and Bioengineering* 100(4), 744–755. DOI: 10.1002/bit.21798.

Himber, C., Dunoyer, P., Moissiard, G., Ritzenthaler, C. and Voinnet, O. (2003) Transitivity-dependent and -independent cell-to-cell movement of RNA silencing. *The EMBO Journal* 22(17), 4523–4533. DOI: 10.1093/emboj/cdg431.

Koch, A., Biedenkopf, D., Furch, A., Weber, L., Rossbach, O. *et al.* (2016) An RNAi-based control of *Fusarium graminearum* infections through spraying of long dsRNAs involves a plant passage and is controlled by the fungal silencing machinery. *PLoS Pathogens* 12(10), e1005901. DOI: 10.1371/journal.ppat.1005901.

Larue, C., Castillo-Michel, H., Sobanska, S., Trcera, N., Sorieul, S. *et al.* (2014) Fate of pristine TiO$_2$ nano-particles and aged paint-containing TiO$_2$ nanoparticles in lettuce crop after foliar exposure. *Journal of Hazardous Materials* 273, 17–26. DOI: 10.1016/j.jhazmat.2014.03.014.

Lau, S.E., Schwarzacher, T., Othman, R.Y. and Harikrishna, J.A. (2015) dsRNA silencing of an R2R3-MYB transcription factor affects flower cell shape in a *Dendrobium* hybrid. *BMC Plant Biology* 15(1), 194. DOI: 10.1186/s12870-015-0577-3.

Li, H., Guan, R., Guo, H. and Miao, X. (2015) New insights into an RNAi approach for plant defence against piercing-sucking and stem-borer insect pests. *Plant, Cell & Environment* 38(11), 2277–2285. DOI: 10.1111/pce.12546.

Liu, Y., Schiff, M. and Dinesh-Kumar, S.P. (2002) Virus-induced gene silencing in tomato. *Plant Journal* 31(6), 777–786. DOI: 10.1046/j.1365-313X.2002.01394.x.

Luo, Z. and Chen, Z. (2007) Improperly terminated, unpolyadenylated mRNA of sense transgenes is tar-geted by RDR6-mediated RNA silencing in *Arabidopsis*. *The Plant Cell* 19(3), 943–958. DOI: 10.1105/tpc.106.045724.

Luo, Q., Kang, Q., Song, W.X., Luu, H.H., Luo, X. *et al.* (2007) Selection and validation of optimal siRNA target sites for RNAi-mediated gene silencing. *Gene* 395(1-2), 160–169. DOI: 10.1016/j.gene.2007.02.030.

Martí, C., Orzáez, D., Ellul, P., Moreno, V., Carbonell, J. *et al.* (2007) Silencing of DELLA in-duces facultative parthenocarpy in tomato fruits. *The Plant Journal* 52(5), 865–876. DOI: 10.1111/j.1365-313X.2007.03282.x.

Martínez de Alba, A.E., Elvira-Matelot, E. and Vaucheret, H. (2013) Gene silencing in plants: a diversity of pathways. *Biochimica et Biophysica Acta (BBA) - Gene Regulatory Mechanisms* 1829(12), 1300–1308. DOI: 10.1016/j.bbagrm.2013.10.005.

McGinnis, K.M. (2010) RNAi for functional genomics in plants. *Briefings in Functional Genomics* 9(2), 111–117. DOI: 10.1093/bfgp/elp052.

Melnyk, C.W., Molnar, A. and Baulcombe, D.C. (2011) Intercellular and systemic movement of RNA silenc-ing signals. *The EMBO Journal* 30(17), 3553–3563. DOI: 10.1038/emboj.2011.274.

Mitter, N., Worrall, E.A., Robinson, K.E., Li, P., Jain, R.G. *et al.* (2017) Clay nanosheets for topical delivery of RNAi for sustained protection against plant viruses. *Nature Plants* 3(2), 16207. DOI: 10.1038/nplants.2016.207.

Molesini, B., Rotino, G.L., Spena, A. and Pandolfini, T. (2009) Expression profile analysis of early fruit development in iaaM-parthenocarpic tomato plants. *BMC Research Notes* 2(1), 143. DOI: 10.1186/1756-0500-2-143.

Molesini, B., Mennella, G., Martini, F., Francese, G. and Pandolfini, T. (2015) Involvement of the putative N-acetylornithine deacetylase from *Arabidopsis thaliana* in flowering and fruit development. *Plant and Cell Physiology* 56(6), 1084–1096. DOI: 10.1093/pcp/pcv030.

Mounet, F., Moing, A., Kowalczyk, M., Rohrmann, J., Petit, J. *et al.* (2012) Down-regulation of a single aux-in efflux transport protein in tomato induces precocious fruit development. *Journal of Experimental Botany* 63(13), 4901–4917. DOI: 10.1093/jxb/ers167.

Nitsch, J.P. (1950) Growth and morphogenesis of the strawberry as related to auxin. *American Journal of Botany* 37(3), 211–215. DOI: 10.1002/j.1537-2197.1950.tb12183.x.

Numata, K., Ohtani, M., Yoshizumi, T., Demura, T. and Kodama, Y. (2014) Local gene silencing in plants via synthetic dsRNA and carrier peptide. *Plant Biotechnology Journal* 12(8), 1027–1034. DOI: 10.1111/pbi.12208.

Okabe, Y., Yamaoka, T., Ariizumi, T., Ushijima, K., Kojima, M. *et al.* (2019) Aberrant stamen development is associated with parthenocarpic fruit set through up-regulation of gibberellin biosynthesis in tomato. *Plant and Cell Physiology* 60(1), 38–51. DOI: 10.1093/pcp/pcy184.

Orzaez, D., Mirabel, S., Wieland, W.H. and Granell, A. (2006) Agroinjection of tomato fruits. A tool for rapid functional analysis of transgenes directly in fruit. *Plant Physiology* 140(1), 3–11. DOI: 10.1104/pp.105.068221.

Orzaez, D., Medina, A., Torre, S., Fernández-Moreno, J.P., Rambla, J.L. *et al.* (2009) A visual reporter system for virus-induced gene silencing in tomato fruit based on anthocyanin accumulation. *Plant Physiology* 150(3), 1122–1134. DOI: 10.1104/pp.109.139006.

Osorio, S., Alba, R., Nikoloski, Z., Kochevenko, A., Fernie, A.R. *et al.* (2012) Integrative comparative analyses of transcript and metabolite profiles from pepper and tomato ripening and development stages uncovers species-specific patterns of network regulatory behavior. *Plant Physiology* 159(4), 1713–1729. DOI: 10.1104/pp.112.199711.

Pandolfini, T., Molesini, B. and Spena, A. (2010) Parthenocarpy in crop plants. In: Ostergaard, L. (ed.) *Fruit Development and Seed Dispersal*. Wiley-Blackwell, Oxford, UK, pp. 326–345.

Pinosa, F., Begheldo, M., Pasternak, T., Zermiani, M., Paponov, I.A. *et al.* (2013) The *Arabidopsis thaliana* Mob1A gene is required for organ growth and correct tissue patterning of the root tip. *Annals of Botany* 112(9), 1803–1814. DOI: 10.1093/aob/mct235.

Sammons, R., Ivashuta, S., Liu, H., Wang, D., Feng, P. *et al.* (2011) Polynucleotide molecules for gene regulation in plants. *US Patent 2011/* 0296556, A1.

Sanzari, I., Leone, A. and Ambrosone, A. (2019) Nanotechnology in plant science: to make a long story short. *Frontiers in Bioengineering and Biotechnology* 7, 120. DOI: 10.3389/fbioe.2019.00120.

Takei, H., Shinozaki, Y., Yano, R., Kashojiya, S., Hernould, M. *et al.* (2019) Loss-of-function of a tomato receptor-like kinase impairs male fertility and induces parthenocarpic fruit set. *Frontiers in Plant Science* 10, 403. DOI: 10.3389/fpls.2019.00403.

Ueta, R., Abe, C., Watanabe, T., Sugano, S.S., Ishihara, R. *et al.* (2017) Rapid breeding of parthenocarpic tomato plants using CRISPR/Cas9. *Scientific Reports* 7(1), 507. DOI: 10.1038/s41598-017-00501-4.

Wang, C. and Fu, D. (2018) Virus-induced gene silencing of the eggplant chalcone synthase gene during fruit ripening modifies epidermal cells and gravitropism. *Journal of Agricultural and Food Chemistry* 66(11), 2623–2629. DOI: 10.1021/acs.jafc.7b05617.

Wang, H., Jones, B., Li, Z., Frasse, P., Delalande, C. *et al.* (2005) The tomato Aux/*IAA* transcription factor IAA9 is involved in fruit development and leaf morphogenesis. *The Plant Cell* 17(10), 2676–2692. DOI: 10.1105/tpc.105.033415.

Wang, M., Weiberg, A., Lin, F.M., Thomma, B.P.H.J., Huang, H.D. *et al.* (2016) Bidirectional cross-kingdom RNAi and fungal uptake of external RNAs confer plant protection. *Nature Plants* 2(10), 16151. DOI: 10.1038/nplants.2016.151.

Wang, M., Thomas, N. and Jin, H. (2017) Cross-kingdom RNA trafficking and environmental RNAi for powerful innovative pre- and post-harvest plant protection. *Current Opinion in Plant Biology* 38, 133–141. DOI: 10.1016/j.pbi.2017.05.003.

4 The 'Trojan Horse' Approach for Successful RNA Interference in Insects

Dimitrios Kontogiannatos*, Anna Kolliopoulou and Luc Swevers

Institute of Biosciences & Applications, National Centre for Scientific Research 'Demokritos', Aghia Paraskevi, Greece

Abstract

Since the discovery of RNA interference in 1998 as a potent molecular tool for the selective down-regulation of gene expression in almost all eukaryotes, increasing research is being performed in order to discover applications that are useful for the pharmaceutical and chemical industry. The ease of use of double-stranded RNA for targeted in vivo gene silencing in animal cells and tissues gave birth to a massive interest from industry in order to discover biotechnological applications for human health and plant protection. For insects, RNAi became the 'Holy Grail' of pesticide manufacturing, because this technology is a promising species-specific environmentally friendly approach to killing natural enemies of cultured plants and farmed animals. The general idea to use RNAi as a pest-control agent originated with the realization that dsRNAs that target developmentally or physiologically important insect genes can cause lethal phenotypes as a result of the specific gene downregulation. Most importantly to achieve this, dsRNA is not required to be constitutively expressed via a transgene in the targeted insect but it can be administrated orally after direct spraying on the infested plants. Similarly, dsRNAs can be administered to pests after constitutive expression as a hairpin in plants or bacteria via stable transgenesis. Ideally, this technology could have already been applied in integrated pest management (IPM) if improvements were not essential in order to achieve higher insecticidal effects. There are many limitations that decrease RNAi efficiency in insects, which arise from the biochemical nature of the insect gut as well as from deficiencies in the RNAi core machinery, a common phenomenon mostly observed in lepidopteran species. To overcome these obstacles, new technologies should be assessed to ascertain that the dsRNA will be transferred intact, stable and in high amounts to the targeted insect cells. In this chapter we will review a wide range of recent discoveries that address the delivery issues of dsRNAs in insect cells, with a focus on the most prominent and efficient technologies. We will also review the upcoming and novel use of viral molecular components for the successful and efficient delivery of dsRNA to the insect cell.

4.1 Introduction

In 1998 researchers first discovered that double-stranded RNA (dsRNA), instead of antisense RNA, was substantially more potent at producing RNA interference (RNAi) (Fire *et al.*, 1998). These researchers showed that injection of *Caenorhabditis elegans* adults with purified antisense or sense

*Corresponding author: dim_kontogiannatos@yahoo.gr

© CAB International 2021. *RNAi for Plant Improvement and Protection*
(eds B. Mezzetti *et al.*)
DOI: 10.1079/9781789248890.0004

RNAs targeting a wide range of selected genes had, at most, a mild effect, while on the contrary dsRNA mixtures acted with stronger efficiency and specificity against their targets (Fire et al., 1998). Only low doses of injected dsRNA were sufficient per affected cell, thus depicting that stoichiometric interference with endogenous mRNA was not necessary and suggesting the involvement of a component that catalyses or amplifies the silencing effect (Fire et al., 1998).

Follow-up studies showed that a specific nuclease activity is associated with dsRNA-mediated RNAi in Drosophila melanogaster S2 cells and that this activity is responsible for the degradation of endogenous transcripts homologous to the transfected dsRNA (Hammond et al., 2000). The nuclease was speculated to contain an essential RNA component after the discovery of small RNA species that could act as specificity agents through homology to the substrate mRNAs (Hammond et al., 2000).

Nowadays we know that RNAi is initiated by ribonucleases that generate small interfering RNAs (siRNAs) from long dsRNAs and mature microRNAs (miRNAs) from primary transcripts (Hammond, 2005). This is accomplished by the action of two RNase III enzymes: Dicer (Fig. 4.1) and Drosha. Class III RNAse enzymes contain two RNase III catalytic domains, a helicase domain and a Piwi/Argonaute/Zwille (PAZ) domain. This last domain is also present in Argonaute family proteins that are essential in later steps of RNAi (Hammond, 2005), while especially in Dicer it is responsible for the recognition of the dsRNA substrate (Lau et al., 2012). Dicer cleaves the substrate at ~22 nucleotides (nt) from the open helicoid end (Lau et al., 2012). In the Dicer enzyme of the protozoan Giardia lamblia a 'platform' domain has been observed that separates the PAZ domain from the RNase III catalytic site, thereby providing structural insights for the production of small RNAs of 25–27 nt in length (Lau et al., 2012). Other eukaryotic

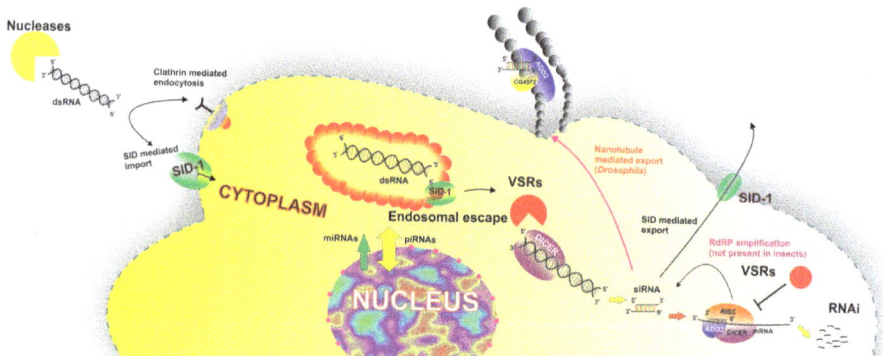

Fig. 4.1. The RNAi 'decathlon'. Insects possess all three RNAi (miRNA, siRNA and piRNA) machineries. Exogenously applied dsRNAs must overcome a series of cellular barriers in order to be processed by the RNAi core machinery. In insects, extracellular and intracellular nucleases degrade dsRNAs before and after entering cells. Intracellular uptake is being promoted by SID-1-like transporters and clathrin-mediated endocytosis. Both pathways might act in parallel in some insects, e.g. *Leptinotarsa decemlineata*. When taken up by endocytosis, dsRNA must undergo endosomal escape in order to interact with the RNAi core machinery. SID-1-like proteins may be responsible for endosomal escape of dsRNAs in some insects. In lepidopteran species, dsRNA may accumulate in endosomes because of inefficiency of endosomal escape. In the cytoplasm, the dsRNA is being processed by Dicer protein to 20–22 nt siRNAs. The complementary siRNA strand is then introduced into the RISC complex and mRNA degradation is initiated. In plants and worms, but not insects, initial RNAi triggers can be amplified by RdRP proteins. Viral suppressors of RNAi (VSRs) may antagonize Dicer and Ago proteins either by protein–protein interactions or dsRNA/siRNA sequestration as an antiviral defence mechanism. Extracellular transportation of siRNAs may be mediated by SID-1-like proteins and intercellular transport can be carried out by nanotubes (in *Drosophila*).

Dicers use a similar mechanism of molecular ruler although their products are 4 nt shorter (Lau *et al.*, 2012). Small RNA products of Dicer are incorporated into large multiprotein processing complexes termed RNA-induced silencing complexes (RISCs) (Fig. 4.1) (Lau *et al.*, 2012). RISC selects one strand of the small RNA duplex, such as miRNA or siRNA (Filipowicz *et al.*, 2008). The single strand acts as a guide for RISC to recognize complementary sequences in mRNAs (Fig. 4.1). One of the proteins in RISC, called Argonaute (Ago), exhibits slicer activity and cleaves the mRNA. A complex that consists of an Ago and a single guide strand is referred to as 'mature RISC' or simply 'RISC', while the same complex is also called 'RISC core' in the context of considering RISC as a huge complex that includes many other components to achieve silencing, e.g. by translational repression and de-adenylation (Fig. 4.1) (Nakanishi, 2016).

In insects, three main small RNA-based silencing pathways are observed: the miRNA, siRNA and PIWI-interacting RNA (piRNA) pathways (Fig. 4.1) (Mongelli and Saleh, 2016). Although all three pathways use small RNAs (from 18 to 33 nt) to guide the sequence-specific recognition of target sequences by an Ago effector protein, the small RNAs in each pathway differ according to their biogenesis, the nature and fate of their targets, and their biological function (Mongelli and Saleh, 2016). Despite the fact that the main RNAi pathways' operation strategies are highly conserved among several organisms, they can involve different proteins and operate via different mechanisms (Terenius *et al.*, 2011). The primary example is the amplification of the RNAi effect in nematodes, plants and fungi through the action of a cellular RNA-dependent RNA polymerase (RdRP) that generates target gene-derived secondary siRNAs (Terenius *et al.*, 2011). It is highly probable that RDRP is responsible for the robust effect of dsRNA-mediated RNAi in these organisms. Homologues of cellular RdRPs do not exist in insect genomes, although they have been identified in genomes of basal arthropods such as ticks (Fig. 4.1) (Terenius *et al.*, 2011).

RNAi efficiency is relatively low in lepidopteran insects compared with many other insect species (Terenius *et al.*, 2011; Guan *et al.*, 2018). dsRNA degradation as well as inefficient cellular uptake and transport seem to be crucial main factors that determine the various levels of RNAi efficiency among insects (Guan *et al.*, 2018). Previous research demonstrated that dsRNA may remain stable for much longer periods after uptake in many species of Coleoptera compared with most lepidopteran species (Terenius *et al.*, 2011; Shukla *et al.*, 2016; Guan *et al.*, 2018). New studies demonstrated that Lepidoptera contain a specific nuclease (REase) which is responsible for the digestion of dsRNA before its processing by Dicer and therefore can negatively affect the RNAi efficiency (Guan *et al.*, 2018). In addition, degradation of dsRNA in the lumen of the gut and in the haemocoel is considered to be an important factor responsible for differences in the efficacy of RNAi between coleopteran and lepidopteran species (Shukla *et al.*, 2016). Shukla *et al.* (2016) demonstrated that intracellular transport of dsRNA can be a major factor affecting the differential efficacy of RNAi that is observed between a lepidopteran (*Heliothis virescens*) and a coleopteran (*Leptinotarsa decemlineata*) species. Moreover, in sharp contrast to coleopteran species, it was observed that Lepidoptera do not efficiently process plant-originated long dsRNAs to 21 bp siRNAs (Ivashuta *et al.*, 2015).

Numerous molecular and physiological processes may be responsible for the insufficient response of RNAi in particular insect orders such as Lepidoptera. Thus, the variability of RNAi in insects is a phenomenon that has to be addressed in order for this technique to become a widely valuable tool in efficient pest control strategies. In this chapter, we will focus on the most important causes of RNAi deficiency and will review scientific and technical methodologies to overcome them.

4.2   dsRNA Uptake in Insects: Molecular Mechanisms and Endosomal Escape

dsRNAs can penetrate the insect's cells via several routes (Fig. 4.1). It has been demonstrated that two inhibitors (chlorpromazine and bafilomycin-A1) of clathrin-dependent endocytosis (Fig. 4.1) can nearly abolish or significantly diminish RNAi of the *Lethal giant larvae* (*TcLgl*) gene in *Tribolium castaneum* (Coleoptera) whereas methyl-β-cyclodextrin and cytochalasin-D, substances that are known to have inhibitory action on other endocytic pathways, showed no effect (Xiao *et al.*, 2015).

In addition to clathrin-mediated endocytosis, transport by the SID-1-like transmembrane channel is also considered as a major pathway for dsRNA uptake in insects (Fig. 4.1) (Cappelle *et al.*, 2016). SID-1-like genes have been identified in the genomes of many insects, with the notable exception of Diptera. While three genes similar to *sid-1* were identified in the genome of *T. castaneum*, none of these genes seem to be indispensable for systemic RNAi in this species (Yoon *et al.*, 2017). On the other hand, two recently identified *sid-1*-like genes in the Colorado potato beetle *Leptinotarsa decemlineata* are necessary for an efficient RNAi response in the *L. decemlineata* cell line Lepd-SL1 (Yoon *et al.*, 2017). In *L. decemlineata*, therefore, both endocytosis and SID-1-like pathways are involved in dsRNA uptake (Cappelle *et al.*, 2016).

Studies in lepidopteran Sf9 cells also suggest that the clathrin-mediated pathway through endosomes is used as the major route for transport of dsRNA into and within these cells (Yoon *et al.*, 2017). However, despite efficient uptake of dsRNA, no silencing effects are observed, which seems to be caused by the ability of dsRNA to escape from the endosomes. Overexpression of *Caenorhabditis elegans* SID-1 improved RNAi efficiency in Sf9 and *Bombyx mori* cells, which could be related to a stimulation of endosomal escape by dsRNA. On the other hand, overexpression of the SID-1 homologue of the migratory locust was not effective (Yoon *et al.*, 2017), indicating that not all SID-1-like homologues are involved in dsRNA transport. Interestingly, in mammals, SIDT1 and SIDT2, closely related members of the SID-1 transmembrane family, are required to transport internalized dsRNA molecules from endosomes to the cytosol to activate the innate immune response (Nguyen *et al.*, 2019). The role for SID-1-like transmembrane channels in the regulation of endosomal escape of dsRNA molecules needs to be further investigated in the future.

In dipteran insects, fatty acid biosynthesis and metabolism may play important roles in the regulation of RNAi efficiency (Dong *et al.*, 2017). Prior exposure to dsRNA ('dsRNA priming') in the fly *Bactrocera dorsalis* resulted in changes in the ratio between linoleic acid (LA) to arachidonic acid (AA) in the haemolymph and inhibition of endocytosis of dsRNA into the midgut cells. Interestingly, injection of AA resulted in an increase in the uptake of ingested dsRNA in *Drosophila melanogaster* and a facilitation of RNAi effects (Dong *et al.*, 2017).

4.3 Physiological and Cellular Mechanisms that Affect RNAi Efficiency

The cotton boll weevil (*Anthonomus grandis*) is a coleopteran insect for which reports on RNAi-mediated gene silencing showed that it does not function efficiently when dsRNA feeding is used (Almeida Garcia *et al.*, 2017). Three nucleases of the DNA/RNA non-specific endonuclease family were identified in the cotton boll weevil transcriptome (AgraNuc1, AgraNuc2, AgraNuc3) and were found to be mainly expressed in the posterior midgut region of the insect (Almeida Garcia *et al.*, 2017). Gene silencing of *AgraNuc1-2-3* showed that *A. grandis* midgut nucleases are one of the main barriers to dsRNA delivery (Almeida Garcia *et al.*, 2017).

Consistent with the above result, a major observation is that insects of different orders express different levels of dsRNA-degrading enzymes in both haemolymph and midgut tissues (Wang *et al.*, 2016). In comparative RNAi studies among species that belonged to four different insect orders, the cockroach *Periplaneta americana* exhibited the best silencing response followed by the coleopteran *Zophobas atratus*, the orthopteran *Locusta migratoria* and the lepidopteran *Spodoptera litura* (Wang *et al.*, 2016). This variability in RNAi response was correlated with the enzymatic degradation of dsRNA, which functions as a key factor that determines the effective dosage duration of inner target exposure (Wang *et al.*, 2016).

Expression of the core RNAi machinery can vary among different developmental stages of insects (Guo *et al.*, 2015). Developmental and growth defects associated with the silencing of the *S-adenosyl-L-homocysteine hydrolase* (*LdSAHase*) gene of *L. decemlineata* occurred with different levels of penetrance depending on the stage of the larva at which dsRNA was administered (Guo *et al.*, 2015). In young larvae the expression levels of *LdDcr2a*, *LdDcr2b*, *LdAgo2a* and *LdAgo2b* (encoding Dicer-2 and Argonaute-2 proteins in the siRNA pathway, respectively) were higher, which affected RNAi efficiency in *L. decemlineata* (Guo *et al.*, 2015).

In some organisms, mRNA suppression by dsRNA is observed in many tissues throughout the body as the RNAi signal spreads between tissues, with this non-cell-autonomous RNAi response being referred to as systemic RNAi (Fig. 4.1) (Cooper *et al.*, 2018). In the nematode model, *C. elegans*, siRNAs generated by Dicer are transported from cell to cell and their abundance is amplified by RdRP (Cooper *et al.*, 2018). Both the recipient and donor cells must possess SID-1 channels so that cell-to-cell transport can occur (Fig. 4.1) (Cooper *et al.*, 2018). However, vesicle transport as well as endocytosis are also involved in the systemic RNAi response (discussed above). A similar mechanism as in nematodes was hypothesized to exist in insects; however, all insect genomes lack RdRP genes and some, like dipterans, also lack SID-1 homologs (Cooper *et al.*, 2018). In *D. melanogaster*, it was reported that nanotube-like structures can establish a systemic RNAi response that functions as an antiviral mechanism in different cell types (Karlikow *et al.*, 2016). The nanotubules are composed of actin and tubulin, and associate with components of the RNAi machinery, including Ago-2, dsRNA and CG4572 (Fig. 4.1) (Karlikow *et al.*, 2016).

4.4   Viral RNAi Suppressors

RNAi is a mechanism that is necessary for antiviral defence in insects, including vectors of human viral diseases such as mosquitoes. Viruses have evolved to escape this antiviral defence system by encoding suppressors of RNAi that function as obstacles for the elimination of viral RNAs, thus contributing to efficient viral replication (Fareh *et al.*, 2018). It was shown that viral suppressors of RNAi (VSRs) from *Drosophila* RNA viruses antagonize Dcr-2 enzyme by safeguarding viral RNA molecules (Fareh *et al.*, 2018). VSR proteins such as VP3 of *Drosophila* X virus and *Culex* Y virus (both of genus *Entomobirnavirus*, family *Birnaviridae*) and 1A protein of *Drosophila* C virus (genus *Cripavirus*, family *Dicistroviridae*) directly bind to dsRNA molecules and prevent the recognition by Dcr-2 in an irreversible manner (Fig. 4.1) (Fareh *et al.*, 2018).

RNAi suppressors can demonstrate host-specific activities (van Mierlo *et al.*, 2014). VP1 of *D. melanogaster* Nora virus (DmelNV) suppresses Ago-2-mediated target RNA cleavage to antagonize antiviral RNAi (Fig. 4.1) (van Mierlo *et al.*, 2014). Ago-2 antagonists of divergent Nora-like viruses in natural populations of *D. immigrans* (DimmNV) and *D. subobscura* (DsubNV), however, cannot suppress RNAi in *D. melanogaster* S2 cells. RNAi suppressor activity of DimmNV VP1 is restricted to its natural host species, *D. immigrans* (van Mierlo *et al.*, 2014). While *DimmNV* VP1 interacts with *D. immigrans* Ago-2 by suppressing slicer activity in embryo lysates from the same species, it does not interact with *D. melanogaster* Ago-2, thus presenting no suppressive effect in this species lysates (van Mierlo *et al.*, 2014).

The possible role of RNA virus infections in inhibiting RNAi in lepidopteran insects has been investigated (Swevers *et al.*, 2016). Several lepidopteran cell lines were found to be persistently infected by the RNA viruses Flock house virus (FHV; *Nodaviridae*) and Macula-like virus (MLV; related to plant viruses of the family *Tymoviridae*) without any apparent pathogenic effects. RNAi reporter assays failed to detect a significant interference with gene silencing in Sf21 and Hi5-SF cells that were persistently infected, when compared with virus-free cells. In Hi5 cells, FHV could be easily eliminated through the expression of an RNA hairpin specific for its VSR gene, confirming that the RNAi mechanism was not inhibited (Swevers *et al.*, 2016). Despite the above-mentioned results, functional tests indicated that the B2 gene of FHV coding for an RNAi inhibitor exhibited RNAi suppressor activity, indicating that protection against RNAi is essential for virus survival (Swevers *et al.*, 2016). In another study using lepidopteran cell lines, overexpression of Dcr-2 and Ago-2 could delay the progression of pathogenic infection by Cricket paralysis virus (*Dicistroviridae*) while knockdown of these RNAi factors resulted in an increase in the levels of persistent infections of FHV and MLV (Santos *et al.*, 2018). The impact of persistent virus infections on the performance of the RNAi machinery requires further study, because reports have revealed the ubiquitous presence of viruses in many insects after the applications of high-throughput sequencing techniques (Bolling *et al.*, 2015).

Also, DNA viruses can express VSR proteins. Baculovirus (*Autographa californica* multiple nucleopolyhedrovirus, AcMNPV) infection induces an RNAi response in *Spodoptera frugiperda* cells, as documented by the detection of a large number of viral siRNAs (Mehrabadi *et al.*, 2015). The p35 gene in the AcMNPV genome, an established inhibitor of apoptosis, was also found to have VSR activity when tested in RNAi reporter assays that employed diverse insect and mammalian cell lines. VSR activity of p35 was not due to the inhibition of dsRNA cleavage by Dicer-2, but because of a downstream action in the RNAi pathway (Mehrabadi *et al.*, 2015).

4.5 Improvement of RNAi

RNAi can be improved by identifying methodologies that overcome the biochemical, molecular and physical boundaries imposed by insect cells. Many technologies have been developed in order to confront these limitations by focusing on the improvement of dsRNA stability and penetrative ability in insect cells.

4.5.1 Nanoparticle-mediated dsRNA encapsulation

The cationic polymer chitosan is able to form stable nanoparticles with anionic nucleic acids (dsRNAs) via electrostatic interactions (Fig. 4.2) that can be observed by atomic force microscopy (Ramesh Kumar *et al.*, 2016). Chitosan/dsRNA-mediated knockdown of a reporter gene was first demonstrated in the lepidopteran Sf21 insect cell line (Ramesh Kumar *et al.*, 2016). In subsequent studies, chitosan/dsRNA nanoparticles targeting the *vestigial* gene in the mosquito *Aedes egypti* were able to cause significant mortality, adult wing-malformation and delayed growth development (Ramesh Kumar *et al.*, 2016). Moreover, a comparative study of nanoparticles that complexed dsRNA with chitosan, carbon quantum dot (CQD) or silica showed that CQD was the most efficient carrier for dsRNA retention, delivery and concomitant gene silencing and mortality in *Ae. aegypti* (Fig. 4.2) (Das *et al.*, 2015). Aerosolization of siRNA–nanoparticle complexes was described and used as a delivery method in three aphid species (*Acyrthosiphon pisum*, *Aphis glycines* and *Schizaphis graminum*) to target genes involved in pigmentation and amino acid metabolism and it was

concluded that the nanoparticle emulsion significantly increased the efficacy of gene knockdown (Thairu *et al.*, 2017).

Particle replication in non-wetting templates (PRINT) technology (Fig. 4.2) has been investigated to be used as an alternative dsRNA-carrying technology for mosquito control (Phanse *et al.*, 2015). Phanse *et al.* (2015) fabricated fluorescently labelled polyethylene glycol-based nano-complexes of specific sizes, shapes and charges and evaluated their properties both *in vitro* in mosquito cell culture and *in vivo* in *Anopheles gambiae* larvae following injection and feeding. Following direct administration into the larval body, the bio-distribution of positively and negatively charged PRINT nanoparticles of each size and shape was similar and accumulation was mainly observed in the thoracic and abdominal regions of the larvae. Positively charged nanoparticles were more likely to be associated with the gastric caeca in the gastrointestinal tract. Negatively charged nanoparticles could have been persisting through metamorphosis and were localized in adult insect organs such as head, body and ovaries (Phanse *et al.*, 2015). During *in vitro* experiments, positively charged nanoparticles were more efficiently internalized into the cells and trafficked to the cytosol, while negatively charged nanoparticles accumulated in lysosomes (Phanse *et al.*, 2015). No cytotoxic effects were observed for any of the tested nanoparticles (Phanse *et al.*, 2015). The authors finally concluded that the excellent low cell and larval toxicity profiles, efficient internalization and widespread bio-distribution of PRINT nanoparticles rendered them as attractive candidates for dsRNA delivery in mosquitoes.

Liposomes have also been examined as a potential dsRNA delivery system (Fig. 4.2) in the German cockroach (*Blattela germanica*). Injection of non-complexed dsRNA into the abdomen of *B. germanica* caused dramatic depletion of the essential *α-tubulin* gene and associated mortality (Lin *et al.*, 2016). In contrast, when the naked dsRNA was orally delivered, lower RNAi efficiency was observed, which was accounted for by the rapid degradation of the dsRNA in the midgut of *B. germanica* (Lin *et al.*, 2016). On the other hand, continuous ingestion of dsRNA-containing lipoplexes was potent with respect to slowing down the degradation of dsRNA in the midgut and to increasing the mortality of the German cockroach

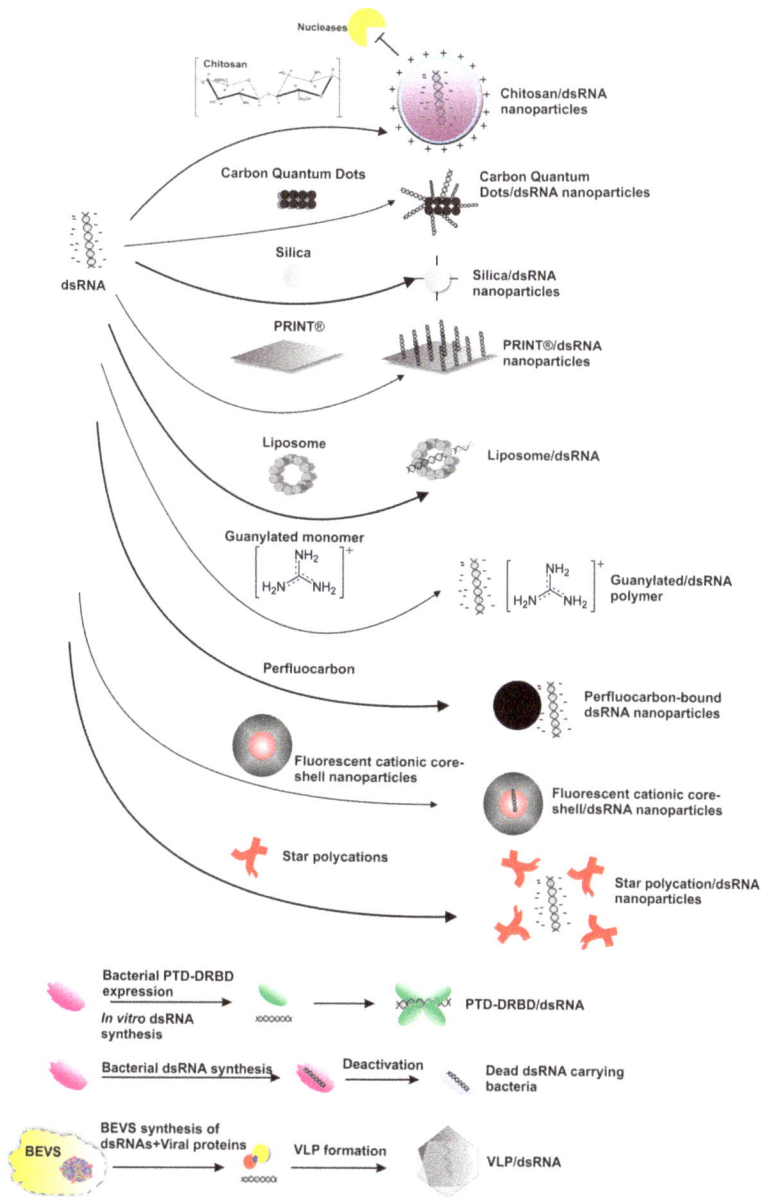

Fig. 4.2. Molecular 'Trojan horses' of dsRNA delivery. Chitosan, carbon quantum dots, silica, PRINT®-based, perfluocarbon-bound, fluorescent cationic core-shell and star polycation nanoparticles as well as liposomes, guanylated polymers and PTD–DRBD peptides have been used for efficient dsRNA delivery in insects. Mechanisms include stabilization and protection of dsRNAs from nuclease degradation and enhancement of dsRNA uptake and endosomal escape. Bacteria-mediated dsRNA delivery is a low-cost method for RNAi delivery but improvements can be made in efficiency of dsRNA synthesis. VLPs loaded with dsRNAs is an alternative methodology to create potent insecticidal dsRNAs. VLP components and dsRNAs could be co-expressed in biotechnological platforms such as the baculovirus expression vector system (BEVS) for efficient packaging.

following inhibition of α-tubulin expression in the midgut (Lin *et al.*, 2016).

In a recent work (Christiaens *et al.*, 2018) researchers used guanidine-containing polymers to protect dsRNA against degradation in *Spodoptera exigua*, which has a very alkaline gut environment while it is also characterized by a strong intestinal nucleic acid-degrading activity. In this research, it was shown that polymers with high guanidine content (Fig. 4.2) proved to be highly protective against dsRNA degradation at pH 11, while the shielding effect lasted for up to 30 h (Christiaens *et al.*, 2018). Moreover, the use of this polymer enhanced the cellular uptake in the lepidopteran CF203 midgut cells. Additionally, a synthetic cationic polymer, poly-[N-(3-guanidinopropyl)methacrylamide] (pGPMA), which mimics arginine-rich cell penetrating peptides, was found to be efficiently taken up by Sf9 cells and to drive highly efficient gene knockdown and moderate larval mortality in *Spodoptera frugiperda* (Parsons *et al.*, 2018).

As well as the above-mentioned experimental cases, a wide range of nanoparticles of different compositions have been fabricated resulting in improved insecticidal properties. For example, perfluocarbon-bound siRNA nanoparticles have been administered by aerosol to aphids (Thairu *et al.*, 2017) in order to stabilize the RNA trigger and deliver it to internal organs via the tracheoles. Using this technology, the aerosolized perfluocarbon-bound siRNA nanoparticles were transferred through tracheoles to the gut and to the haemolymph of aphids. Aerosolization of naked RNAs improved RNAi efficiency, which was even more profound using the perfluocarbon-bound siRNA nanoparticles (Thairu *et al.*, 2017). Moreover, fluorescent cationic core-shell nanoparticles, which consisted of a fluorescent core of perylene-3,4,9,10-tetracarboxydiimide chromophore (PDI) in the centre and polymer shells terminating with multiple amino groups, efficiently entered into live cells presenting low cytotoxicity as well as high gene delivery efficacy (He *et al.*, 2013). Using this technology, researchers have efficiently silenced *CHT10*, a midgut-specific chitinase gene expressed in the gut peritrophic membrane of the Asian corn borer, *Ostrinia furnacalis* (Lepidoptera: Crambidae), leading to severe defects in larval growth and consequently to death (He *et al.*, 2013).

Manufacturing of dsRNA-bound nanoparticles should be cost effective and convenient for industrialization. Recently, researchers have developed a star polycation (SPc) as an efficient but low-cost gene carrier for pest management (Li *et al.*, 2019). As emphasized by the authors, the chemical sources of SPc are cheap and easily available. The chemical structure of SPc, containing four arms in one core with a compact tertiary amine, confers high gene transfection efficiency. The nanoparticles can deliver dsRNAs to knock down insect gene expression and inhibit pest growth (Li *et al.*, 2019).

4.5.2 Bacterial delivery

Bacterial dsRNA administration (Fig. 4.2) was pioneered by Timmons and Fire (1998) who showed that ingestion of bacterially expressed dsRNAs could be effective in the production of specific and potent genetic interference in *C. elegans*. This approach uses an RNase III-deficient *Escherichia coli* strain known as HT115 (DE3) (Timmons and Fire, 1998; Kourti *et al.*, 2017). Following this methodology, cloning of the gene of interest takes place between two T7 promoters on the special RNAi plasmid L4440. For the transformation, HT115 cells are used and dsRNA is produced upon induction of T7 RNA polymerase. Following induction of dsRNA production, the cells are introduced in the worm's growth medium and RNAi happens after a short period of incubation. In a similar way in insects, dsRNA-producing bacteria are incorporated in their artificial diets or are sprayed on plant organs that are the food for the insects, while again RNAi occurs following a period of continuous feeding. Successful bacteria-mediated RNAi has been reported in many insect species (Tian *et al.*, 2009; Zhu *et al.*, 2010; Kontogiannatos *et al.*, 2013; Zhang *et al.*, 2013; Li *et al.*, 2014) with various results, mostly reflecting efficiency issues. In *S. exigua* a high dosage of dsRNA is required to efficiently kill late-instar stages because of high activity of RNases in the midgut lumen (Vatanparast and Kim, 2017). It was observed that sonication of bacterial cells before oral administration minimizes dsRNA release and causes higher larval mortality (Vatanparast and Kim, 2017). Moreover, targeting of young larvae that possessed weak RNase activity in the midgut lumen led to significant enhancement of RNAi efficiency and insecticidal activity against *S. exigua* (Vatanparast and Kim, 2017).

A prominent approach for continuous dsRNA delivery via bacterial expression was published by Whitten *et al.* (2016). In this work, researchers genetically modified symbiotic bacteria of the blood-sucking bug *Rhodnius prolixus* and the western flower thrips *Frankliniella occidentalis* in order to constitutively express dsRNAs. When the modified bacteria were ingested, they colonized the insects, while they also successfully competed with the wild-type microflora and sustainably mediated systemic knockdown phenotypes that could be horizontally transmitted (Whitten *et al.*, 2016).

4.5.3 Ribonucleoprotein delivery

siRNA-based therapeutics are receiving much attention because of their promising impact in human cancer therapy. As happens in insects, the siRNA size and anionic charge are limiting factors that do not facilitate efficient penetration of the dsRNAs into mammalian cells. An efficient siRNA delivery approach was reported for the first time by Eguchi *et al.* (2009), where a peptide transduction domain–dsRNA-binding domain (PTD–DRBD) fusion protein was used. In this report it was shown that DRBDs bind to siRNAs strongly, thus bypassing the siRNAs' negative charge and allowing PTD-mediated cellular uptake. The RNAi response was quickly induced by PTD–DRBD-delivered siRNA in many types of primary and transformed cells (Eguchi *et al.*, 2009).

The use of PTD–DRBD peptides to improve insect RNAi (Fig. 4.2) was investigated in the coleopteran *Anthonomus grandis* almost 8 years after the discovery of PTD–DRBDs (Gillet *et al.*, 2017). As could reasonably be expected, the chimeric PTD–DRBD protein combined with dsRNA formed a ribonucleoprotein particle that improved the effectiveness of the RNAi mechanism in this insect. The same authors reported that the complex slows down nuclease activity in the gut of *A. grandis* and that PTD-mediated internalization in insect gut cells is achieved within minutes after plasma-membrane contact, limiting the exposure time to gut nucleases.

Most importantly, the efficiency of insect gene silencing upon oral delivery presented an approximately twofold increase when PTD–DRBDs were used as a delivery method compared with naked dsRNA (Gillet *et al.*, 2017).

4.6 Viral Components as 'Trojan Horses' of Insect RNAi

4.6.1 dsRNA viruses

dsRNA viruses comprise a diverse group that infect a wide range of hosts from animal, plant, fungal and bacterial kingdoms. Their genome is organized in segments numbered from 1 to 12 and their virions are all non-enveloped and possess icosahedral capsids that are differentiated by T-number, capsid layer and turret forms. The dsRNA viruses are segregated into 12 families: *Amalgaviridae, Birnaviridae, Chrysoviridae, Cystoviridae, Endornaviridae, Hypoviridae, Megabirnaviridae, Partitiviridae, Picobirnaviridae, Quadriviridae, Reoviridae* and *Totiviridae*. Of these, *Reoviridae* is the largest and most diverse family with respect to host range, with the most important members in this group being rotaviruses that cause gastroenteritis in young children and bluetongue virus, an economically important pathogen of cattle and sheep that is transmitted by mosquitoes (Louten and Reynolds, 2016).

In contrast to DNA viruses, RNA viruses typically do not penetrate an infected cell nucleus. Since they do not form a DNA intermediate, they do not need any of the host enzymes to replicate their RNA genome. However, RNA viruses still need to transcribe their mRNAs to allow host ribosomes to translate viral proteins and to form new virions. Because cells do not contain the enzymes required to transcribe mRNA from an RNA template, all RNA viruses therefore must carry and encode their own RdRP enzyme to transcribe viral mRNA. dsRNA viruses therefore typically code for and contain an RdRP that is carried into the cell within the virion (Louten and Reynolds, 2016).

dsRNA replication occurs in the cytoplasm for all dsRNA viruses that have been investigated. Transcription, known as the synthesis of a dsRNA template's viral positive strands, occurs in viral particles or core particles. Usually, the new positive strands are extruded from the viral particles and translated into viral proteins. The same positive strands are then packed to produce new particles or subviral particles. Negative strand synthesis on the positive strand template (replication) completes the creation of new dsRNA once the new particles or cores have been formed. For dsRNA viruses with more complicated structures, addition of new layers of protein and/or membrane completes the virus reproduction cycle (Wickner, 1993).

4.6.2   The dsRNA virus replication machinery

dsRNA viruses have evolved sophisticated mechanisms for cellular entry. Because they are carriers of dsRNA molecules (genome segments), it is interesting to analyse their life cycle in some detail and get an idea of how the dsRNA fragments are replicated and shielded from degradation or interaction with the RNAi machinery. Two dsRNA viruses with relatively well-characterized life cycles are discussed.

Bluetongue virus (BTV) (genus: *Orbivirus*; family: *Reoviridae*) is an arthropod-borne virus (arbovirus) that is transmitted by midges (*Culicoides* sp.). The typical dsRNA virus replication machinery resembles that of the BTV. As described by Lourenco and Roy (2011) and Sung *et al.* (2019), the BTV particle has two capsids, an outer capsid and an inner capsid, the latter of which is also called the core. The outer capsid contains proteins VP2 and VP5 to facilitate virus entry through the cellular membrane and the release of the core into the cytoplasm. The core particles do not further disassemble and are capable, using the encapsidated dsRNA genome segments as template, of producing positive strand RNA molecules that are transported to the cytoplasm through channels in the core particles. The icosahedral-shaped core principally comprises two proteins, VP7 and VP3, which are arranged in two layers. The VP3 layer encloses the viral genome of ten dsRNA segments (S1–S10). In addition, the core contains three minor proteins: the polymerase (VP1), the capping enzyme (VP4) and VP6, an essential structural protein of 36 kDa with RNA and ATP

binding activity. VP6 is unique in the *Orbivirus* genus within the *Reoviridae* family. Upon entry, core particles become transcriptionally active, producing and extruding single-stranded positive sense RNAs (ssRNA) through the local channels at the fivefold axis, without further disassembly. These ssRNAs then act as mRNAs for viral protein synthesis and as templates for genomic RNA synthesis. The ten newly synthesized ssRNA segments are first combined via specific intersegment RNA–RNA interactions to form RNA complexes of all ten segments and then packaged together with VP1, VP4 and VP6 into the assembling VP3 capsid layer. Genomic dsRNA molecules are subsequently synthesized within this assembled particle (known as the 'subcore'), prior to encapsidation by the VP7 layer, leading to robust core particle formation (Lourenco and Roy, 2011; Sung *et al.*, 2019).

Cypoviruses (cytoplasmic polyhedrosis viruses (CPVs); genus: *Cypovirus*; family: *Reoviridae*) are widespread pathogens of insects. The type species is Cypovirus 1, which specifically infects silkworms (*Bombyx mori*) and negatively affects the sericulture industry (Cao *et al.*, 2012; He *et al.*, 2017; Zhao *et al.*, 2019). In contrast to all other reoviruses, CPV virions consist of one layer of capsid that corresponds to the core particle of other reoviruses (discussed above for BTV). A distinctive feature of CPVs, which is shared by the DNA viruses, baculoviruses, is the production of polyhedra or occlusion bodies that protect encapsulated virions against damage and enhance viral survival in the environment (and actually can be regarded as a replacement for the outer capsid layer in other reoviruses). After feeding, CPV polyhedra are lysed because of the high pH in the lepidopteran midgut, which results in the infection of the midgut epithelium by the released virions. Interestingly, electron microscope images show that CPV virions can directly penetrate the plasma membrane of the microvilli that are localized at the apical sides of the enterocytes during midgut infection of silkworm larvae (Cao *et al.*, 2012; He *et al.*, 2017; Zhao *et al.*, 2019). Functional studies also show the involvement of clathrin-mediated endocytosis for uptake of CPV virions in silkworm-derived BmN cells and the midgut epithelium (Tan *et al.*, 2003). In the cytoplasm, CPV virions undergo activation and become capable of RNA transcription using similar mechanisms as described for core particles of BTV (see previous section). Besides an RdRP, CPV virions contain additional processing activities that form a cap-like structure, mGpppAmpGp, to protect viral mRNAs from degradation by exonuclease enzymes. All of the dsRNA segments in the genome of Cypovirus 1 contain the conserved sequence (GUUAA......GUUAGCC) at their ends which likely function as recognition signals for the RdRP complex to initiate transcription or replication but may also have a role in the binding of the mRNAs to the ribosomes or the interaction with viral structural proteins (Tan *et al.*, 2003; Cao *et al.*, 2012; He *et al.*, 2017; Chen *et al.*, 2018).

4.6.3 The use of unique viral components as molecular tools to achieve improved RNAi efficiency in insects

The unique mechanism of dsRNA virus replication is an evolutionary conserved molecular adaptation that aims to protect viral dsRNA genomes from the hostile cellular environment of their hosts. For efficient infection, dsRNA viruses must be able to: (i) penetrate efficiently the host cell's cytoplasmic membrane; (ii) protect their dsRNA genome from RNA degradation and the RNAi response; (iii) translate the non-structural and structural proteins encoded by their genome; and (iv) multiply and construct new virions. In order to use RNAi as a potent tool for efficient pest control, applied dsRNAs must have two characteristics in common with those found in dsRNA virus infections: (i) penetration efficiency; and (ii) RNA degradation resistance. The question then arises as to whether one could mimic the molecular components of viral infection in order to improve RNAi in insects.

In the pharmaceutical industry, viruses provide an ideal basis for the development of targeted drug delivery vehicles (Yildiz *et al.*, 2011). Interest in the exploitation of virus-based nanoparticles (VNPs) and virus-like particles (VLPs) has united efforts among researchers in different fields such as biology, chemistry, engineering and medicine (Fig. 4.2). VLPs are the genome-free counterparts of virions and are valuable because of their biocompatibility and biodegradability. Plant and bacterial VLPs have the additional advantage of being non-infectious and non-hazardous in humans and other mammals (Yildiz *et al.*, 2011). VNPs are well-characterized, monodisperse structures that can be produced in large quantities, which also enables solving their structures at atomic resolution. VNPs have highly symmetrical structures and can be considered as one of the most advanced and flexible nanomaterials. Furthermore, the basic VNP structure can be 'programmed' for loading with drug molecules, imaging reagents, quantum dots and other nanoparticles, while its external surface can be changed to reveal targeting ligands that allow cell-specific delivery (Yildiz *et al.*, 2011).

The use of VLPs for biotechnological applications in agriculture remains unexplored so far. However, the potential of RNA viruses for triggering of gene silencing and concomitant lethal effects was underscored in a recent study that employed recombinant FHV that was engineered to package foreign RNA sequences (Taning *et al.*, 2018). Nonetheless, in this case viruses with replicating genetic material were used that can be classified as genetically modified organisms (GMOs). Because GMOs are associated with strong public and political opposition and require lengthy evaluation procedures, the approach of VLPs with inert genetic cargo (non-replicating dsRNAs) may be considered safer and more feasible from a regulatory viewpoint. For transport and delivery of dsRNAs, VLPs based on dsRNA viruses are more suitable than those of ssRNA viruses that may not package efficiently long dsRNA molecules that form strong secondary structures (Zhao *et al.*, 2018). However, packaging of short RNA hairpins is possible, as was illustrated for VLPs based on the bacteriophage Qβ that naturally packages a positive strand ssRNA genome. In this case, RNAi scaffolds consisted of fusions of the 29 nt Qβ RNA hairpin packaging signal with a miRNA-based stem loop of 59 nt (Fang *et al.*, 2016). When co-expressed in bacteria, the Qβ capsid protein and the RNAi scaffold become spontaneously assembled in VLP-RNAi particles. While this example illustrates that delivery of short RNA hairpins with VLPs of ssRNA viruses is possible, VLPs of dsRNA viruses may have the advantage of packaging long dsRNAs that may have more potent silencing and insecticidal effects (Fang *et al.*, 2016; Zhao *et al.*, 2018).

Among dsRNA viruses, cypoviruses may constitute the basis for the development of a biotechnological platform in agriculture for production of VLPs that carry long dsRNAs as cargo (RNAi-VLPs). Cryo-electron microscope studies established that the CPV virion consists of three major capsid proteins: (i) the capsid shell protein (also known as VP1) that spontaneously forms a thin icosahedral capsid shell of 66 nm; (ii) the turret protein (also known as VP3); and (iii) 'large protruding protein' (also known as VP5) (Cheng *et al.*, 2011; Fang *et al.*, 2016). The mature virion also contains a small number of copies of the A-spike protein (also known as VP2) that could be involved in cell attachment, and the transcription enzyme complex consisting of the RdRP and VP4 (Hagiwara *et al.*, 2002; Cheng *et al.*, 2011; Fang *et al.*, 2016). The well-known structure of CPV virions permits the rational design of VLPs with enhanced properties such as increased stability and facilitated cell penetration. For delivery of RNAi, methods for efficient incorporation of (long) dsRNA molecules also need to be devised (Kolliopoulou *et al.*, 2017; Zhao *et al.*, 2018).

4.7 Conclusions

RNAi technology is one of the most appealing trends in the field of crop protection and has major advantages in comparison with chemical insecticides that are currently in use. In RNAi applications, the requirement of specific base-pairing almost guarantees the precise targeting of the intended pest with minimal repercussions on non-target species. In comparison with the non-specific detrimental effects of chemical insecticides on non-target organisms (pollinators, parasitoids, predators and vertebrates), this can be considered as a major asset. However, RNAi technology suffers from issues with efficiency and speed of killing and more research efforts are required to improve the methodology.

Currently many laboratories are investigating different dsRNA delivery methods in order to achieve better performances of RNAi in insects. Chemically synthesized nanoparticles, ribonucleoproteins, specialized bacterial strains and VLPs underline the continuous effort that the scientific community is currently taking to produce more efficient but also safer and more environmentally friendly RNAi-based pesticides. Considering the amount of basic knowledge that still needs to be acquired, RNAi research remains an ongoing process whose valuable applications will likely not be shown sooner than the ending of the coming decade.

Acknowledgements

The research work was supported by the Hellenic Foundation for Research and Innovation (HFRI) under the 'First Call for HFRI Research Projects to support Faculty members and Researchers and the procurement of high-cost research equipment grant' (Project Number: 785).

References

Almeida Garcia, R., Lima Pepino Macedo, L., Cabral do Nascimento, D., Gillet, F.-X., Moreira-Pinto, C.E. *et al.* (2017) Nucleases as a barrier to gene silencing in the cotton boll weevil, *Anthonomus grandis*. *PLoS ONE* 12(12), e0189600. DOI: 10.1371/journal.pone.0189600.

Bolling, B.G., Weaver, S.C., Tesh, R.B. and Vasilakis, N. (2015) Insect-specific virus discovery: significance for the arbovirus community. *Viruses* 7(9), 4911–4928. DOI: 10.3390/v7092851.

Cao, G., Meng, X., Xue, R., Zhu, Y., Zhang, X. *et al.* (2012) Characterization of the complete genome segments from BmCPV-SZ, a novel *Bombyx mori* cypovirus 1 isolate. *Canadian Journal of Microbiology* 58(7), 872–883. DOI: 10.1139/w2012-064.

Cappelle, K., de Oliveira, C.F.R., Van Eynde, B., Christiaens, O. and Smagghe, G. (2016) The involvement of clathrin-mediated endocytosis and two Sid-1-like transmembrane proteins in double-stranded

RNA uptake in the Colorado potato beetle midgut. *Insect Molecular Biology* 25(3), 315–323. DOI: 10.1111/imb.12222.

Chen, F., Zhu, L., Zhang, Y., Kumar, D., Cao, G. *et al.* (2018) Clathrin-mediated endocytosis is a candidate entry sorting mechanism for *Bombyx mori* cypovirus. *Scientific Reports* 8(1), 7268. DOI: 10.1038/s41598-018-25677-1.

Cheng, L., Sun, J., Zhang, K., Mou, Z., Huang, X. *et al.* (2011) Atomic model of a cypovirus built from cryo-EM structure provides insight into the mechanism of mRNA capping. *Proceedings of the National Academy of Sciences of the United States of America* 108(4), 1373–1378. DOI: 10.1073/pnas.1014995108.

Christiaens, O., Tardajos, M.G., Martinez Reyna, Z.L., Dash, M., Dubruel, P. *et al.* (2018) Increased RNAi efficacy in *Spodoptera exigua* via the formulation of dsRNA with guanylated polymers. *Frontiers in Physiology* 9, 316. DOI: 10.3389/fphys.2018.00316.

Cooper, A.M., Silver, K., Zhang, J., Park, Y. and Zhu, K.Y. (2018) Molecular mechanisms influencing efficiency of RNA interference in insects. *Pest Management Science* 75(1), 18–28. DOI: 10.1002/ps.5126.

Das, S., Debnath, N., Cui, Y., Unrine, J. and Palli, S.R. (2015) Chitosan, carbon quantum dot, and silica nanoparticle mediated dsRNA delivery for gene silencing in *Aedes aegypti*: a comparative analysis. *ACS Applied Materials & Interfaces* 7(35), 19530–19535. DOI: 10.1021/acsami.5b05232.

Dong, X., Li, X., Li, Q., Jia, H. and Zhang, H. (2017) The inducible blockage of RNAi reveals a role for polyunsaturated fatty acids in the regulation of dsRNA-endocytic capacity in *Bactrocera dorsalis*. *Scientific Reports* 7(1), 5584. DOI: 10.1038/s41598-017-05971-0.

Eguchi, A., Meade, B.R., Chang, Y.-C., Fredrickson, C.T., Willert, K. *et al.* (2009) Efficient siRNA delivery into primary cells by a peptide transduction domain-dsRNA binding domain fusion protein. *Nature Biotechnology* 27(6), 567–571. DOI: 10.1038/nbt.1541.

Fang, P.-Y., Gómez Ramos, L. M., Holguin, S.Y., Hsiao, C., Bowman, J.C. *et al.* (2016) Functional RNAs: combined assembly and packaging in VLPs. *Nucleic Acids Research* 45(6), 3519–3527. DOI: 10.1093/nar/gkw1154.

Fareh, M., van Lopik, J., Katechis, I., Bronkhorst, A.W., Haagsma, A.C. *et al.* (2018) Viral suppressors of RNAi employ a rapid screening mode to discriminate viral RNA from cellular small RNA. *Nucleic Acids Research* 46(6), 3257–3257. DOI: 10.1093/nar/gky042.

Filipowicz, W., Bhattacharyya, S.N. and Sonenberg, N. (2008) Mechanisms of post-transcriptional regulation by microRNAs: are the answers in sight? *Nature Reviews Genetics* 9(2), 102–114. DOI: 10.1038/nrg2290.

Fire, A., Xu, S., Montgomery, M.K., Kostas, S.A., Driver, S.E. *et al.* (1998) Potent and specific genetic interference by double-stranded RNA in *Caenorhabditis elegans*. *Nature* 391(6669), 806–811. DOI: 10.1038/35888.

Gillet, F.-X., Garcia, R.A., Macedo, L.L.P., Albuquerque, E.V.S., Silva, M.C.M. *et al.* (2017) Investigating engineered ribonucleoprotein particles to improve oral RNAi delivery in crop insect pests. *Frontiers in Physiology* 8, 256. DOI: 10.3389/fphys.2017.00256.

Guan, R.-B., Li, H.-C., Fan, Y.-J., Hu, S.-R., Christiaens, O. *et al.* (2018) A nuclease specific to lepidopteran insects suppresses RNAi. *Journal of Biological Chemistry* 293(16), 6011–6021. DOI: 10.1074/jbc.RA117.001553.

Guo, W.-C., Fu, K.-Y., Yang, S., Li, X.-X. and Li, G.-Q. (2015) Instar-dependent systemic RNA interference response in *Leptinotarsa decemlineata* larvae. *Pesticide Biochemistry and Physiology* 123, 64–73. DOI: 10.1016/j.pestbp.2015.03.006.

Hagiwara, K., Rao, S., Scott, S.W. and Carner, G.R. (2002) Nucleotide sequences of segments 1, 3 and 4 of the genome of *Bombyx mori* cypovirus 1 encoding putative capsid proteins VP1, VP3 and VP4, respectively. *The Journal of General Virology* 83(6), 1477–1482. DOI: 10.1099/0022-1317-83-6-1477.

Hammond, S.M. (2005) Dicing and slicing: the core machinery of the RNA interference pathway. *FEBS Letters* 579(26), 5822–5829. DOI: 10.1016/j.febslet.2005.08.079.

Hammond, S.M., Bernstein, E., Beach, D. and Hannon, G.J. (2000) An RNA-directed nuclease mediates post-transcriptional gene silencing in *Drosophila* cells. *Nature* 404(6775), 293–296. DOI: 10.1038/35005107.

He, B., Chu, Y., Yin, M., Müllen, K., An, C. *et al.* (2013) Fluorescent nanoparticle delivered dsRNA toward genetic control of insect pests. *Advanced Materials* 25(33), 4580–4584. DOI: 10.1002/adma.201301201.

He, L., Hu, X., Zhu, M., Liang, Z., Chen, F. *et al.* (2017) Identification and characterization of vp7 gene in *Bombyx mori* cytoplasmic polyhedrosis virus. *Gene* 627, 343–350. DOI: 10.1016/j.gene.2017.06.048.

Ivashuta, S., Zhang, Y., Wiggins, B.E., Ramaseshadri, P., Segers, G.C. *et al.* (2015) Environmental RNAi in herbivorous insects. *RNA* 21(5), 840–850. DOI: 10.1261/rna.048116.114.

Karlikow, M., Goic, B., Mongelli, V., Salles, A., Schmitt, C. *et al.* (2016) *Drosophila* cells use nanotube-like structures to transfer dsRNA and RNAi machinery between cells. *Scientific Reports* 6(1), 346–350. DOI: 10.1038/srep27085.

Kolliopoulou, A., Taning, C.N.T., Smagghe, G. and Swevers, L. (2017) Viral delivery of dsRNA for control of insect agricultural pests and vectors of human disease: prospects and challenges. *Frontiers in Physiology* 8, 399. DOI: 10.3389/fphys.2017.00399.

Kontogiannatos, D., Swevers, L., Maenaka, K., Park, E.Y., Iatrou, K. *et al.* (2013) Functional characterization of a juvenile hormone esterase related gene in the moth *Sesamia nonagrioides* through RNA Interference. *PLoS ONE* 8(9), e73834. DOI: 10.1371/journal.pone.0073834.

Kourti, A., Swevers, L. and Kontogiannatos, D. (2017) In search of new methodologies for efficient insect pest control: the RNAi 'movement'. In: Shields, V. (ed.) *Biological Control of Pest and Vector Insects*, 1st edn. IntechOpen, Rijeka, pp. 1–27.

Lau, P.-W., Guiley, K.Z., De, N., Potter, C.S., Carragher, B. *et al.* (2012) The molecular architecture of human Dicer. *Nature Structural & Molecular Biology* 19(4), 436–440. DOI: 10.1038/nsmb.2268.

Li, J., Li, X., Bai, R., Shi, Y., Tang, Q. *et al.* (2014) RNA interference of the P450 CYP6CM1 gene has different efficacy in B and Q biotypes of *Bemisia tabaci*. *Pest Management Science* 71(8), 1175–1181. DOI: 10.1002/ps.3903.

Li, J., Qian, J., Xu, Y., Yan, S., Shen, J. *et al.* (2019) A facile-synthesized star polycation constructed as a highly efficient gene vector in pest management. *ACS Sustainable Chemistry & Engineering* 7(6), 6316–-6322. DOI: 10.1021/acssuschemeng.9b00004.

Lin, Y.-H., Huang, J.-H., Liu, Y., Belles, X. and Lee, H.-J. (2016) Oral delivery of dsRNA lipoplexes to German cockroach protects dsRNA from degradation and induces RNAi response. *Pest Management Science* 73(5), 960–966. DOI: 10.1002/ps.4407.

Lourenco, S. and Roy, P. (2011) *In vitro* reconstitution of Bluetongue virus infectious cores. *Proceedings of the National Academy of Sciences* 108(33), 13746–13751. DOI: 10.1073/pnas.1108667108.

Louten, J. and Reynolds, N. (2016) *Essential Human Virology*. Elsevier, Amsterdam, Paris, 344.

Mehrabadi, M., Hussain, M., Matindoost, L. and Asgari, S. (2015) The baculovirus antiapoptotic p35 protein functions as an inhibitor of the host RNA interference antiviral response. *Journal of Virology* 89(16), 8182–8192. DOI: 10.1128/JVI.00802-15.

Mongelli, V. and Saleh, M.-C. (2016) Bugs are not to be silenced: small RNA pathways and antiviral responses in insects. *Annual Review of Virology* 3(1), 573–589. DOI: 10.1146/annurev-virology-110615-042447.

Nakanishi, K. (2016) Anatomy of RISC: how do small RNAs and chaperones activate Argonaute proteins? *Wiley Interdisciplinary Reviews: RNA* 7(5), 637–660. DOI: 10.1002/wrna.1356.

Nguyen, T.A., Smith, B.R.C., Elgass, K.D., Creed, S.J., Cheung, S. *et al.* (2019) SIDT1 localizes to endolysosomes and mediates double-stranded RNA transport into the cytoplasm. *The Journal of Immunology* 202(12), 3483–3492. DOI: 10.4049/jimmunol.1801369.

Parsons, K.H., Mondal, M.H., McCormick, C.L. and Flynt, A.S. (2018) Guanidinium-functionalized interpolyelectrolyte complexes enabling RNAi in resistant insect pests. *Biomacromolecules* 19(4), 1111–1117. DOI: 10.1021/acs.biomac.7b01717.

Phanse, Y., Dunphy, B.M., Perry, J.L., Airs, P.M., Paquette, C.C.H. *et al.* (2015) Biodistribution and toxicity studies of PRINT hydrogel nanoparticles in mosquito larvae and cells. *PLoS Neglected Tropical Diseases* 9(5), e0003735. DOI: 10.1371/journal.pntd.0003735.

Ramesh Kumar, D., Saravana Kumar, P., Gandhi, M.R., Al-Dhabi, N.A., Paulraj, M.G. *et al.* (2016) Delivery of chitosan/dsRNA nanoparticles for silencing of wing development vestigial (vg) gene in *Aedes aegypti* mosquitoes. *International Journal of Biological Macromolecules* 86, 89–95. DOI: 10.1016/j.ijbiomac.2016.01.030.

Santos, D., Wynant, N., Van den Brande, S., Verdonckt, T.-W., Mingels, L. *et al.* (2018) Insights into RNAi-based antiviral immunity in *Lepidoptera*: acute and persistent infections in *Bombyx mori* and *Trichoplusia ni* cell lines. *Scientific Reports* 8(1), 2423. DOI: 10.1038/s41598-018-20848-6.

Shukla, J.N., Kalsi, M., Sethi, A., Narva, K.E., Fishilevich, E. *et al.* (2016) Reduced stability and intracellular transport of dsRNA contribute to poor RNAi response in lepidopteran insects. *RNA Biology* 13(7), 656–669. DOI: 10.1080/15476286.2016.1191728.

Sung, P.-Y., Vaughan, R., Rahman, S.K., Yi, G., Kerviel, A. *et al.* (2019) The interaction of Bluetongue virus VP6 and genomic RNA is essential for genome packaging. *Journal of Virology* 93(5), e02023–18. DOI: 10.1128/JVI.02023-18.

Swevers, L., Ioannidis, K., Kolovou, M., Zografidis, A., Labropoulou, V. *et al.* (2016) Persistent RNA virus infection of lepidopteran cell lines: interactions with the RNAi machinery. *Journal of Insect Physiology* 93-94, 81–93. DOI: 10.1016/j.jinsphys.2016.09.001.

Tan, Y.-R., Sun, J.-C., Lu, X.-Y., Su, D.-M. and Zhang, J.-Q. (2003) Entry of *Bombyx mori* cypovirus 1 into midgut cells *in vivo*. *Journal of Electron Microscopy* 52(5), 485–489. DOI: 10.1093/jmicro/52.5.485.

Taning, C.N.T., Christiaens, O., Li, X., Swevers, L., Casteels, H. *et al.* (2018) Engineered flock house virus for targeted gene suppression through RNAi in fruit flies (*Drosophila melanogaster*) *in vitro* and *in vivo*. *Frontiers in Physiology* 9, 805. DOI: 10.3389/fphys.2018.00805.

Terenius, O., Papanicolaou, A., Garbutt, J.S., Eleftherianos, I., Huvenne, H. *et al.* (2011) RNA interference in Lepidoptera: an overview of successful and unsuccessful studies and implications for experimental design. *Journal of Insect Physiology* 57(2), 231–245. DOI: 10.1016/j.jinsphys.2010.11.006.

Thairu, M.W., Skidmore, I.H., Bansal, R., Nováková, E., Hansen, T.E. *et al.* (2017) Efficacy of RNA interference knockdown using aerosolized short interfering RNAs bound to nanoparticles in three diverse aphid species. *Insect Molecular Biology* 26(3), 356–368. DOI: 10.1111/imb.12301.

Tian, H., Peng, H., Yao, Q., Chen, H., Xie, Q. *et al.* (2009) Developmental control of a lepidopteran pest *Spodoptera exigua* by ingestion of bacteria expressing dsRNA of a non-midgut gene. *PLoS ONE* 4(7), e6225. DOI: 10.1371/journal.pone.0006225.

Timmons, L. and Fire, A. (1998) Specific interference by ingested dsRNA. *Nature* 395(6705), 854. DOI: 10.1038/27579.

van Mierlo, J.T., Overheul, G.J., Obadia, B., van Cleef, K.W.R., Webster, C.L. *et al.* (2014) Novel *Drosophila* viruses encode host-specific suppressors of RNAi. *PLoS Pathogens* 10(7), e1004256. DOI: 10.1371/journal.ppat.1004256.

Vatanparast, M. and Kim, Y. (2017) Optimization of recombinant bacteria expressing dsRNA to enhance insecticidal activity against a lepidopteran insect, *Spodoptera exigua*. *PLoS ONE* 12(8), e0183054. DOI: 10.1371/journal.pone.0183054.

Wang, K., Peng, Y., Pu, J., Fu, W., Wang, J. *et al.* (2016) Variation in RNAi efficacy among insect species is attributable to dsRNA degradation *in vivo*. *Insect Biochemistry and Molecular Biology* 77, 1–9. DOI: 10.1016/j.ibmb.2016.07.007.

Whitten, M.M.A., Facey, P.D., Del Sol, R., Fernández-Martínez, L.T., Evans, M.C. *et al.* (2016) Symbiont-mediated RNA interference in insects. *Proceedings of the Royal Society B: Biological Sciences* 283(1825), 20160042. DOI: 10.1098/rspb.2016.0042.

Wickner, R.B. (1993) Double-stranded RNA virus replication and packaging. *The Journal of Biological Chemistry* 268(6), 3797–3800.

Xiao, D., Gao, X., Xu, J., Liang, X., Li, Q. *et al.* (2015) Clathrin-dependent endocytosis plays a predominant role in cellular uptake of double-stranded RNA in the red flour beetle. *Insect Biochemistry and Molecular Biology* 60, 68–77. DOI: 10.1016/j.ibmb.2015.03.009.

Yildiz, I., Shukla, S. and Steinmetz, N.F. (2011) Applications of viral nanoparticles in medicine. *Current Opinion in Biotechnology* 22(6), 901–908. DOI: 10.1016/j.copbio.2011.04.020.

Yoon, J.-S., Gurusamy, D. and Palli, S.R. (2017) Accumulation of dsRNA in endosomes contributes to inefficient RNA interference in the fall armyworm, *Spodoptera frugiperda*. *Insect Biochemistry and Molecular Biology* 90, 53–60. DOI: 10.1016/j.ibmb.2017.09.011.

Zhang, X., Liu, X., Ma, J. and Zhao, J. (2013) Silencing of cytochrome P450 CYP6B6 gene of cotton bollworm (*Helicoverpa armigera*) by RNAi. *Bulletin of Entomological Research* 103(5), 584–591. DOI: 10.1017/S0007485313000151.

Zhao, Y., Sun, J., Labropoulou, V. and Swevers, L. (2018) Beyond baculoviruses: additional biotechnological platforms based on insect RNA viruses. *Advances in Insect Physiology* 55, 123–162.

Zhao, Y., Kolliopoulou, A., Ren, F., Lu, Q., Labropoulou, V. *et al.* (2019) Transcriptional response of immune-related genes after endogenous expression of VP1 and exogenous exposure to VP1-based VLPs and CPV virions in lepidopteran cell lines. *Molecular Genetics and Genomics* 294(4), 887–899. DOI: 10.1007/s00438-019-01551-1.

Zhu, F., Xu, J., Palli, R., Ferguson, J. and Palli, S.R. (2010) Ingested RNA interference for managing the populations of the Colorado potato beetle, *Leptinotarsa decemlineata*. *Pest Management Science* 67(2), 175–182. DOI: 10.1002/ps.2048.

5 Biogenesis and Functional RNAi in Fruit Trees

Michel Ravelonandro* and Pascal Briard

Unite Mixte de Recherches 1332, INRAE-Bordeaux; Villenave d'Ornon, France

Abstract

In plants, genome expression is linked to the transcribed mRNAs that are synthesized by RNA polymerase. Following its move to the cytoplasm, the generated mRNA is briefly translated to the encoded protein. If transcription and translation are dependent on the family of RNA polymerase, these two phenomena could be interfered with through the process designated as gene regulation. Thus, large molecules of RNA (single-stranded or double-stranded) consequently sliced into small molecules produce nascent small interfering RNA ranging from 21 to 27 nucleotides. This chapter revisits the biogenesis of these two types of RNAi, miRNA and siRNA, and notably their involvement in plant gene regulation. Following their sequential transcription and their specific involvement, we will consider the sources and roles of RNA interference in plants and we will look at their detection in fruit crops. We discuss their applications and the risk assessment studies in fruit crops.

5.1 Brief Report about Biogenesis of RNAi in Plants

Recent progress with plant genome sequencing has increasingly led to the rapid development of gene regulation studies. In parallel with better knowledge about cellular components, knowledge has increased about plant promoters (Chow *et al.*, 2016) and how plant genes can interact in leading to either gene knockout or an overexpression of plant phenotypes (Baulcombe, 2004). The development of many satellite studies on gene interference has been successfully performed and among these were studies of the silencing of changes to coloured petals of petunias (Napoli *et al.*, 1990; Jorgensen *et al.*, 1996). These studies showed that knockout or overexpression was due to small RNA molecules that interfere with the homologous nucleotide sequences of the encoding genes. Interfering sequences and consequent phenotypes were closely dependent (Hamilton and Baulcombe, 1999), showing that a genetic character can be either reverted or exclusively fixed (Zotti *et al.*, 2018). Two types of RNA interference (RNAi) can be involved in plants: microRNA (miRNA) and small interfering RNA (siRNA). Whereas miRNA is single-stranded (Bartel, 2004; Carthew and Sontheimer, 2009), siRNA functions as a small dsRNA (21–27 nt) produced from cleavage of a larger double-stranded RNA (dsRNA), designated as a precursor (Nakahara and Carthew, 2004). Levels of miRNA are variable in plant cells, because the binding of the miRNA to its complementary endogenous mRNA is specifically occurring,

---

*Corresponding author: michel.ravelonandro@wanadoo.fr

© CAB International 2021. *RNAi for Plant Improvement and Protection*
(eds B. Mezzetti *et al.*)
DOI: 10.1079/9781789248890.0005

and they also change either in different tissues or in some developmental stages (Carrington and Ambros, 2003; Carthew and Sontheimer, 2009). It has been shown that miRNA can bind to a specific peptide, conferring their stable function in cells through miRNA encoded peptides (miPEPs) (Couzigou *et al.*, 2015). However, siRNA can interact with the plant enzyme RDR6 in amplifying the molecule ratios in cells that cause a robust gene interference that increasingly confers a strong phenotype (Fahlgren *et al.*, 2006). While a high population of RNAi is synthesized in cells, the lifespan of these molecules results in a long chain of gene regulation and interfering interactions that can be affected either spatially or temporally. It is known that a population of miRNA can be involved in regulation of several genes; however, an siRNA response is specific (Carthew and Sontheimer, 2009).

5.2 Diversity of RNAi in Fruit Crops

Studies of new technologies in fruit crops have lagged behind those in the annual crops that have a high economic value as foods (Dempewolf *et al.*, 2017). In perennial fruit crops, traditional plant breeding has been the main technology for crop improvement (Peña and Séguin, 2001). This has limitations as fruit trees grow slowly and need to achieve maturity for fruiting, so that genetic improvement required patience and good management (good knowledge and a high expertise about tree physiology and cultivation). Molecular breeding as an approach developed by scientists led to the strategic discovery of new genes of interest (Wei *et al.*, 2015). New traits that were discovered conferred virus resistance (Zuriaga *et al.*, 2018) and plant transformation technologies permitted introduction of these new traits (Petri *et al.*, 2011). Genetic engineering of *Prunus*, *Citrus*, *Malus*, *Vitis* and other crops brought variable benefits for fruit-tree breeders, including resistance to viruses (Scorza *et al.*, 1994; Ravelonandro *et al.*, 1997; Reyes *et al.*, 2011; Scorza *et al.*, 2013; Rubio *et al.*, 2015). Papaya is a promising example that reflects the success of RNAi against papaya ringspot virus (PRSV) (Gonsalves, 2006). This has led to growing interest in the discovery of new gene(s) and the subsequent exploitation of gene silencing.

Many plant genes are known to control the metabolic chain of any enzymatic complex in the Krebs cycle (Senthil-Kumar and Mysore, 2010). Among the challenges was the feasibility of investigating the role of genes encoding the enzymes expressed in different plant tissue. Physical studies that focused on agronomic and horticultural traits of fruit tree species have revealed that mapping different genes helped complement full genome sequencing (Iwata *et al.*, 2016). Hence, the ability to transform a fruit crop facilitates the introduction and regulation of engineered gene(s) in order to achieve expression of specific traits (Petri *et al.*, 2011).

In juvenile fruit trees, the immature age of the fruit trees did not allow verification of the RNAi effects until a few years later. Once introduced into plant genomes, new engineered sequences can be verified through sequencing and molecular hybridization (Ravelonandro *et al.*, 2019). RNAi occurs in any cellular compartment so has a key role in regulating plant life. The role of RNAi is threefold: (i) to regulate the endogenous genes; (ii) to specifically interfere with the targeted sequences; and (iii) to convert such molecular interactions in plant phenotype (Zotti *et al.*, 2018). Focusing on fruit, RNAi appears as part of molecules enabling specific interference with genes expressed either endogenously or exogenously. The models supporting these phenomena are the apple 'Arctic', the 'HoneySweet' plum, the 'Rainbow' papaya and the activity against certain pests of grapevines (Nandety *et al.*, 2015; Taning *et al.*, 2016) (Figs 5.1 and 5.2).

5.3 Detection and Application in Fruit Crops

The relevance of the efficiency of RNAi and the targeted virus RNA was significantly highlighted in plum (Scorza *et al.*, 2013) and other fruit. In 'Arctic' apple, four genes are silenced that control polyphenol oxidase (PPO) production (Armstrong and Lane, 2009), which causes the production of brown melanin due to oxidation following fruit damage. Consequently, the 'Arctic' apple differs from conventional fruit in that its flesh does not turn brown after slicing (Fig. 5.2).

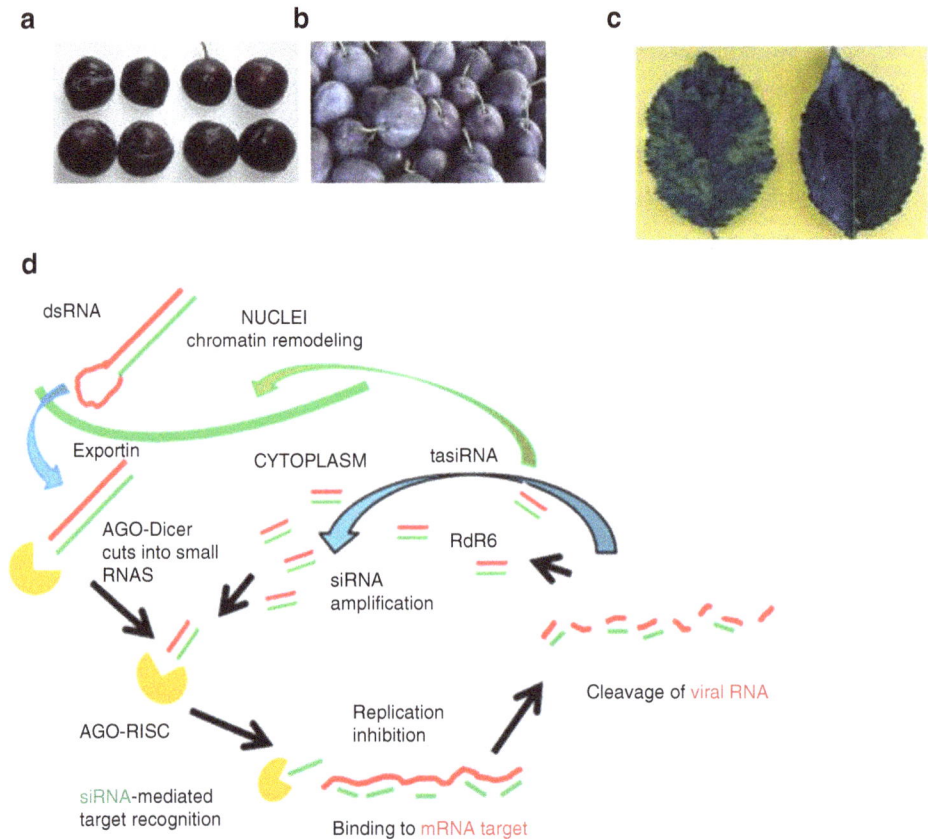

Fig. 5.1. (a) Deformed fruits of susceptible conventional plums. (b) 'HoneySweet' fruits. (c) Diseased leaf of susceptible plum (left) and symptomless leaf of the resistant 'HoneySweet' plum (right). (d) The production of specific small RNAs targeting and silencing the viral mRNA in a cell of 'HoneySweet' plum.

Studies of healthy plums sampled from trees challenged with plum pox virus (PPV) and 'Arctic' apple have shown that the engineered RNAi construct functioned in whole plants (Figs 5.1 and 5.2). Studies of any other effects of the introduced gene showed homology of 'HoneySweet' fruits with those sampled from healthy conventional cultivars (either 'Stanley' or 'Reinclod') (Bobis et al., 2019). Enzymes involved in fruit storage and maturation function similarly and there is homologous fruit composition (Ravelonandro et al., 2013; Callahan et al., 2019). The genetic engineering of the PPV coat protein (CP) gene as an introduced sequence did not lead to any change in plum tree traits, apart from PPV resistance, showing that the RNAi is restrictively expressed to target only PPV RNA (Fig. 5.1c,d). Resistance traits against either PPV ('HoneySweet' plum parent) or PRSV ('Rainbow' papaya parent) share the same properties as those transferred in hybrids and conserved similar effects.

5.4   Biosafe Use of RNAi

This interference strategy occurring in plants can be exploited to change metabolic paths to optimize desired characteristics that could not

a

Regular **Arctic**

b

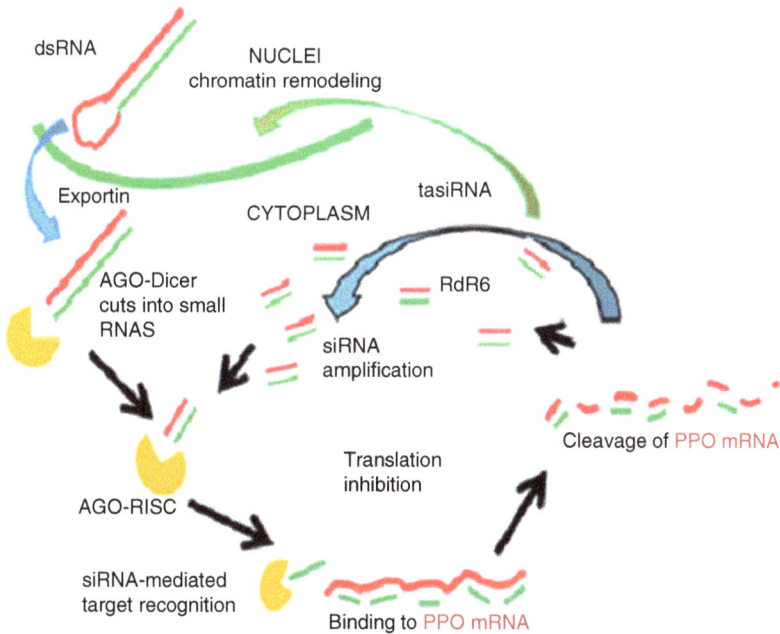

Fig. 5.2. (a) Sliced fruits of conventional apple showing browning (left) compared with 'Arctic' apple (right). (Figure courtesy of Okanagan Company, Summerland, Canada.) (b) Molecular mechanism of the silencing of the mRNA encoding the polyphenol oxidase (PPO) so that browning does not occur in 'Arctic' apple when sliced.

be achieved with classical breeding. However, in common with the use of new genetic techniques, the introduction of this new RNAi technology raised safety concerns, although the highly selective nature of RNA activity reduces the likelihood of off-target and non-target effects. This has been supported by the genetic and bioinformatic information obtained through next-generation sequencing (NGS), high-throughput sequencing (HTS) and other techniques (Shendure and Ji, 2008). The sources, paths and compartmental cell storage are also relevant. A number of risk assessment studies of engineered fruit trees have been conducted and have not shown any unexpected or adverse effects (Yien et al., 2011; Scorza et al., 2019). For example, oral feeding of mammals with PPV-resistant plum in experimental models showed no adverse effects on mice and no allergenic reactions. The RNAi modified papaya did not reveal any genotoxicity in any analysed gastro-organs in rats (Yien et al., 2011). These results suggest that RNAi does not elicit any unexpected toxic reactions and does not represent any bio-risk to mammals (Scorza et al., 2019).

5.5 Conclusions

The aim of this chapter has been to provide information on RNAi that can be either endogenously produced by plant cells and then accumulated in fruits (Figs 5.1 and 5.2) or exogenously applied on fruits in order to protect them against parasites (Taning et al., 2016). For RNAi technology to be firmly and clearly appreciated by consumers, it is important for us to deliver honest and relevant communications, especially in EU countries. First, during the period of development of genetically modified organisms for these past four decades, the writing of laws and rules concerning the use and release of modified plants in the environment has been a dominant factor (DeFrancesco, 2013). Secondly, the emergence of new technologies, ranging from RNAi to gene editing (Shan et al., 2013), provides a powerful platform for gene regulation that should offer potential resources in crop improvement. Further improvements in metabolomics, genetics and bioinformatics will provide further evidence (Shameer et al., 2014) to support the acceptance of RNAi.

Acknowledgements

The authors would like to thank Dr Ralph Scorza and colleagues (ARS-USDA, Kearneysville, West Virginia, USA) and other European collaborators who have contributed to the research and development of 'HoneySweet' plum. Special thanks to Mrs Angela Tipton, Communications Manager of Okanagan company (Canada), who shared her expertise with 'Arctic' apple by supplying Fig. 5.2a. We also thank Professor Mezzetti, who reviewed this chapter.

References

Armstrong, J.D. and Lane, W.D. (2009) Genetically modified reduced-browning fruit-producing plant and produced fruit thereof, and method of obtaining such. US Patent Application Number 12/ 919, 735.

Bartel, D.P. (2004) MicroRNAs: genomics, biogenesis, mechanism, and function. Cell 116(2), 281–297. DOI: 10.1016/s0092-8674(04)00045-5.

Baulcombe, D. (2004) RNA silencing in plants. Nature 431(7006), 356–363. DOI: 10.1038/nature02874.

Bobis, O., Bonta, V., Marghitas, L.A., Dezmirean, D.S., Pasca, C. et al. (2019) Does genetic engineering influence the nutritional value of plums? Case study on two conventional and one genetic engineering plum fruits. Bulletin of University of Agricultural Sciences and Veterinary Medicine Cluj-Napoca. Animal Science and Biotechnologies 76(1), 51–56. DOI: 10.15835/buasvmcn-asb:2019.0003.

Callahan, A.M., Dardick, C.D. and Scorza, R. (2019) Multilocation comparison of fruit composition for 'HoneySweet', an RNAi based plum pox virus resistant plum. PLoS ONE 14(3), e0213993. DOI: 10.1371/journal.pone.0213993.

Carrington, J.C. and Ambros, V. (2003) Role of microRNAs in plant and animal development. *Science* 301(5631), 336–338. DOI: 10.1126/science.1085242.

Carthew, R.W. and Sontheimer, E.J. (2009) Origins and mechanisms of miRNAs and siRNAs. *Cell* 136(4), 642–655. DOI: 10.1016/j.cell.2009.01.035.

Chow, C.-N., Zheng, H.-Q., Wu, N.-Y., Chien, C.-H., Huang, H.-D. *et al.* (2016) Plant PAN 2.0: an update of plant promoter analysis navigator for reconstructing transcriptional regulatory networks in plants. *Nucleic Acids Research* 44(D1), D1154–D1160. DOI: 10.1093/nar/gkv1035.

Couzigou, J.-M., Lauressergues, D., Bécard, G. and Combier, J.-P. (2015) miRNA-encoded peptides (mi-PEPs): a new tool to analyze the roles of miRNAs in plant biology. *RNA Biology* 12(11), 1178–1180. DOI: 10.1080/15476286.2015.1094601.

DeFrancesco, L. (2013) How safe does transgenic food need to be? *Nature Biotechnology* 31(9), 794–802. DOI: 10.1038/nbt.2686.

Dempewolf, H., Baute, G., Anderson, J., Kilian, B., Smith, C. *et al.* (2017) Past and future use of wild relatives in crop breeding. *Crop Science* 57(3), 1070–1082. DOI: 10.2135/cropsci2016.10.0885.

Fahlgren, N., Montgomery, T.A., Howell, M.D., Allen, E., Dvorak, S.K. *et al.* (2006) Regulation of AUXIN RESPONSE FACTOR3 by TAS3 ta-siRNA affects developmental timing and patterning in *Arabidopsis*. *Current Biology* 16(9), 939–944. DOI: 10.1016/j.cub.2006.03.065.

Gonsalves, D. (2006) Transgenic papaya: development, release, impact and challenges. *Advanced Virus Research* 67, 317–354. DOI: 10.1016/S0065-3527(06)67009-7.

Hamilton, A.J. and Baulcombe, D.C. (1999) A species of small antisense RNA in posttranscriptional gene silencing in plants. *Science* 286(5441), 950–952. DOI: 10.1126/science.286.5441.950.

Iwata, H., Minamikawa, M.F., Kajiya-Kanegae, H., Ishimori, M. and Hayashi, T. (2016) Genomics-assisted breeding in fruit trees. *Breeding Science* 66(1), 100–115. DOI: 10.1270/jsbbs.66.100.

Jorgensen, R.A., Cluster, P.D., English, J., Que, Q. and Napoli, C.A. (1996) Chalcone synthase cosuppression phenotypes in petunia flowers: comparison of sense vs. antisense constructs and single-copy vs. complex T-DNA sequences. *Plant Molecular Biology* 31(5), 957–973. DOI: 10.1007/BF00040715.

Nakahara, K. and Carthew, R.W. (2004) Expanding roles for miRNAs and siRNAs in cell regulation. *Current Opinion in Cell Biology* 16(2), 127–133. DOI: 10.1016/j.ceb.2004.02.006.

Nandety, R.S., Kuo, Y.-W., Nouri, S. and Falk, B.W. (2015) Emerging strategies for RNA interference (RNAi) applications in insects. *Bioengineered* 6(1), 8–19. DOI: 10.4161/21655979.2014.979701.

Napoli, C., Lemieux, C. and Jorgensen, R. (1990) Introduction of a chimeric chalcone synthase gene into Petunia results in reversible co-suppression of homologous genes in trans. *The Plant Cell* 2(4), 279–289. DOI: 10.2307/3869076.

Petri, C., Hily, J.-M., Vann, C., Dardick, C. and Scorza, R. (2011) A high-throughput transformation system allows the regeneration of marker-free plum plants (*Prunus domestica*). *Annals of Applied Biology* 159(2), 302–315. DOI: 10.1111/j.1744-7348.2011.00499.x.

Peña, L. and Séguin, A. (2001) Recent advances in the genetic transformation of trees. *Trends in Biotechnology* 19(12), 500–506. DOI: 10.1016/S0167-7799(01)01815-7.

Ravelonandro, M., Scorza, R., Bachelier, J.C., Labonne, G., Levy, L. *et al.* (1997) Resistance of transgenic *Prunus domestica* to Plum pox virus infection. *Plant Disease* 81(11), 1231–1235. DOI: 10.1094/PDIS.1997.81.11.1231.

Ravelonandro, M., Scorza, R., Polak, J., Callahan, A., Krška, B. *et al.* (2013) 'HoneySweet' plum – a valuable genetically engineered fruit-tree cultivar. *Food Nutrition Science* 4, No.6A, 45–49.

Ravelonandro, M., Scorza, R. and Briard, P. (2019) Innovative RNAi strategies and tactics to tackle plum pox virus (PPV) genome in *Prunus domestica*-plum. *Plants* 8(12), 565. DOI: 10.3390/plants8120565.

Reyes, C.A., De Francesco, A., Peña, E.J., Costa, N., Plata, M.I. *et al.* (2011) Resistance to citrus psorosis virus in transgenic sweet orange plants is triggered by coat protein-RNA silencing. *Journal of Biotechnology* 151(1), 151–158. DOI: 10.1016/j.jbiotec.2010.11.007.

Rubio, M., Ballester, A.R., Olivares, P.M., Castro de Moura, M., Dicenta, F. *et al.* (2015) Gene expression analysis of plum pox virus (sharka) susceptibility/resistance in apricot (*Prunus armeniaca* L.). *PLoS ONE* 10(12), e0144670. DOI: 10.1371/journal.pone.0144670.

Scorza, R., Ravelonandro, M., Callahan, A.M., Cordts, J.M., Fuchs, M. *et al.* (1994) Transgenic plums (*Prunus domestica* L.) express the plum pox virus coat protein gene. *Plant Cell Reports* 14(1), 18–22. DOI: 10.1007/BF00233291.

Scorza, R., Callahan, A., Dardick, C., Ravelonandro, M., Polak, J. *et al.* (2013) Genetic engineering of Plum pox virus resistance: 'HoneySweet' plum – from concept to product. *Plant Cell, Tissue and Organ Culture* 115(1), 1–12. DOI: 10.1007/s11240-013-0339-6.

Scorza, R., Ravelonandro, M., Cambra, M., Capote, N., Badenes, M. *et al.* (2019) RNA profiling and compositional studies of HoneySweet plum for regulatory approvals. *Crop Innovations and Regulations (CIR) Conference*, 10–12 September 2019. Crowne Plaza Barcelona – Fira Center, Barcelona.

Senthil-Kumar, M. and Mysore, K.S. (2010) RNAi in plants: recent developments and applications in agriculture. In: Catalano, A.J. (ed.) *Gene Silencing: Theory, Techniques and Applications*. Nova Science Publishers, New York, pp. 183–199.

Shameer, K., Naika, M.B., Mathew, O.K. and Sowdhamini, R. (2014) POEAS: automated plant phenomic analysis using plant ontology. *Bioinformatics and Biology Insights* 8, 209–214. DOI: 10.4137/BBI.S19057.

Shan, Q., Wang, Y., Li, J., Zhang, Y., Chen, K. *et al.* (2013) Targeted genome modification of crop plants using a CRISPR-Cas system. *Nature Biotechnology* 31(8), 686–688. DOI: 10.1038/nbt.2650.

Shendure, J. and Ji, H. (2008) Next-generation DNA sequencing. *Nature Biotechnology* 26(10), 1135–1145. DOI: 10.1038/nbt1486.

Taning, C.N.T., Christiaens, O., Berkvens, N., Casteels, H., Maes, M. *et al.* (2016) Oral RNAi to control *Drosophila suzukii*: laboratory testing against larval and adult stages. *Journal of Pest Science* 89(3), 803–814. DOI: 10.1007/s10340-016-0736-9.

Wei, H., Chen, X., Zong, X., Shu, H., Gao, D. *et al.* (2015) Comparative transcriptome analysis of genes involved in anthocyanin biosynthesis in the red and yellow fruits of sweet cherry (*Prunus avium* L.). *PLoS ONE* 10(3), e0121164. DOI: 10.1371/journal.pone.0121164.

Yien, G.-C., Lin, H.-T., Cheng, Y.-H., Lin, Y.-J., Chang, H.-C. *et al.* (2011) Food safety evaluation of papaya fruits resistant to papaya ring spot virus. *Journal of Food and Drug Analysis* 19(3), 269–280.

Zotti, M., dos Santos, E.A., Cagliari, D., Christiaens, O., Taning, C.N.T. *et al.* (2018) RNA interference technology in crop protection against arthropod pests, pathogens and nematodes. *Pest Management Science* 74(6), 1239–1250. DOI: 10.1002/ps.4813.

Zuriaga, E., Romero, C., Blanca, J.M. and Badenes, M.L. (2018) Resistance to plum pox virus (PPV) in apricot (*Prunus armeniaca* L.) is associated with down-regulation of two MATHd genes. *BMC Plant Biology* 18(1), 25. DOI: 10.1186/s12870-018-1237-1.

6 Gene Silencing or Gene Editing: the Pros and Cons

Huw D. Jones*

IBERS, Aberystwyth University, UK

Abstract

Research into plant genetics often requires the suppression or complete knockout of gene expression to scientifically validate gene function. In addition, the phenotypes obtained from gene suppression can occasionally have commercial value for plant breeders. Until recently, the methodological choices to achieve these goals fell into two broad types: either some form of RNA-based gene silencing; or the screening of large numbers of natural or induced random genomic mutations. The more recent invention of gene editing as a tool for targeted mutation potentially gives researchers and plant breeders another route to block gene function. RNAi is widely used in animal and plant research and functions to silence gene expression by degrading the target gene transcript. Although RNAi offers unique advantages over genomic mutations, it often leads to the formation of a genetically modified organism (GMO), which for commercial activities has major regulatory and acceptance issues in some regions of the world. Traditional methods of generating genomic mutations are more laborious and uncertain to achieve the desired goals but possess a distinct advantage of not being governed by GMO regulations. Gene editing (GE) technologies have some of the advantages of both RNAi and classical mutation breeding in that they can be designed to give simple knockouts or to modulate gene expression more subtly. GE also has a more complex regulatory position, with some countries treating it as another conventional breeding method whilst the EU defines GE as a technique of genetic modification and applies the normal GMO authorization procedures. This chapter explores the pros and cons of RNAi alongside other methods of modulating gene function.

6.1 Introduction

Blocking the expression of a (candidate) gene has long been an experimental tool for research that aims to define the cellular function of specific DNA sequences. Alongside other methods, it can provide strong evidence to support a hypothesis on the role of a gene. It can also provide novel phenotypes with useful characteristics for commercial products (see elsewhere in this text). Until recently, the options for doing this were restricted to screening individuals possessing a knockout phenotype due to random natural or induced mutations in the genome, or gene silencing that resulted in reduced protein synthesis from the gene under investigation. These two fundamentally different approaches have coexisted in research and commercial arenas over the past few decades but have various pros and cons, which are discussed below.

*hdj2@aber.ac.uk

© CAB International 2021. *RNAi for Plant Improvement and Protection*
(eds B. Mezzetti *et al.*)
DOI: 10.1079/9781789248890.0006

The invention of gene editing offered researchers another, potentially more powerful, approach to gene suppression. CRISPR (clustered regularly interspaced short palindromic repeats) (Jinek *et al.*, 2012; Cong *et al.*, 2013; Mali *et al.*, 2013) is proving to be faster, cheaper and easier to use than other existing programmable gene editing technologies, including oligonucleotide-directed mutagenesis (ODM), meganucleases (MN), zinc-finger nucleases (ZFN) and transcription activator-like effector nucleases (TALENs) reviewed by Guha *et al.* (2017). The exponential growth in research outputs using CRISPR is testament to its utility for targeted mutation and gives researchers and plant breeders yet another route to block or fine-tune gene function and expression. This chapter compares and contrasts these various methods of modulating gene expression in plants and highlights the pros and cons of RNAi in particular.

6.2   Random Mutations

The earliest approach was simply to screen wild populations for natural variants that possessed random mutations resulting in a 'knockout' of gene function. While its use as a 'forward genetic' tool (with no control and often no knowledge of the genetic changes made) has been integral to plant and animal breeding for centuries, it requires long time scales and very large plant populations. It is often not a feasible approach to use when specific, pre-defined genetic mutations are sought. One way to overcome this limitation is to substantially increase the cellular mutation rate using chemicals or radiation that randomly damages DNA. Mutation breeding has exploited this approach since the 1930s, although the term 'Mutationszüchtung' (mutation breeding) was not coined until 1944 (Freisleben and Lein, 1944). The reverse genetic tool of TILLING (Targeting Induced Local Lesions in Genomes) also depends on chemically induced or natural mutations and can be used to identify individuals that carry specific genetic regions that may or may not result in gene knockout (McCallum *et al.*, 2000). An advantage of inducing mutations is that it can generate very high numbers of genetic changes in some species. For

example, in a wheat TILLING experiment, after selfing plants germinated from seeds exposed to ethyl methanesulfonates (EMS), it was estimated that individual plants carried an average of 340,000 mutations (Chen *et al.*, 2012). TILLING also exploits high-throughput molecular methods to screen for mutations in specific genes, such as targeted sequencing or mismatch cleavage assays that utilize specific endonucleases such as CEL1 (Kurowska *et al.*, 2011).

An alternative method to generate knockout mutants is via insertional mutagenesis, which utilizes the random integration of transgenes (often by *Agrobacterium* T-DNAs) to interrupt gene function. The main drawback of this method is the low efficiency of generating T-DNA insertions in functional genes. For this reason, insertional mutagenesis is fully applicable only in highly transformable plant species with small genomes, such as *Arabidopsis* and rice (Bolle *et al.*, 2011). However, it also has advantages in that the resulting mutation can be tagged using a reporter gene incorporated within the T-DNA and thus can be modified to identify promoter sequences in genomes. For example, a promoterless GUS construct randomly inserted into the cereal *Tritordeum* identified anther-specific expression patterns (Salgueiro *et al.*, 2002).

All the methods described above generate random changes in genomic DNA where the location and type of mutation cannot be predicted. Of these mutations, only a subset would reside in a target gene and only some of these would result in knockout phenotypes. While whole genome or other sequencing strategies may retrospectively be able to identify the sequence changes generated, there remain two major drawbacks of these methods. Firstly, because it is impossible to target the mutations, very large numbers of individuals, each carrying a very large number of mutations, must be generated. As a consequence of this, considerable cost in terms of time, labour and money must be invested to screen large populations of individuals to identify those carrying any useful mutations. In addition, to 'clean up' the desired mutation from the many unwanted DNA changes in the same individuals, back-crossing to a recurrent parent for many generations is needed.

6.3 Gene Editing

Gene editing is a set of molecular tools developed over the past few decades that aims to precisely change an organism's genome in a targeted fashion. Although many engineered nucleases have been developed to perform this task, CRISPR coupled with a CRISPR-associated protein (CAS), which is based on a natural system found in *Streptococcus pyogenes*, has proved to be the most facile and popular (Martínez-Fortún *et al.*, 2017). Other programmable nuclease systems include MN, ZFN and TALENs but, in general, these require more work to get them functioning optimally for each new target sequence. Two common features of these editing tools are the ability to scan the host genome for the pre-determined short DNA sequence and the subsequent binding of an exonuclease to generate a double-strand break (DSB) at the target site. Plant cells tend to repair DSBs in nuclear DNA using the error-prone, non-homologous end-joining (NHEJ) pathway, which can introduce small insertions and deletions (indels) at the cut site. Although the location of the cut site can be precisely pre-determined by the design of the guide sequence, the exact mutation resulting from erroneous repair cannot. Thus, in practice, many different gene-edited individuals must be generated and screened by sequencing or phenotype for the desired knockout or other endpoint. A more deterministic variation of gene editing is to supply a short additional DNA fragment, which may or may not have homology to the flanking regions of the cut site. This can be incorporated into the host genome in a targeted manner either by the NHEJ pathway or, if sufficient identical sequence overlap is present, by an alternative minor repair pathway, known as homology directed repair (HDR), which can be also exploited to insert a DNA fragment into the DSB site. Like conventional mutation breeding, gene editing to generate mutations can result in knockouts or the synthesis of aberrant proteins. Where these mutations are in the coding regions of genomic DNA, the altered expression will appear in all cell types at all developmental stages. Recent developments have further expanded the capacity of the CRISPR-Cas system to produce, for example, nickases that cut only one DNA strand, methods to edit many targets simultaneously and Cas variants lacking nuclease activity that instead can recruit synthetic enhancers or repressors to alter gene expression. Using specific repressors, it has been possible to achieve heterochromatin-mediated gene silencing (termed CRISPRi) (Gilbert *et al.*, 2013). However, we still lack a full understanding of the rules by which a given guide RNA may engage and be active on a given target site (Boettcher and McManus, 2015).

Thus, while there may be theoretical approaches to use of gene editing for silencing or to give tissue-specific or developmentally regulated alterations of expression, current commercial products under development (of which the author is aware) lack the subtle control of expression possible with RNAi.

6.4 RNA Interference

Post-transcriptional gene silencing via RNAi is a series of molecular interactions that lead to the suppression of target gene translation. There are several pathways of epigenetic regulation of gene expression found in cells but double-stranded RNA (dsRNA) designed to the coding region of an endogenous gene often leads to post-transcriptional degradation of target gene mRNA. To achieve RNAi in plants, dsRNA designed to complement the target sequence is inserted into cells, where it is cleaved by Dicer, incorporated into the RNA-induced silencing complex (RISC) and acts as a guide for Argonaute to degrade mRNAs specific to the target gene (Baulcombe, 2000). Transgene-induced RNAi requires a genetic transformation step and for some species it is relatively straightforward to design the necessary plasmid constructs to produce a dsRNA sequence and to transform plants so that they routinely display silencing of the gene target. However, it is also possible to observe transient silencing in transformed tissues when stable and heritable germline expression of dsRNA molecules is not the intention. For example, RNAi has been demonstrated following high-pressure spray application of siRNA into plant cells (Dalakouras *et al.*, 2016) and by physical rubbing of virus particles on to leaves as in virus-induced gene silencing (VIGS) approaches (reviewed by Robertson, 2004). In addition, feeding or soaking animals such as nematodes and certain insects with dsRNA can induce robust silencing (reviewed in Britton *et al.*, 2012 and Christiaens *et al.*, 2018).

The levels of gene silencing that result from transgene-induced RNAi is highly variable, ranging from no apparent effect to high levels of suppression where expression of the target gene is undetectable. The most common outcome is partial suppression and, even when the same dsRNA cassette is used, different transgenic events can be found with different levels of silencing (Eamens *et al.*, 2008). This can be advantageous if reduced expression rather than complete knockout phenotypes is desired. By choosing tissue-specific, developmentally regulated or inducible promoters to drive the dsRNA cassette, it is possible to direct silencing to specific cells (Tuteja *et al.*, 2004; Rao and Wilkinson, 2006). This is a significant advantage of RNAi over the genomic knockout methods described above. However, silencing has also been observed in transgenic lines where RNAi was not intended, for example in events possessing multiple copies of gene cassettes intended for expressing functional genes. This silencing is difficult to predict, being variable between lines and over time (Howarth *et al.*, 2005).

VIGS is a particularly well used research tool that has significant advantages over other techniques used for reverse genetic analysis of gene function. It is relatively rapid, facile and has low start-up costs. The optimization of several virus vectors for different plant species makes VIGS attractive for research in many monocot and dicot crops. VIGS can be used to rapidly screen many tens or hundreds of candidate genes because it does not need the stable, germline transformation step associated with T-DNA mutagenesis or transgene-induced RNAi.

As described above, RNAi can be readily used to silence native genes to alter the biochemistry or other phenotypic characteristics in the host organism. In addition, RNA silencing has been exploited as a powerful tool for engineering pest resistance into crop plants and the strategies to achieve this via mutation breeding or gene editing are still in their infancy. For example, a range of approaches have been deployed to silence the

expression of viral components in crops such as papaya, squash, banana, plum and common beans (Wang *et al.*, 2012). There are also many examples of plant-derived and sprayable sources of dsRNA being successfully used against pest insects (Bachman *et al.*, 2013). Although there have been attempts to adapt genome editing for control of plant viruses (Mushtaq *et al.*, 2019), it is not clear whether this will ever become a viable alternative to RNAi (Romay and Bragard, 2017).

6.5 Regulatory Considerations

In addition to the scientific rationale for choosing one methodology over another, where the equivalent end point can be achieved by more than one method, there may also be regulatory factors that influence the decision (Jones, 2015).

Applications of plant-derived RNAi necessitate the generation of a genetically modified organism (GMO) and so would be captured by GMO legislation and liable for risk assessment, authorization and, in some countries, product labelling. The exact procedures vary between the competent authorities in different countries but many are costly and have long time-frames. This is a particular issue in the EU, driven by the perverse voting patterns of various member states in advisory committees, which results in considerable uncertainty regarding the outcome. Silencing resulting from dsRNA sprays or other topical applications are not yet commercially available and there is considerable discussion regarding how they will be regulated.

Several countries in North and South America, along with others in Asia, have ruled that products of simple gene editing are not GMOs and would be regulated as any other conventionally bred variety. Prior to the European Court of Justice (ECJ) ruling on mutagenesis in 2018, there was a general expectation that the EU would follow a similar path by treating gene-edited mutations in a similar manner to classical mutagenesis with both being exempted from EU GMO legislation. However, the ECJ ruled that gene editing and other new forms of mutagenesis did not fit the exemption and that even simple mutations generated by gene editing must be regulated as GMOs. Although there is an expectation that the EC will revisit this situation sometime in the future, as of now both plant-derived RNAi and gene editing are GMOs in EU law and have similar expectations in terms of risk assessment and labelling.

New lines produced by mutation breeding, whether incorporating wild mutations found naturally in the gene pool, or induced artificially by radiation or chemical mutagens, are dealt with as any new conventional variety. The exact procedures vary from region to region but in the EU they involve national trials and listing in the plant variety catalogue. Technically, mutation breeding is defined in Directive EC 2001/18 as a technique of genetic modification but exempted from the regulation because it was considered at the time to have a history of safe use. Thus, classical mutation breeding or TILLING, a relatively rapid method to generate characterized mutations in target genes, would be advantageous from a regulatory perspective in that new varieties would not require expensive GMO authorization or labelling.

6.6 Conclusions

Recent years have seen great advances in developing technologies for modulating gene expression in plants. Induced mutation breeding is hampered by its lack of precision, the numbers of mutations per individual and large population sizes required. It also needs to use genetic segregation over multiple generations to remove the undesired mutations from the breeding lines. However, the significant benefit of its non-GMO status in law means that it will always have a place in some breeding programmes that lack accessible variation in target traits.

The CRISPR/Cas9 technology can be used to edit nucleotide sequences of gene coding regions, regulatory elements or other selected genomic loci in plants. Early commercial examples have been simple loss-of-function alleles, because these are straightforward to generate. In the immediate future, commercial gene editing will likely focus on traits under simple genetic control and where the results of modification are already well understood from null alleles in existing gene pools or other knockout or silencing approaches, such as induced mutations or RNA interference (Martínez-Fortún et al., 2017). In regions of the world where simple gene edits are not governed by overburdensome GMO regulations and where food from these plants has broad consumer acceptance, gene editing is likely to displace RNAi approaches for applications where complete knockout phenotypes are desired. However, RNA silencing is now a well-established and easy-to-use technology, which will continue to serve as a useful tool in gene function analysis and crop improvement. Where complete knockout of genes is undesirable or indeed lethal to plants, or where silencing is required in some cells and not others, RNAi is the preferred method. With continuing efforts in further understanding the RNA silencing mechanisms in plants, it can be anticipated that RNA silencing technologies will be further improved to overcome potential limitations, allowing for wider applications in agriculture.

Acknowledgements

The Institute of Biological, Environmental and Rural Sciences (IBERS) receives strategic funding from the Biotechnology and Biological Sciences Research Council (BBSRC) via grant [BBS/E/W/0012843].

References

Bachman, P.M., Bolognesi, R., Moar, W.J., Mueller, G.M., Paradise, M.S. et al. (2013) Characterization of the spectrum of insecticidal activity of a double-stranded RNA with targeted activity against Western Corn Rootworm (Diabrotica virgifera virgifera LeConte). Transgenic Research 22(6), 1207–1222. DOI: 10.1007/s11248-013-9716-5.

Baulcombe, D.C. (2000) Molecular biology. unwinding RNA silencing. *Science* 290(5494), 1108–1109. DOI: 10.1126/science.290.5494.1108.

Boettcher, M. and McManus, M.T. (2015) Choosing the right tool for the job: RNAi, TALEN, or CRISPR. *Molecular Cell* 58(4), 575–585. DOI: 10.1016/j.molcel.2015.04.028.

Bolle, C., Schneider, A. and Leister, D. (2011) Perspectives on systematic analyses of gene function in *Arabidopsis thaliana*: new tools, topics and trends. *Current Genomics* 12(1), 1–14. DOI: 10.2174/138920211794520187.

Britton, C., Samarasinghe, B. and Knox, D.P. (2012) Ups and downs of RNA interference in parasitic nematodes. *Experimental Parasitology* 132(1), 56–61. DOI: 10.1016/j.exppara.2011.08.002.

Chen, L., Huang, L., Min, D., Phillips, A., Wang, S. *et al.* (2012) Development and characterization of a new TILLING population of common bread wheat (*Triticum aestivum* L.). *PLoS ONE* 7, e41570. DOI: 10.1371/journal.pone.0041570.

Christiaens, O., Dzhambazova, T., Kostov, K., Arpaia, S., Reddy Joga, M. *et al.* (2018) *Literature review of baseline information on RNAi to support the environmental risk assessment of RNAi-based GM plants*. EFSA Supporting Publication 15(5). European Food Safety Authority, Parma, Italy, p. 173.

Cong, L., Ran, F.A., Cox, D., Lin, S., Barretto, R. *et al.* (2013) Multiplex genome engineering using CRISPR/Cas systems. *Science* 339(6121), 819–823. DOI: 10.1126/science.1231143.

Dalakouras, A., Wassenegger, M., McMillan, J.N., Cardoza, V., Maegele, I. *et al.* (2016) Induction of silencing in plants by high-pressure spraying of in vitro-synthesized small RNAs. *Frontiers in Plant Science* 7, 1327–1327. DOI: 10.3389/fpls.2016.01327.

Eamens, A., Wang, M.-B., Smith, N.A. and Waterhouse, P.M. (2008) RNA silencing in plants: yesterday, today, and tomorrow. *Plant Physiology* 147(2), 456–468. DOI: 10.1104/pp.108.117275.

Freisleben, R.A. and Lein, A. (1944) Möglichkeiten und praktische Durchführung der Mutationszüchtung. *Kühn-Arhiv* 60, 211–222.

Gilbert, L.A., Larson, M.H., Morsut, L., Liu, Z., Brar, G.A. *et al.* (2013) CRISPR-mediated modular RNA-guided regulation of transcription in eukaryotes. *Cell* 154(2), 442–451. DOI: 10.1016/j.cell.2013.06.044.

Guha, T.K., Wai, A. and Hausner, G. (2017) Programmable genome editing tools and their regulation for efficient genome engineering. *Computational and Structural Biotechnology Journal* 15, 146–160. DOI: 10.1016/j.csbj.2016.12.006.

Howarth, J.R., Jacquet, J.N., Doherty, A., Jones, H.U.W.D. and Cannell, M.E. (2005) Molecular genetic analysis of silencing in two lines of *Triticum aestivum* transformed with the reporter gene construct pAHC25. *Annals of Applied Biology* 146(3), 311–320. DOI: 10.1111/j.1744-7348.2005.040121.x.

Jinek, M., Chylinski, K., Fonfara, I., Hauer, M., Doudna, J.A. *et al.* (2012) A programmable dual-RNA-guided DNA endonuclease in adaptive bacterial immunity. *Science* 337(6096), 816–821. DOI: 10.1126/science.1225829.

Jones, H.D. (2015) Regulatory uncertainty over genome editing. *Nature Plants* 1, 14011. DOI: 10.1038/nplants.2014.11.

Kurowska, M., Daszkowska-Golec, A., Gruszka, D., Marzec, M., Szurman, M. *et al.* (2011) TILLING: a shortcut in functional genomics. *Journal of Applied Genetics* 52(4), 371–390. DOI: 10.1007/s13353-011-0061-1.

Mali, P., Yang, L., Esvelt, K.M., Aach, J., Guell, M. *et al.* (2013) RNA-guided human genome engineering via Cas9. *Science* 339(6121), 823–826. DOI: 10.1126/science.1232033.

Martínez-Fortún, J., Phillips, D.W. and Jones, H.D. (2017) Potential impact of genome editing in world agriculture. *Emerging Topics in Life Sciences* 1, 117.

McCallum, C.M., Comai, L., Greene, E.A. and Henikoff, S. (2000) Targeted screening for induced mutations. *Nature Biotechnology* 18(4), 455–457. DOI: 10.1038/74542.

Mushtaq, M., Sakina, A., Wani, S.H., Shikari, A.B., Tripathi, P. *et al.* (2019) Harnessing genome editing techniques to engineer disease resistance in plants. *Frontiers in Plant Science* 10, 550. DOI: 10.3389/fpls.2019.00550.

Rao, M.K. and Wilkinson, M.F. (2006) Tissue-specific and cell type-specific RNA interference in vivo. *Nature Protocols* 1(3), 1494–1501. DOI: 10.1038/nprot.2006.260.

Robertson, D. (2004) VIGS vectors for gene silencing: many targets, many tools. *Annual Review of Plant Biology* 55, 495–519. DOI: 10.1146/annurev.arplant.55.031903.141803.

Romay, G. and Bragard, C. (2017) Antiviral defenses in plants through genome editing. *Frontiers in Microbiology* 8, 47. DOI: 10.3389/fmicb.2017.00047.

Salgueiro, S., Matthes, M., Gil, J., Steele, S., Savazzini, F. *et al.* (2002) Insertional tagging of regulatory sequences in tritordeum; a hexaploid cereal species. *Theoretical and Applied Genetics* 104(6-7), 916–925. DOI: 10.1007/s00122-001-0836-6.

Tuteja, J.H., Clough, S.J., Chan, W.-C. and Vodkin, L.O. (2004) Tissue-specific gene silencing mediated by a naturally occurring chalcone synthase gene cluster in *Glycine max*. *The Plant Cell* 16(4), 819–835. DOI: 10.1105/tpc.021352.

Wang, M.-B., Masuta, C., Smith, N.A. and Shimura, H. (2012) RNA silencing and plant viral diseases. *Molecular Plant-Microbe Interactions* 25(10), 1275–1285. DOI: 10.1094/MPMI-04-12-0093-CR.

7 Application of RNAi Technology in Forest Trees

Matthias Fladung[1]*, Hely Häggman[2] and Suvi Sutela[3]

[1]*Thuenen-Institute of Forest Genetics, Grosshansdorf, Germany; [2]University of Oulu, Oulu, Finland; [3]Natural Resources Institute Finland, Helsinki, Finland*

Abstract

A diverse set of small RNAs is involved in the regulation of genome organization and gene expression in plants. These regulatory sRNAs play a central role for RNA in evolution and ontogeny in complex organisms, including forest tree species, providers of indispensable ecosystem services. RNA interference is a process that inhibits gene expression by double-stranded RNA and thus causes the degradation of target messenger RNA molecules. Targeted gene silencing by RNAi has been utilized in various crop plants in order to enhance their characteristics. For forest tree species, most of the successful RNAi modification has been conducted in poplar. Over the past 20 years, successful RNAi-mediated suppression of gene expression has been achieved with a variety of economically important traits. Moreover, the stability of RNAi-mediated transgene suppression has been confirmed in field-grown poplars. In this chapter, we describe examples of successful RNAi applications mainly in poplar but also provide some information about application of RNAi in pest control in forest tree species. Advantages and disadvantages of this technology with respect to the particular features of forest tree species will be discussed.

7.1 Introduction

Forests contribute profoundly to human well-being by providing a diverse set of important ecosystem services. These services may be divided into regulating, provisioning (e.g. production of timber and non-timber products) and cultural services. The regulating ecosystem services include various vital processes such as fire-risk prevention and soil erosion control as well as water and climate regulation. Indeed, forests play a major role in the global carbon cycle by absorbing CO_2 and storing it in their biomass through photosynthesis. In contrast, deforestation elevates atmospheric CO_2 levels and it has been estimated that deforestation and forest degradation can account for 26% of the CO_2 emissions since 1870 (Le Quéré *et al.*, 2016). Regardless of the well-recognized importance of forests, the global forest land area continues to decline as forests are converted to other land uses (FAO, 2016), predominantly commercial and subsistence agriculture (Whiteman, 2014). The loss of natural intact forests (primary forest, see Box 7.1.) is alarming, as these forests have greater capability to adapt to environmental changes and short-term climatic anomalies than forests that have been under human influence (Watson *et al.*, 2018). Furthermore, intact forests support globally significant environmental values such

---

*Corresponding author: matthias.fladung@thuenen.de

© CAB International 2021. *RNAi for Plant Improvement and Protection*
(eds B. Mezzetti *et al.*)
DOI: 10.1079/9781789248890.0007

Box 7.1. How primary and planted forests can be defined

Forest is determined both by the presence of trees in a land area (≥ 0.5 ha) and the absence of other predominant land uses (FAO, 2012).

Primary forest is naturally regenerated forest composed of native tree species with natural forest dynamics, including species composition, occurrence of dead wood, age structure and regeneration processes (FAO, 2012). Furthermore, there should be no clearly visible indications of human activities, the area is large enough to maintain its natural characteristics and the ecological processes are not significantly disturbed.

Planted forest is predominantly composed of trees established through planting and/or deliberate seeding (FAO, 2012). Planted forests include plantation forests and semi-natural forests. **Plantation forests** can be defined as intensively managed planted forest aimed for commercial production of wood and non-wood forest products, or production of specific environmental service (Carle and Holmgren, 2003). **Semi-natural** forest can be defined as a managed forest having some of the principal characteristics and key elements of native ecosystems (e.g. complexity, structure and diversity) and which is predominantly composed of native species (FAO, 2002).

---

as conservation of biodiversity; therefore, extra efforts should be made for their preservation. Halting deforestation and restoring degraded forests are important, as loss of forests threatens sustainable development as well as human well-being (Watson *et al.*, 2018).

Simultaneously, the demand for wood biomass and other bio-based products is increasing with the needs of growing human populations and bio-based economies. Jürgensen *et al.* (2014) estimated that, in the year 2012, natural forests supplied most of industrial roundwood, while production of forest plantations was 33% (562 million m³). Projections on industrial roundwood supply indicate an increase of 67% in plantation wood production over the period 2000 (624 million m³) to 2040 (1043 million m³). The area of planted forests (Box 7.1) has increased since 1990, with an average annual rate of 3.2 million ha for the period 2010–2015 (FAO, 2016). However, it is likely that climate change and food production pressures will restrict land availability for planted forests, and thus create a need for more intensive management regimes for existing forests, including improved health management (Payn *et al.*, 2015).

Wood production can be enhanced with improved plantation management, which includes soil preparation, weed and pathogen control and fertilization in addition to utilization of improved tree varieties (Häggman *et al.*, 2013). Production of tree varieties with improved traits (growth, stem characteristics, abiotic and biotic resistance) has

been the goal of modern tree breeding programmes launched in the 1950s. Due to the characteristics of forest tree species (Fig. 7.1), tree breeding is a slow and costly process. The tree breeding cycle typically consists of selection, field testing, controlled crossings and progeny/clonal testing. As tree species have a long juvenile phase, one must wait for years before trees flower. Moreover, it also takes years to be able to assess the phenotype of the progeny: DNA-based molecular markers have not yet enabled early selection of material, because of the complex patterns of inheritance of desired tree traits (Häggman *et al.*, 2014). For instance, the breeding cycle for Scots pine was estimated to take 40 years (Ruotsalainen, 2014) or, if progeny tests were omitted, less than 30 years (Rosvall and Mullin, 2013). Genomic selection utilizing single nucleotide polymorphism (SNP) as genome-wide markers in predicting phenotypes may speed tree domestication by accelerating breeding cycles, increasing selection intensity and improving the accuracy of breeding values (Grattapaglia *et al.*, 2018). Isik and McKeand (2019) reported on the fourth cycle of loblolly pine breeding of the Cooperative Tree Improvement Program at North Carolina State University, initiated in 1960. The authors were positive that high-quality SNP markers and SNP array available for loblolly pine will be a major advantage and that the predictive power of SNP markers will be verified in the near future.

Forest tree breeding can also be accelerated by using genetic engineering. Genetic

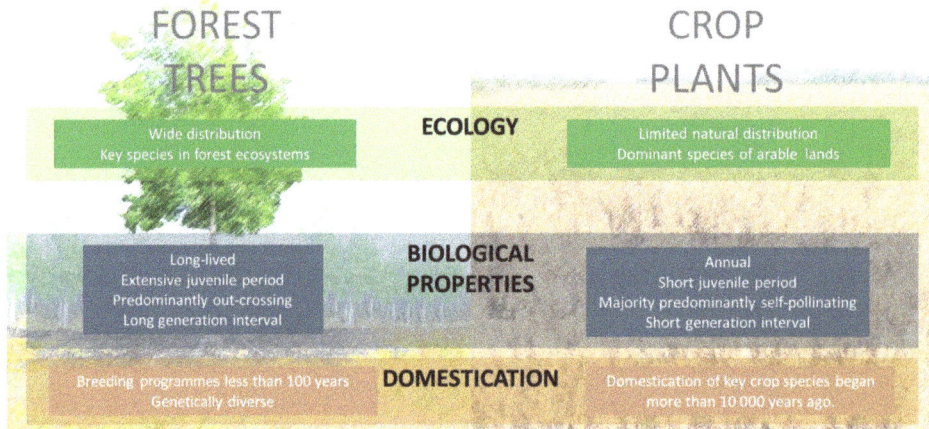

Fig. 7.1. The characteristics of forest trees species differ from annual crop plants and thus also the degree of domestication and breeding practices. Today, improved forest tree varieties produced in breeding programmes can be considered undomesticated if compared with crop plants.

engineering enables expression or repression/ silencing of targeted genes (recombinant or endogenous) at certain developmental stages, in different tissues or by specific environmental cues (Hernandez-Garcia and Finer, 2014). In general, the economically most significant tree characteristics have been successfully modified with genetic engineering, including wood properties and productivity as well as abiotic and biotic resistance of trees (e.g. Häggman *et al.*, 2013, 2014; Séguin *et al.*, 2014; Chang *et al.*, 2018). The stability of genetic modification, in addition to issues related to flowering onset and fertility, has been demonstrated as trustworthy in greenhouse and field studies (Häggman *et al.*, 2013, 2016). Several different genetic engineering approaches repress the expression of the target gene via the RNA interference (RNAi) process present naturally in cells containing a nucleus. RNAi inhibits gene expression by double-stranded RNA (dsRNA) and thus causes the degradation of target messenger RNA (mRNA) molecules. Moreover, RNAi includes the suppression of the transcription of the target gene and also inhibition of the target gene translation. In this chapter we cover examples of successful RNAi-mediated genetic modifications conducted with dsRNA producing gene

constructs and poplar, a model woody tree species in plant science.

7.2 Discovery of RNAi

For a long time, RNA was believed to only act as a messenger between DNA and protein; however, discoveries in the past ten years suggest that RNA is also involved in the regulation of genome organization and gene expression. Evidence has been obtained that regulatory RNA molecules play a central role for RNA in evolution and ontogeny in complex organisms, including tree species. Regulatory RNA comprises all types of small RNA molecules (sRNAs), including micro- and small interfering RNAs (miRNAs, siRNAs) that mediate the silencing effect of RNA interference (RNAi), an antiviral defence system discovered by Andrew Fire and Craig Mello.

Andrew Fire started collaboration with Craig Mello at the Carnegie Institution in Baltimore, Maryland, in 1986. The pioneering gene expression studies were done using *Caenorhabditis elegans* worms and injecting mRNA (sense RNA) of a gene encoding for muscle protein production (*unc-22*); however, no responses from the worms were found, neither did injecting the worms with antisense RNA cause twitching movements typical for reduction of *unc-22*. Twitching movements from the worms were only detected when both the sense and the antisense RNA were applied, indicating silencing of the worm *unc-22* gene. These findings, published in 1998 (Fire *et al.*, 1998), led to the Nobel Prize in Physiology or Medicine in 2006 being awarded jointly to Andrew Z. Fire and Craig C. Mello 'for their discovery of RNA interference – gene silencing by double-stranded RNA'. It turned out that the role of introns in DNA was to code for RNAi elements. These early discoveries of RNAi technology were groundbreaking for all the applications presented in this chapter.

Before the universal mechanisms of RNAi were revealed, the RNAi phenomenon had been observed 30 years ago in plants when attempts were made to overexpress chalcone synthase (*CHS*) in petunia in order to make the flower colour more purple (Jorgensen, 1990; Napoli *et al.*, 1990). However, instead of *CHS* overexpression, the gene was suppressed in varying levels, resulting in white-purple variegated and even white-coloured petunia flowers. Since its discovery, RNAi has been found to be common in almost all organisms as a basic biological process serving protection against viral infections and disabling the spread of transposable elements within a genome. RNAi induced silencing has been used widely in basic and applied research to functionally characterize gene-of-interest by loss of function and the RNAi mechanism has also been extensively used in crop protection platforms. So far, RNAi approaches have been conventionally based on the use of transgenic plants expressing dsRNAs against selected targets. However, the use of transgenes and genetically modified (GM) organisms has raised considerable scientific and public concerns; hence the need for alternative approaches has emerged, as underlined by Dalakouras *et al.* (2020).

7.3   RNAi in Plants

RNAi enables regulation of gene expression at transcriptional and post-transcriptional level mediated via target mRNA cleavage and/or translation inhibition. RNAi-mediated post-transcriptional gene silencing (PTGS) can be achieved with different genetic engineering approaches, including artificial/synthetic miRNA-induced gene silencing (MIGS) as well as virus- and host-induced gene silencing (VIGS, HIGS), while siRNAs can be exploited in transcriptional gene silencing (TGS).

In the RNAi process, 'exogenous' dsRNAs, originating from viral replication, transgenes or transposons are first recognized and cleaved in a cell by Dicer-like endonucleases into 21–24 nt short siRNAs. In plants, there are several Dicer-like endonucleases producing siRNAs with characteristic 3′ and 5′ termini (Bologna and Voinnet, 2014; Borges and Martienssen, 2015). The RNase III family enzyme Dicer-like1 generates miRNA/miRNA duplexes by processing imperfect primary hairpin RNAs (pri-miRNAs) encoded by the plant microRNA (*MIR*) genes. Both siRNAs and miRNAs possess 3′ overhangs of 2 nt which are stabilized by methylation by Hue enhancer1 (Yu *et al.*, 2005; Yang *et al.*, 2006). The 5′ terminal nucleotide of siRNAs and miRNAs will determine which one of the two sRNAs is loaded on to Argonaute protein, a core component of RNA-induced silencing complex (RISC).

The loaded sRNA acts as a guide for Argonaute in a search for complementary transcripts (mRNA) that are degraded (Hamilton and Baulcombe, 1999; Mi *et al.*, 2008). In addition, Argonaute, together with sRNAs of 22 nt, will direct RNA-directed RNA polymerase to the 3′ of the target RNA, which leads to transcription of the target and generation of dsRNA. Subsequently generated secondary siRNAs, defined as transitive RNAi, cause systemic genetic interference. Moreover, Argonaute 24-nt sRNA complexes guide DNA methyltransferases to cognate DNA or its nascent transcript, leading to methylation of cytosines in both DNA strands (RNA-directed DNA methylation) (RdDM) and suppression of transcription (Wassenegger *et al.*, 1994; Chan *et al.*, 2004). In plants, the RNA silencing signal spreads to adjacent cells and long-distance through the vascular system, creating systemic signalling. The exact mechanism of RNAi signal movement is still undetermined. Cell-to-cell transportation is likely to occur through plasmodesmata, while long-distance transport has been shown to involve siRNA and miRNAs 21–24 nt long (Dunoyer *et al.*, 2010; Mermigka *et al.*, 2016).

7.4 Functioning of RNAi Vectors in Poplar

RNAi involves the silencing of a target gene by introduction of dsRNA corresponding to the sequences within the target gene to be silenced. Different software can be used in the design of RNAi constructs. The efficient prediction of long dsRNA RNAi constructs can be conducted, for instance, with siDirect (Naito *et al.*, 2009) or siRNA-Finder (Si-Fi) (Lück *et al.*, 2019), the latter incorporated with a tool specifically intended for VIGS, HIGS and miRNA. To make dsRNA, one needs to transcribe both sense and antisense strands of RNA from a complementary DNA (cDNA) and allow them to anneal. This is achieved by utilizing a construct (vector) containing a partial sequence of the target gene which is subsequently expressed in the plant cell. There are several different plasmid vectors available for this purpose; the target sequence may be cloned to both sides of intron-containing hairpin RNA (ihpRNA) vector in antisense and sense orientation; or the target sequence may be surrounded by two promoters, as presented below.

A set of RNAi vectors was constructed and transferred to poplar by Meyer *et al.* (2004). To address the question of silencing, the *GUS* reporter gene was applied as a test system. The functionality of these dsRNA-forming vectors was then proofed in *GUS*-transgenic poplar in both transient assays by transforming protoplasts with the RNAi constructs and in stably transformed *GUS*-expressing poplar (Meyer *et al.*, 2004). Based on the observation that the RNAi:*GUS* construct with the Intron290 spacer showed the strongest downregulation of the reporter gene, the authors concluded that the RNAi vectors are functional in poplar.

A novel RNAi approach without spacer but with two promoters flanking the gene to be silenced has been proposed by DNA Cloning Service (Hamburg, Germany) (Fig. 7.2). The advantage of this approach is that no cloning of sense and antisense sequences of the gene to be silenced is needed. The approach was first tested in transgenic poplar constitutively expressing the *GUS* gene under the cauliflower mosaic virus 35S promoter.

A modified RNAi construct carrying the *GUS* gene flanked by two 35S-promoters (Fig. 7.3A) was transferred to stably *GUS*-expressing poplar (M. Fladung, unpublished results). Silencing of the *GUS* gene was validated in GUS-staining experiments of chlorophyll-less leaf discs harvested from different independent *GUS*:RNAi-35::*GUS*-transgenic poplar lines. The GUS-stains ranged from slightly decreased blue to nearly completely white leaf discs (M. Fladung, unpublished results, Fig. 7.3B).

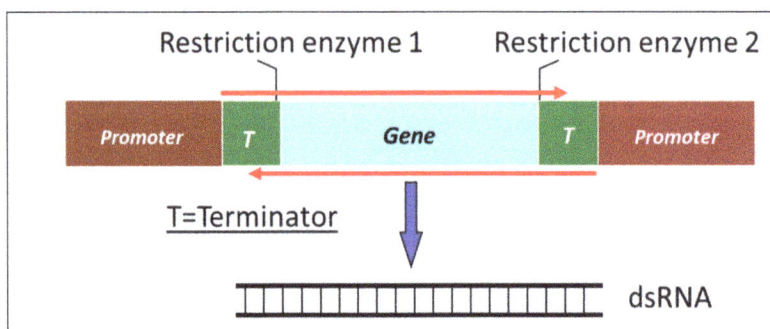

Fig. 7.2. Schematic representation of the novel RNAi approach without intron spacer but with two promoters flanking the gene to be silenced. Promoters can be constitutive, inducible, or tissue and developmental specific. T = terminator, dsRNA = double-stranded RNA. Red arrows indicate the transcription directions. (Source: DNA Cloning Service, Hamburg, Germany).

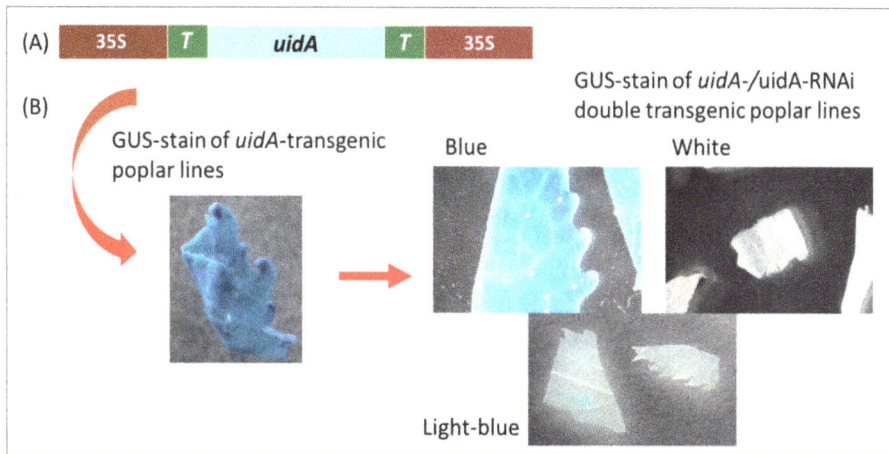

Fig. 7.3. (A) Modified RNAi construct carrying the *uidA* gene (encoding β-glucuronidase, GUS enzyme) flanked by two 35S-promoters.
T = terminator. (B) GUS-staining experiments. Left: blue-stained leaf disc from 35::*GUS*-transgenic poplar. Right: different GUS-staining intensities in different independent *GUS*:RNAi-35::*GUS*-transgenic poplar lines. The blue colour indicates the activation of the reporter gene and, thus, the functioning of the RNAi construct. (Source: DNA Cloning Service, Hamburg, Germany).

7.4.1 Flowering time genes and genetic containment

Flowering onset is very important with respect to yield in many plant species. Unravelling the interactions of genes involved in flowering time is, therefore, of high interest for crop breeders. In addition, a variety of growth factors, secondary metabolites and exogenous compounds have been shown to influence flowering time in annuals and perennial plants (Ionescu *et al.*, 2016). The role of genes controlling these processes or the identification of genes inducing flowering has mostly been studied in *Arabidopsis* at the beginning of this century; however, some of the genes are being analysed in the *Populus* tree model system by applying RNAi suppression.

Tylewicz *et al.* (2015) studied the possible participation of poplar homologues of the evolutionarily conserved basic-leucine zipper (bZIP) domain transcription factor *FD* and the *Arabidopsis FLOWERING LOCUS T* (*FT*) on floral transition by using gain of function and RNAi-suppressed *FD* transgenic plants. Following the identification of two *FD*-like homologues (*FDL1* and *FDL2*) in *Populus*, the authors studied the role of both *FDL* genes by RNAi suppression. In

addition to being primarily involved in flowering induction in combination with *FT*, it seems that, independently from *FT*, *FD* has dual roles in the photoperiodic control of seasonal growth and stress tolerance in trees (Tylewicz *et al.*, 2015).

From another well-known flowering *Arabidopsis* time regulator, *GIGANTEA* (*GI*), which connects networks involved in developmental stage transitions and environmental stress responses, only a little is known about its role in poplar (Ke *et al.*, 2017). The authors identified three *GI*-like genes in poplar, and following overexpression and RNAi suppression of these genes, *Arabidopsis GI* functions seemed to be conserved in poplar. Downregulation of the poplar *GI*-like genes by RNAi led to vigorous growth, higher biomass and enhanced salt stress tolerance in transgenic poplar plants (Ke *et al.*, 2017).

So far, the function of the floral homeotic genes *AGAMOUS* (*AG*) and *SEEDSTICK* (*STK*) in the development of poplar catkins have been studied by Lu *et al.* (2019). RNAi co-suppression of both the two *AG* and the two *STK* paralogues led to modifications in poplar floral phenotypes, e.g. carpel-inside-carpel phenotypes, complete disruption of seed production, or sterile anther-like organs (Lu *et al.*, 2019), but without changes in biomass growth or leaf morphology. Lu *et al.* (2019) concluded that *AG* and *STK* gene functions are strongly conserved during poplar catkin development.

Genetically modified forest trees, including poplar, eucalyptus and pine, have been produced in many laboratories in the world and safety has been tested in the field (Walter *et al.*, 2010). A few GM forest trees have been commercialized in China and the USA, but in Europe, market introduction is impeded by environmental concerns and political and social interference with the EU regulatory system (Custers *et al.*, 2016). One biosafety concern regarding commercialization of GM forest trees is possible transgene flow into wild tree populations. Reducing flower fertility or induction of complete flower sterility is a containment strategy and will probably be necessary before most commercial uses of GM trees are possible.

As early as in 2001, Meilan *et al.* (2001) applied RNAi to downregulate genes involved in flowering to engineer sterility in poplar. Although some genes were successfully downregulated, it could not be confirmed that RNAi can be applied as a long-term containment measure that is stable in the field under natural environmental conditions, because of expression changes during tree maturation. Stability of RNAi in the field has been investigated by Li *et al.* (2008) by testing 56 independent poplar RNAi transgenic events over 2 years (over winter-to-summer seasonal cycles). Here, the *BAR* resistance transgene was targeted with two different RNAi constructs. Although the degree of RNAi suppression varied widely, the authors found that it was highly stable in each event over the two years (Meilan *et al.*, 2001). The authors concluded that RNAi is highly effective for functional genomics and biotechnology of perennial plants. An effective containment strategy in transgenic trees was postulated by Klocko *et al.* (2016, 2018). Targeting of the poplar homologue of *LEAFY* (*LFY*) via RNAi resulted in a decrease in catkin size and loss of functional sexual organ development in field-grown poplar plants (Klocko *et al.*, 2016). RNAi silencing has also been successfully used in *Populus tremula* × *tremuloides* trees to engineer sterility with constructs targeting the *LFY* and *AGAMOUS* (*AG*) flowering genes (Klocko *et al.*, 2018).

Stability of RNAi in the field over several years has already been indicated by Meilan *et al.* (2001), Mohamed *et al.* (2010) and Klocko *et al.* (2016). A comprehensive study on stability of catkin sterility by testing over 3300 genetically engineered (GE) poplar trees and 948 transformation events in a single, 3.6 ha field trial was performed by Klocko *et al.* (2018). The goal was to assess modified RNA expression or protein function of floral regulatory genes, including *LFY*, *AG*, *APETALA1* (*AP1*), *SHORT VEGETATIVE PHASE* (*SVP*) and *FLOWERING LOCUS T* (*FT*) for seven growing seasons in the field. All modifications induced by the RNAi or overexpression constructs revealed stability over three to five flowering seasons (Klocko *et al.*, 2018). No somaclonal variation and no floral modification that was not related to the added transgene could be observed. This study has shown that RNAi-based sterility of catkins is stable and could be one successful containment option for transgenic forest trees (Klocko *et al.*, 2018).

7.5 Secondary Cell Wall Formation

Modifications of lignin and/or cellulose biosynthesis have been one of the major goals of tree biotechnology and molecular biology for more than 25 years to improve biofuel production from woody biomass, because its energy largely resides in plant cell walls. However, wood is composed of 40–50% cellulose, 15–20% hemicellulose and 25–30% lignin. The complex structure of lignified cell walls makes wood largely inaccessible to cellulases for cellulose degradation and breakdown into sugars (Hisano *et al.*, 2009). Because the presence of lignin is responsible for wood hardiness, downregulation of major lignin genes could lead to reduced lignin content and, therefore, increased accessibility of celluloses for cellulose degradation. On the other hand, modification of cellulose content is also of interest to improve wood quality and strength.

Modification of lignin biosynthesis by RNAi suppression of *4-coumaroyl-CoA 3'-hydroxylase* (*C3'H*) has been investigated by Coleman *et al.* (2008). *C3'H* catalyses the hydroxylation of 4-coumaroyl shikimate and 4-coumaroyl quinate. When downregulated, *C3'H* becomes a rate-limiting step in lignin biosynthesis. RNAi suppression of *C3'H* led to a significant decrease in total lignin content and to a significant shift in lignin monomer composition in the accumulation of phenylpropanoid glycosides (Coleman *et al.*, 2008). In another study, both *C3'H* and *SHIKIMATE HYDROXYCINNAMOYL TRANSFERASE* (*HCT*) were RNAi downregulated in transgenic poplar (Zhou *et al.*, 2018). Wood analyses revealed that lignin content was lower in the *C3'H/HCT* double RNAi transgenic poplar than in the non-modified control plants. In addition, wood anatomical characteristics like cell wall thickness, diameter of fibre cells and mechanical properties were changed in the transgenic poplars (Zhou *et al.*, 2018). Transgenic up- and downregulation of *4-coumarate:coenzyme A ligase 1* (*4CL1*) altered lignin content and composition in transgenic poplars (Tian *et al.*, 2013). *4CL1* ligates 4-coumarate with CoA. There were no negative effects on growth of the transgenic plants but an enhanced growth performance could be observed. The results suggest that *4CL1* is a traffic control gene in monolignol biosynthesis in poplar (Tian *et al.*, 2013). In another study, orthologues of *cinnamyl alcohol dehydrogenase* (*CAD*) and *cinnamoyl-CoA reductase* (*CCR*) were RNAi downregulated in *Populus trichocarpa* (Yan *et al.*, 2019). Suppression of *PtrCAD1* in transgenics led to reduced CCR protein activity in the stem-differentiating xylem, while downregulation of *PtrCCR2* caused a lower CAD protein activity. The results provide evidence for the formation of PtrCAD1/PtrCCR2 protein complexes in monolignol biosynthesis *in planta* (Yan *et al.*, 2019).

In plant cell walls, members of the cellulose synthase A gene family (*CesAs*) control cellulose biosynthesis in plant cell walls. To understand the functional role of single *CesA* genes in the complex pathway in *P. trichocarpa*, Abbas *et al.* (2020) RNAi downregulated *PtrCesA4*, *PtrCesA7-A/B* and *PtrCesA8-A/B* during wood formation. RNAi knockdown of *CesA* led to a dramatic decrease in cellulose content, possibly responsible for changes in phenotype, physiology and wood characteristics. *CesA*:RNAi poplar revealed stunted growth and narrow leaves, and the reduced mechanical strength may be due to thinner fibre cell walls (Abbas *et al.*, 2020). Xylem vessels in the *CesA*:RNAi poplar were collapsed, indicating that water transport in xylem

may be affected and thus causing early necrosis in leaves. The authors conclude that *PtrCesA4*, *PtrCesA7-A/B* and *PtrCesA8-A/B* are not only involved in wood formation but also trigger pleiotropic effects of their perturbations on wood formation (Abbas *et al.*, 2020).

The transcription factor *WOX4* regulates cell divisions in the cambium in *Arabidopsis*. *WOX4* is a key target of the *CLAVATA3* (*CLV3*)/*EMBRYO SURROUNDING REGION* (*ESR*)-*RELATED 41* (*CLE41*) signalling pathway. The functions of homologues of both genes during secondary growth were studied in *P. tremula* × *P. tremuloides* (Kucukoglu *et al.*, 2017). In *Populus*, *WOX4* homologues are specifically expressed only in the cambial region during vegetative growth but not after growth cessation and during dormancy (Kucukoglu *et al.*, 2017). Transgenic trees with RNAi downregulated poplar *WOX4* revealed unchanged primary growth; however, secondary growth was reduced. Further, the poplar *CLE41* homologues positively regulate the poplar *WOX4* homologues, indicating that regulation of vascular cambium activity between angiosperm and gymnosperm tree species is evolutionarily conserved (Kucukoglu *et al.*, 2017).

The functional role of secretory carrier-associated membrane proteins (SCAMPs) in wood formation of *Populus* has been studied by Obudulu *et al.* (2018). SCAMPs are highly conserved 32–38 kDa proteins that are involved in membrane trafficking. An RNAi vector to downregulate *SCAMP3* was constructed and transferred to *Populus tremula* × *tremuloides* (Obudulu *et al.*, 2018). Wood harvested from *SCAMP3* downregulated transgenic trees revealed increased amounts of both polysaccharides and lignin oligomers, indicating that SCAMP proteins influence accumulation of secondary cell wall components. Indeed, secondary cell walls from *SCAMP3*:RNAi transgenic trees deposited higher amounts of both carbohydrate and lignin (Obudulu *et al.*, 2018).

A very important component of hemicelluloses is xylan. Xylan is abundant in plant biomass, occurs mainly in all cell walls of grasses and is the second most abundant polysaccharide in secondary cell walls of dicot wood (Lee *et al.*, 2011; Mellerowicz and Gorshkova, 2011; Rennie and Scheller, 2014). Molecular dissection of xylan biosynthesis was performed by Li *et al.* (2011) through RNAi knockdown of several candidate genes. Members of glycosyl-transferase protein families GT8, GT43 and GT47 have been identified to be involved in the biosynthesis of xylan in the secondary cell walls of *Arabidopsis*. However, their functional role in xylan biosynthesis in poplar was largely unknown. Knockdown of poplar *GT8* homologues (*PtrGT8D1* and *PtrGT8D2*) through RNAi resulted in 29–36% reduction in stem wood xylan content (Li *et al.*, 2011). Interestingly, xylan reduction in poplar wood had essentially no effect on cellulose quantity but caused an 11–25% increase in lignin and anatomically an increased vessel diameter and thinner fibre cell walls (Li *et al.*, 2011). For GT43, five genes were shown to be highly expressed in the developing wood in the genome of poplar (Lee *et al.*, 2011). Downregulation of both *GT43B* and *PoGT8D* by RNAi in hybrid poplar led to smaller cell walls and lower xylan content in wood, indicating that both genes are involved in xylan biosynthesis in poplar wood (Lee *et al.*, 2011).

7.6 Seasonal Growth, Tree Architecture and Yield

Tree growth and architecture play an important role for biomass production (Teichmann and Muhr, 2015). Determinants of productivity are, among others, large leaves, sylleptic branching, narrow crown architecture, adapted activity of stomata and a compact root system. Improved trees could show tolerance towards desiccation, anaerobic conditions and high temperatures, and resistance to insect damage and diseases. Reports have been published over the past 50 years describing individuals from different tree species showing modified plant architecture named, e.g. *dwarf, nana, erecta, fastigiata, pyramidalis* and *columnar*. Induction of mutations has been a key element of mutation breeding for many plant species, including trees, for more than 70 years.

Enhanced shoot and root growth in poplar by RNAi suppression of the poplar homologues of the *Short Internodes* (*SHI*) and its closely related gene *STYLISH1* (*STY1*) from *Arabidopsis* has been described by Zawaski *et al.* (2011). The *SHI* gene belongs to a gene family that includes important developmental regulators. In addition,

increased fibre length and modified proportion of xylem tissue were found, indicating that both genes play an important role in the regulation of vegetative growth and wood formation (Zawaski *et al.*, 2011). To study the role of bioactive gibberellic acid (GA) concentrations on above- and below-ground biomass growth, Gou *et al.* (2011) overexpressed and RNAi suppressed both paralogues of the gibberellin 2-oxidase (GA2ox) gene in poplar. *PtGA2ox4* and its paralogue *PtGA2ox5* are primarily expressed in aerial organs. Overexpression of *PtGA2ox5* produced a strong dwarfing phenotype, while RNAi suppression of both paralogues promoted leaf growth and led to changes in wood development and to a decrease of root biomass, but did not modify the overall plant phenotype (Gou *et al.*, 2011).

An interesting study described the effects following RNAi downregulation of central circadian clock components in *P. tremula* × *P. tremuloides* trees (Edwards *et al.*, 2018). The circadian clock is a biochemical oscillator that regulates and coordinates physiological and biochemical factors in roughly 24 h cycles. Transgenic trees with reduced expression of two *late elongated hypocotyl* genes (*LHY1* and *LHY2*) revealed reduced growth and lower biomass production than wild-type trees. Analysis of the activity of genes involved in growth regulation showed arrhythmic and misaligned expression, indicating that impaired circadian clock function leads to misregulation of cell division genes (Edwards *et al.*, 2018).

Double knockout mutations of the flowering genes *SUPPRESSOR OF CONSTANS1* (*SOC1*) and *FRUITFULL* (*FUL*) in *Arabidopsis* led to plants revealing wood formation and perennial growth (Melzer *et al.*, 2008). Double overexpression of both *SOC1* and *FUL* genes in poplar led to stunted plants and changes in leaf morphology. RNAi suppression of the closest poplar *SOC1* and *FUL* homologues yielded plants with unchanged plant phenotype and wood formation (Bruegmann and Fladung, 2019). However, due to salicoid genome duplication, possibly additional paralogues with redundant function exist and not all of these paralogues were RNAi knocked out, i.e. in *P. trichocarpa*, three paralogues of *SOC1* and two paralogues of *FUL* were found.

Onset of flowering, axillary meristem identity and dormancy release were studied by Mohamed *et al.* (2010) by modifying the expression of

CENTRORADIALIS (CEN) and *MOTHER OF FT AND TFL1 (TERMINAL FLOWER 1) (MFT)* for 6 years in the field. Members of these subfamilies control shoot meristem identity; and loss-of-function mutations in herbaceous plants result in dramatic changes in plant architecture. RNAi downregulation of *PopCEN1* and its close paralogue, *PopCEN2*, yielded precocious first flowering with higher number of inflorescences and changed proportion of short shoots (Mohamed *et al.*, 2010). Strikingly, terminal vegetative meristems did not develop inflorescences, indicating that the flowering signal is transported to axillary meristems rather than the shoot apex. Thus, *PopCEN1/PopCEN2* genes are involved in shoot developmental transitions correlated with age (e.g. catkin formation on adult trees).

7.7   Abiotic Stress Tolerance

Trees are exposed to a number of environmental stresses throughout their entire lifespan. Besides biotic interactions, in particular, abiotic stresses such as drought, high soil salinity, heat, cold, oxidative stress and heavy metal toxicity are the most harmful environmental conditions that affect and limit crop productivity worldwide. As trees are sessile and long-lived organisms, the responses to occasionally detrimental environmental conditions are crucial for their survival. Therefore, there is a strong need to understand how trees react against high stress severity or when multiple stresses like high temperatures, drought and diseases act on trees. Unfortunately, plant responses to these stresses are mostly very complex; thus holistic, system biology or 'omics' approaches allow the identification of regulatory knots in the complex network of molecular and biochemical interactions (Cramer *et al.*, 2011).

By applying RNAi, research has revealed important information about the role of involved candidate gene families that may help tree breeders to develop abiotic stress-tolerant clones. Downregulation of poplar plasma membrane intrinsic proteins (PIPs) has led to a number of leaf physiology trait changes (Bi *et al.*, 2015). PIPs are a subfamily of aquaporins whose primary function is the transport of water across cell membranes in response to changes in osmotic pressures. RNAi:PIP poplar

leaves indeed revealed wider-opened stomata, leading to higher net CO_2 assimilation and transpiration rates. Possibly the higher transpiration caused a certain level of dehydration in the leaf, implying that leaves of RNAi:PIP plants were at risk of drought stress (Bi *et al.*, 2015). But levels of hormones like abscisic acid (ABA), auxin and brassinosteroids were also altered. In particular, ABA is a well-known regulator of the water status in plants, controlling various abiotic stress responses such as drought. Changes in levels of ABA are therefore expected to affect drought tolerance in plants. Yu *et al.* (2019) overexpressed and downregulated genes involved in ABA stress signalling and photoperiodic regulation in a poplar hybrid. Poplar lines overexpressing *bZIP transcription factor FD like1 (FDL1)* or its close homologue *FDL2* revealed drought sensitivity, whereas RNAi:*FDL* lines showed higher biomass allocation to roots under drought.

Ethylene responsive factors (ERFs) are also very important in responses to abiotic stress. To study the role of an ERF gene from *Betula platyphylla* (birch), *BpERF11* was overexpressed and RNAi downregulated (Zhang *et al.*, 2016). Overexpression of *BpERF11* led to plants with higher electrolyte leakage revealing increased transpiration rates, while downregulation of this gene resulted in increase of genes involved in abiotic stress tolerance. The authors conclude that *BpERF11* is a transcription factor that negatively regulates salt and severe osmotic tolerance by modulating various physiological processes (Zhang *et al.*, 2016). Another group of transcription factor, the zinc-finger proteins (ZFPs), were analysed by Zang *et al.* (2015, 2017) in *Tamarix hispida*. ZFPs are abundant in plants and characterized by a zinc finger domain. First, Zang *et al.* (2015) cloned the *ThZFP1* gene from *T. hispida* and could show that *ThZFP1* responds to abiotic stress and plays a role in improving salt and drought tolerance. In a second study, Zang *et al.* (2017) identified ThDof1.4, a transcriptional regulator of *ThZFP1*, and studied its function by up- and RNAi downregulation. As expected, overexpression of ThDof1.4 increased the transcripts of *ThZFP1* in *T. hispida* and RNAi silencing reduced its expression, indicating that *ThZFP1* and its regulator are involved in responses to salt or drought stress in *T. hispida* (Zang *et al.*, 2017).

Freezing tolerance in poplar was studied by Zhou *et al.* (2010) by up- and downregulation of the *fatty acid desaturase* (*PtFAD2*). Whereas *PtFAD2* overexpressing lines revealed significant higher survival rates of cuttings after freezing treatment compared with controls, the down-regulated lines showed lower survival rates. The results indicate that the level of polyunsaturated fatty acids in plant cells affect freezing in poplar (Zhou *et al.*, 2010).

Isoprene emission has been described in many, but not all, plant species (Sharkey *et al.*, 2008; Monson *et al.*, 2013). Isoprene is the most abundant volatile compound emitted by vegetation (Behnke *et al.*, 2009). Plants that emit isoprene are believed to tolerate sunlight-induced rapid heating of leaves as well as ozone and other reactive oxygen species better than non-emitting plants (Sharkey *et al.*, 2008). On the other hand, emission of isoprene from plants is important as it affects atmospheric chemistry. Isoprene emission has appeared and been lost many times independently during the evolution of plants (Monson *et al.*, 2013). Expression of the isoprene synthase gene can account for control of isoprene emission capacity (Sharkey *et al.*, 2008). To better understand the regulation of isoprene emission and to retrieve new insights into the link between isoprene and enhanced temperature tolerance, Behnke *et al.* (2007, 2009) downregulated the expression of the isoprene synthase gene by RNAi. By applying heat stress to isoprene- and non-isoprene-emitting poplars, the non-isoprene-emitting plants showed reduced net assimilation and photosynthetic electron transport rates, but not in the absence of stress (Behnke *et al.*, 2007). Further, the non-isoprene-emitting poplars were more resistant to ozone, as indicated by less damaged leaf area compared with isoprene-emitting wild-type poplars (Behnke *et al.*, 2009). In the field, growth performance and biomass yield of non-isoprene-emitting poplars revealed no change for two growing seasons (Behnke *et al.*, 2012).

7.8   RNAi in Forest Tree Pest Control

An interesting application of RNAi has been reported regarding insect pest control for several forest tree species. The idea behind it is based on entry of specific dsRNA delivery into the insect cell leading to the subsequent degradation of complementary mRNA of a carefully selected essential target gene, leading to insect mortality (Agrawal *et al.*, 2003; Vogel *et al.*, 2019). The sequence specificity of the small RNAs and the fact that, at least theoretically, any 'mortality gene' can be chosen makes RNAi highly attractive as a species-specific pesticide (Vogel *et al.*, 2019).

For the very dangerous pine wood nematode, *Bursaphelenchus xylophilus*, which was the causal agent of pine wilt disease that killed millions of pine trees in China and the rest of eastern Asia in the past, RNAi was used to downregulate the expression of the *endo-beta-1,4-glucanase* gene of the nematode (Ma *et al.*, 2011). Silencing of this gene led to reduced propagation and dispersal ability of this nematode. Another strategy was applied by Qiu *et al.* (2016) when blocking the function of the *pectate lyase 1* gene in *B. xylophilus* (*Bxpel1*) through RNAi. *B. xylophilus* individuals propagated much less in a solution soaked in dsRNA than in a control solution treatment; thus, application of *Bxpel1* dsRNAi to nematode-infected *Pinus thunbergii* trees resulted in reduced migration speed and reproduction rates of the nematodes (Ma *et al.*, 2011). The authors concluded that *Bxpel1* is a significant pathogenic factor in pine wilt disease break-out which could be the starting point for *B. xylophilus* control.

Also, for the interaction of the emerald ash borer (*Agrilus planipennis*), an invasive and destructive insect pest attacking ash (*Fraxinus* spp.), RNAi provides an alternative approach for insect pest management (Zhao *et al.*, 2015). Following microinjection of the dsRNA of the *AplaScrB-2* gene encoding a ß-fructofuranosidase enzyme into the beetle, the expression levels of *AplaScrB-2* decreased in the following days. The authors could show that RNAi is functional in the emerald ash borer *A. planipennis* causing ash dieback (Zhao *et al.*, 2015). Following targeting of two essential genes, *inhibitor of apoptosis* (*IAP*) or *COPI coatomer, beta subunit* (*COP*) by RNAi, Rodrigues *et al.* (2017) observed insect mortality, providing evidence that RNAi could successfully be applied to counteract the dangerous ash dieback.

7.8.1 Crop protection by topical RNAi by spray-induced gene silencing (SIGS)

Plant pathogens cause serious crop losses worldwide. Studies on the pathogenic fungus *Fusarium graminearum* pathosystem (Koch *et al.*, 2016; Wang and Jin, 2017) revealed that spraying dsRNAs (i.e. 791 nt *CYP3*-dsRNA) targeting the three fungal cytochrome P450 lanosterol C-14α-demethylases, required for biosynthesis of fungal ergosterol, inhibited fungal growth in the directly sprayed (local) as well as the non-sprayed (distal) parts of detached leaves. Moreover, efficient spray-induced control of fungal infections in the distal tissue involved passage of *CYP3*-dsRNA via the plant vascular system and processing into siRNAs by fungal DICER-LIKE 1 (*FgDCL-1*) after uptake by the pathogen. The authors also underlined the use of target-specific dsRNA as an anti-fungal agent offering unprecedented potential as a new plant protection strategy. Song *et al.* (2018) studied the effect of spray-induced gene silencing (SIGS) by targeting dsRNA to *myosin5* gene of *Fusarium asiaticum* and found that the RNAi-induced silencing lasted in *Fusarium* for only 9 h, in contrast to wheat cells with efficient and longer-lasting turnover of dsRNA into secondary siRNA. This might indicate that the RNA-dependent RNA polymerases, required for the production of secondary siRNA in plants, are only transiently functional or non-functional in *Fusarium*. Thus, the authors underlined that the mechanism of SIGS is still unknown and demonstrated that secondary siRNA amplification limits the application of SIGS.

Dubrovina and Kiselev (2019) reviewed the exogenous application of RNAs (dsRNAs, hairpin RNAs and siRNAs) designed to silence important genes of plant pathogenic viruses, fungi, or insects. Plants can uptake and process exogenously applied RNAs, leading to local and systemic spread within the plant and resulting in induction of RNAi-mediated plant pathogen resistance. Furthermore, sRNAs originating from a plant host can subsequently be delivered into fungal pathogens and lead to silencing of fungal genes vital for pathogenicity. The authors summarized the studies reporting on exogenous RNA applications for downregulation of essential fungal and insect genes as well as targeting of plant viruses for increased resistance, and, in addition, reported on the suppression of plant transgenes and endogenes by application of exogenous RNAs.

7.8.2 Clay nanosheets for topical delivery of RNAi for sustained protection against plant viruses

Mitter *et al.* (2017) used topical application of pathogen-specific dsRNA for virus resistance in plants. This is an attractive alternative to transgenic RNAi. However, the instability of naked dsRNA sprayed on plants has been a major challenge towards its practical application. The authors showed that dsRNA can be loaded on designed, non-toxic, degradable, layered double hydroxide (LDH) clay nanosheets. Once loaded, it showed sustained release and could be detected on sprayed leaves even 30 days after application. They found evidence for the degradation of LDH, dsRNA uptake in plant cells and silencing of homologous RNA on topical application. Significantly, a single spray of dsRNA loaded on LDH (BioClay) afforded virus protection for at least 20 days when challenged on sprayed and newly emerged unsprayed leaves. To conclude, nanotechnology can be used in crop protection as an environmentally sustainable and easy-to-adopt topical spray.

7.9 Outlook

Forests provide many benefits and services to society, including clean water and air, recreation, wildlife habitat, carbon storage, climate regulation and a variety of forest products (EPA, 2017). Climate influences the structure and function of forest ecosystems and plays an essential role in forest health. A changing climate will challenge the adaptation capacity of forest tree species and may worsen many of the threats to forests, such as pest outbreaks, fires, overexploitation and drought. Greenhouse gas emissions from human activity and livestock are a significant driver of climate change, trapping heat in the earth's atmosphere and triggering global warming. The United Nations 2030 Agenda for Sustainable Development is a commitment made by countries to tackle the complex challenges we face, from ending poverty and hunger and responding to climate change to building resilient communities, achieving inclusive growth and sustainably managing the Earth's natural resources. The Agenda's 17 Sustainable Development Goals lay out specific objectives for countries to meet within a given timeframe, with achievements monitored periodically to measure progress. Universally relevant, they call for comprehensive and participatory approaches (FAO, 2017, 2018).

In addition to human activities, forests are threatened by invasive exotic pathogens either due to climate change and/or due to long-distance trade of host- or pathogen-containing goods and climatic extremes, i.e. wildfires, droughts and storms. As an example, Finland's Ministry of Agriculture and Forestry indicates that, at present, the health of Finnish forests is good, but climate change and immigrant species increase the risk for damage (MMM Finland, 2020). European spruce bark beetle (*Ips typographus*) caused serious damage in spruce forests in southern and south-eastern Finland during 2010–2013 and in Germany from 2015 to 2018. Spruce bark beetle benefits from dry, hot summers.

The rapid pace of climate change may exceed the ability of many species to adapt in place or migrate to suitable habitats and this fundamental mismatch raises the possibility of extinction or local extirpation. Assisted migration (AM), i.e. human-assisted movement of species in response to climate change, is one management option that is available to address this challenge (e.g. Ste-Marie *et al.*, 2011). Winder *et al.* (2011) discussed the ecological constraints and consequences of AM and options for their mitigation at three scales: translocation over long distances (assisted long-distance migration), translocation just beyond the range limit (assisted range expansion) and translocation of genotypes within the existing range (assisted population migration). They concluded that, from an ecological perspective, AM is a feasible management option for tree species. However, AM needs honest considerations in each case to evaluate its potential benefits and threats for future forestry, as species may have potential to become harmful in new locations or transmit diseases to new areas (Ricciardi and Simberloff, 2009). The US Forest Service offers a comprehensive online search engine for literature about climate change and AM (US Forest Service, 2020).

In 2012 an international group of experts in silviculture, forest tree breeding, forest biotechnology and environmental risk assessment (ERA) met to examine how the ERA paradigm used for GE plants may be applied to GE trees for use in plantation forests (Häggman *et al.*, 2013). The group pointed out that intensively managed, highly productive forestry incorporating the most advanced methods for tree breeding and application of genetic engineering has tremendous potential for producing more wood on less land. Furthermore, they emphasized the need to differentiate between ERA for confined field trials of GE trees, compared with ERA for unconfined or commercial-scale releases. In the latter case, attention should be paid to characteristics of forest trees distinguishing them from shorter-lived plant species, the temporal and spatial scale of forests, and the biodiversity of the plantation forest as a receiving environment (Häggman *et al.*, 2013). Yet, the deployment of GE trees in plantation forests is still a controversial topic even though no indications of any risk to the environment or human health have been found in hundreds of field trials conducted with GE forest trees (Walter *et al.*, 2010).

Klocko *et al.* (2018) published a paper on phenotypic expression and stability in a large-scale field study of GE poplars with sexual containment transgenes. They tested over 3300 GE poplar trees and 948 transformation events in a single 3.6 ha field trial for seven growing seasons. The trial is the largest field-based study of GE forest trees in the world. The goal was to assess a diversity of approaches for obtaining bisexual sterility by modifying RNA expression or protein function of floral regulatory genes. Modified floral traits were stable over three to five flowering seasons and they identified RNAi or overexpression constructs that either postponed floral onset or led to sterile flowers. No detectable somaclonal variation and no trees with vegetative or floral modifications were related to the transgene added. Thus, GE containment traits can be obtained which are effective, stable and not associated with vegetative abnormalities or somaclonal variation.

Regardless of the promising RNAi plants generated and potential of genetic engineering to aid the adaptation of future reforestation material to climate change, the annual number of confined field trials (CFTs) conducted with

GM plants was shown to decrease in a review by Smets and Rüdelsheim (2018). The observed decrease in CFTs during the study period 2014–2017 was especially drastic in North America and Europe, while only a slight decrease was found in Latin America. Public research institutes, i.e. not-for-profit research organizations, such as universities and government-owned institutes accounted for only 4.2% of all CFTs; in contrast, industry accounted for 95.5% of all CFTs. Three categories of trees in CFTs were discerned: poplar/aspen and eucalyptus (for timber and biofuel); fruit trees; and ornamental trees. Generally, 88% expressed marker genes, 29% virus resistance, 28% nematode resistance and 28% product quality traits. During the study period the number of recorded CFTs conducted with tree species was 216, which comprised less than 1% of the total number of CFTs (Smets and Rüdelsheim, 2018). The low number of tree CFTs may be partly explained by the strict regulation on the containment of GE material during excessive laboratory, greenhouse and field testing (Strauss *et al.*, 2015).

References

Abbas, M., Peszlen, I., Shi, R., Kim, H., Katahira, R. *et al.* (2020) Involvement of *CesA4, CesA7-A/B* and *CesA8-A/B* in secondary wall formation in *Populus trichocarpa* wood. *Tree Physiology* 40(1), 73–89. DOI: 10.1093/treephys/tpz020.

Agrawal, N., Dasaradhi, P.V.N., Mohmmed, A., Malhotra, P., Bhatnagar, R.K. *et al.* (2003) RNA interference: biology, mechanism, and applications. *Microbiology and Molecular Biology Reviews* 67(4), 657–685. DOI: 10.1128/MMBR.67.4.657-685.2003.

Behnke, K., Ehlting, B., Teuber, M., Bauerfeind, M., Louis, S. *et al.* (2007) Transgenic, non-isoprene emitting poplars don't like it hot. *The Plant Journal* 51(3), 485–499. DOI: 10.1111/j.1365-313X.2007.03157.x.

Behnke, K., Kleist, E., Uerlings, R., Wildt, J., Rennenberg, H. *et al.* (2009) RNAi-mediated suppression of isoprene biosynthesis in hybrid poplar impacts ozone tolerance. *Tree Physiology* 29(5), 725–736. DOI: 10.1093/treephys/tpp009.

Behnke, K., Grote, R., Brüggemann, N., Zimmer, I., Zhou, G. *et al.* (2012) Isoprene emission-free poplars - a chance to reduce the impact from poplar plantations on the atmosphere. *New Phytologist* 194(1), 70–82. DOI: 10.1111/j.1469-8137.2011.03979.x.

Bi, Z., Merl-Pham, J., Uehlein, N., Zimmer, I., Mühlhans, S. *et al.* (2015) RNAi-mediated downregulation of poplar plasma membrane intrinsic proteins (PIPs) changes plasma membrane proteome composition and affects leaf physiology. *Journal of Proteomics* 128, 321–332. DOI: 10.1016/j.jprot.2015.07.029.

Bologna, N.G. and Voinnet, O. (2014) The diversity, biogenesis, and activities of endogenous silencing small RNAs in *Arabidopsis*. *Annual Review of Plant Biology* 65(1), 473–503. DOI: 10.1146/annurev-arplant-050213-035728.

Borges, F. and Martienssen, R.A. (2015) The expanding world of small RNAs in plants. *Nature Reviews Molecular Cell Biology* 16(12), 727–741. DOI: 10.1038/nrm4085.

Bruegmann, T. and Fladung, M. (2019) Overexpression of both flowering time genes *AtSOC1* and *SaFUL* revealed huge influence onto plant habitus in poplar. *Tree Genetics & Genomes* 15(2), 20. DOI: 10.1007/s11295-019-1326-9.

Carle, J. and Holmgren, P. (2003) Definitions related to planted forests. Food and Agriculture Organization of the United Nations, Rome. Available at: www.fao.org/forestry/25853-0d4f50dd8626f4bd6248009f c68f892fb.pdf (accessed 6 November 2020).

Chan, S.W.-L., Zilberman, D., Xie, Z., Johansen, L.K., Carrington, J.C. *et al.* (2004) RNA silencing genes control *de novo* DNA methylation. *Science* 303(5662), 1336. DOI: 10.1126/science.1095989.

Chang, S., Mahon, E.L., MacKay, H.A., Rottmann, W.H., Strauss, S.H. *et al.* (2018) Genetic engineering of trees: progress and new horizons. *In vitro Cellular & Developmental Biology* 54(4), 341–376. DOI: 10.1007/s11627-018-9914-1.

Coleman, H.D., Samuels, A.L., Guy, R.D. and Mansfield, S.D. (2008) Perturbed lignification impacts tree growth in hybrid poplar – a function of sink strength, vascular integrity, and photosynthetic assimilation. *Plant Physiology* 148(3), 1229–1237. DOI: 10.1104/pp.108.125500.

Cramer, G.R., Urano, K., Delrot, S., Pezzotti, M. and Shinozaki, K. (2011) Effects of abiotic stress on plants: a systems biology perspective. *BMC Plant Biology* 11(1), 163. DOI: 10.1186/1471-2229-11-163.

Custers, R., Bartsch, D., Fladung, M., Nilsson, O., Pilate, G. *et al.* (2016) EU regulations impede market introduction of GM forest trees. *Trends in Plant Science* 21(4), 283–285. DOI: 10.1016/j.tplants.2016.01.015.

Dalakouras, A., Wassenegger, M., Dadami, E., Ganopoulos, I. and Pappas, M.L. (2020) GMO-free RNAi: exogenous application of RNA molecules in plants. *Plant physiology* 182, 38–50. DOI: 10.1104/pp.19.00570.

Dubrovina, A.S. and Kiselev, K.V. (2019) Exogenous RNAs for gene regulation and plant resistance. *International Journal of Molecular Sciences* 20(9), 2282. DOI: 10.3390/ijms20092282.

Dunoyer, P., Schott, G., Himber, C., Meyer, D., Takeda, A. *et al.* (2010) Small RNA duplexes function as mobile silencing signals between plant cells. *Science* 328(5980), 912–916. DOI: 10.1126/science.1185880.

Edwards, K.D., Takata, N., Johansson, M., Jurca, M., Novák, O. *et al.* (2018) Circadian clock components control daily growth activities by modulating cytokinin levels and cell division-associated gene expression in *Populus* trees. *Plant, Cell & Environment* 41(6), 1468–1482. DOI: 10.1111/pce.13185.

EPA (2017) Climate Impacts on Forests. US Environmental Protection Agency, Washington, DC. Available at: https://19january2017snapshot.epa.gov/climate-impacts/climate-impacts-forests_.html#Overview (accessed 6 November 2020).

FAO (2002) *Proceedings, Second Expert Meeting on Harmonizing Forest-related Definitions for Use by Various Stakeholders*. Food and Agriculture Organization of the United Nations, Rome.

FAO (2012) *FRA 2015 Terms and Definitions. Forest Resources Assessment working paper 180*. Food and Agriculture Organization of the United Nations, Rome.

FAO (2016) *Global Forest Resources Assessment 2015. How are the world's forests changing?* 2nd edn. Food and Agriculture Organization of the United Nations, Rome.

FAO (2017) Strategy on Climate Change. Available at: http://www.fao.org/3/a-i7175e.pdf (accessed 6 November 2020).

FAO (2018) The State of the World's Forests 2018 – Forest pathways to sustainable development. Food and Agriculture Organization of the United Nations, Rome. Available at: http://www.fao.org/climate-change/en/ (accessed 6 November 2020).

Fire, A., Xu, S., Montgomery, M.K., Kostas, S.A., Driver, S.E. *et al.* (1998) Potent and specific genetic interference by double-stranded RNA in *Caenorhabditis elegans*. *Nature* 391(6669), 806–811. DOI: 10.1038/35888.

Gou, J., Ma, C., Kadmiel, M., Gai, Y., Strauss, S. *et al.* (2011) Tissue-specific expression of *Populus* C$_{19}$ GA 2-oxidases differentially regulate above- and below-ground biomass growth through control of bioactive GA concentrations. *New Phytologist* 192(3), 626–639. DOI: 10.1111/j.1469-8137.2011.03837.x.

Grattapaglia, D., Silva-Junior, O.B., Resende, R.T., Cappa, E.P., Müller, B.S.F. *et al.* (2018) Quantitative genetics and genomics converge to accelerate forest tree breeding. *Frontiers in Plant Science* 9, 1693. DOI: 10.3389/fpls.2018.01693.

Häggman, H., Raybould, A., Borem, A., Fox, T., Handley, L. *et al.* (2013) Genetically engineered trees for plantation forests: key considerations for environmental risk assessment. *Plant Biotechnology Journal* 11(7), 785–798. DOI: 10.1111/pbi.12100.

Häggman, H., Sutela, S., Walter, C. and Fladung, M. (2014) Biosafety considerations in the context of deployment of Ge trees. In: Fenning, T. (ed.) *Challenges and Opportunities for the World's Forests in the 21st Century. Forestry Sciences series, no. 81*. Dordrecht, Springer Netherlands, pp. 491–524.

Häggman, H., Sutela, S. and Fladung, M. (2016) Genetic engineering – contribution to forest tree breeding efforts. In: Vettori, C., Fladung, M., Häggman, H., Pilate, G. and Gallardo, F., *et al.* (eds) *Biosafety of Forest Transgenic Trees: Improving the scientific basis for safe tree development and implementation of EU policy directives. Forestry Sciences series, no. 82*. Dordrecht, Springer Netherlands, pp. 11–30.

Hamilton, A.J. and Baulcombe, D.C. (1999) A species of small antisense RNA in posttranscriptional gene silencing in plants. *Science* 286(5441), 950–952. DOI: 10.1126/science.286.5441.950.

Hernandez-Garcia, C.M. and Finer, J.J. (2014) Identification and validation of promoters and *cis*-acting regulatory elements. *Plant Science* 217-218, 109–119. DOI: 10.1016/j.plantsci.2013.12.007.

Hisano, H., Nandakumar, R. and Wang, Z.-Y. (2009) Genetic modification of lignin biosynthesis for improved biofuel production. *In Vitro Cellular & Developmental Biology* 45(3), 306–313. DOI: 10.1007/s11627-009-9219-5.

Ionescu, I.A., Møller, B.L. and Sánchez-Pérez, R. (2016) Chemical control of flowering time. *Journal of Experimental Botany* 74, 369–382. DOI: 10.1093/jxb/erw427.

Isik, F. and McKeand, S.E. (2019) Fourth cycle breeding and testing strategy for *Pinus taeda* in the NC State University Cooperative Tree Improvement Program. *Tree Genetics & Genomes* 15(5), 1–12. DOI: 10.1007/s11295-019-1377-y.

Jorgensen, R. (1990) Altered gene expression in plants due to *trans* interactions between homologous genes. *Trends in Biotechnology* 8, 340–344. DOI: 10.1016/0167-7799(90)90220-R.

Jürgensen, C., Kollert, W. and Lebedys, A. (2014) Assessment of industrial roundwood production from planted forests. FAO Planted Forests and Trees working paper FP/48/E. Food and Agriculture Organization of the United Nations, Rome. Available at: www.fao.org/forestry/plantedforests/67508@170537/en/ (accessed 6 November 2020).

Ke, Q., Kim, H.S., Wang, Z., Ji, C.Y., Jeong, J.C. et al. (2017) Down-regulation of *GIGANTEA - like* genes increases plant growth and salt stress tolerance in poplar. *Plant Biotechnology Journal* 15(3), 331–343. DOI: 10.1111/pbi.12628.

Klocko, A.L., Brunner, A.M., Huang, J., Meilan, R., Lu, H. et al. (2016) Containment of transgenic trees by suppression of *LEAFY*. *Nature Biotechnology* 34(9), 918–922. DOI: 10.1038/nbt.3636.

Klocko, A.L., Lu, H., Magnuson, A., Brunner, A.M., Ma, C. et al. (2018) Phenotypic expression and stability in a large-scale field study of genetically engineered poplars containing sexual containment transgenes. *Frontiers in Bioengineering and Biotechnology* 6, 100. DOI: 10.3389/fbioe.2018.00100.

Koch, A., Biedenkopf, D., Furch, A., Weber, L., Rossbach, O. et al. (2016) An RNAi-based control of *Fusarium graminearum* infections through spraying of long dsRNAs involves a plant passage and is controlled by the fungal silencing machinery. *PLoS Pathogens* 12(10), e1005901. DOI: 10.1371/journal.ppat.1005901.

Kucukoglu, M., Nilsson, J., Zheng, B., Chaabouni, S. and Nilsson, O. (2017) *WUSCHEL-RELATED HOMEOBOX4 (WOX4)* -like genes regulate cambial cell division activity and secondary growth in *Populus* trees. *New Phytologist* 215(2), 642–657. DOI: 10.1111/nph.14631.

Le Quéré, C., Andrew, R.M., Canadell, J.G., Sitch, S., Korsbakken, J.I. et al. (2016) Global carbon budget 2016. *Earth System Science Data* 8(2), 605–649. DOI: 10.5194/essd-8-605-2016.

Lee, C., Teng, Q., Zhong, R. and Ye, Z.-H. (2011) Molecular dissection of xylan biosynthesis during wood formation in poplar. *Molecular Plant* 4, 730–747. DOI: 10.1093/mp/ssr035.

Li, J., Brunner, A.M., Shevchenko, O., Meilan, R., Ma, C. et al. (2008) Efficient and stable transgene suppression via RNAi in field-grown poplars. *Transgenic Research* 17(4), 679–694. DOI: 10.1007/s11248-007-9148-1.

Li, Q., Min, D., Wang, J.P.-Y., Peszlen, I., Horvath, L. et al. (2011) Down-regulation of glycosyltransferase 8D genes in *Populus trichocarpa* caused reduced mechanical strength and xylan content in wood. *Tree Physiology* 31(2), 226–236. DOI: 10.1093/treephys/tpr008.

Lu, H., Klocko, A.L., Brunner, A.M., Ma, C., Magnuson, A.C. et al. (2019) RNA interference suppression of *AGAMOUS* and *SEEDSTICK* alters floral organ identity and impairs floral organ determinacy, ovule differentiation, and seed-hair development in *Populus*. *New Phytologist* 222(2), 923–937. DOI: 10.1111/nph.15648.

Lück, S., Kreszies, T., Strickert, M., Schweizer, P., Kuhlmann, M. et al. (2019) siRNA-Finder (si-Fi) software for RNAi-target design and off-target prediction. *Frontiers in Plant Science* 10, 1023. DOI: 10.3389/fpls.2019.01023.

Ma, H.B., Lu, Q., Liang, J. and Zhang, X.Y. (2011) Functional analysis of the cellulose gene of the pine wood nematode, *Bursaphelenchus xylophilus*, using RNA interference. *Genetics and Molecular Research* 10(3), 1931–1941. DOI: 10.4238/vol10-3gmr1367.

Meilan, R., Brunner, A.M., Skinnera, J.S. and Strauss, S.H. (2001) Modification of flowering in transgenic trees. *Progress in biotechnology* 18, 247–256. DOI: 10.1016/S0921-0423(01)80079-4.

Mellerowicz, E.J. and Gorshkova, T.A. (2011) Tensional stress generation in gelatinous fibres: a review and possible mechanism based on cell-wall structure and composition. *Journal of Experimental Botany* 63(2), 551–565. DOI: 10.1093/jxb/err339.

Melzer, S., Lens, F., Gennen, J., Vanneste, S., Rohde, A. et al. (2008) Flowering-time genes modulate meristem determinacy and growth form in *Arabidopsis thaliana*. *Nature Genetics* 40(12), 1489–1492. DOI: 10.1038/ng.253.

Mermigka, G., Verret, F. and Kalantidis, K. (2016) RNA silencing movement in plants. *Journal of Integrative Plant Biology* 58(4), 328–342. DOI: 10.1111/jipb.12423.

Meyer, S., Nowak, K., Sharma, V.K., Schulze, J. and Mendel, R.R. (2004) Vectors for RNAi technology in poplar. *Plant biology* 7, 100–104. DOI: 10.1055/s-2004-815729.

Mi, S., Cai, T., Hu, Y., Chen, Y., Hodges, E. *et al.* (2008) Sorting of small RNAs into *Arabidopsis* argo-naute complexes is directed by the 5′ terminal nucleotide. *Cell* 133(1), 116–127. DOI: 10.1016/j. cell.2008.02.034.

Mitter, N., Worrall, E.A., Robinson, K.E., Li, P., Jain, R.G. *et al.* (2017) Clay nanosheets for topical delivery of RNAi for sustained protection against plant viruses. *Nature Plants* 3(2), 16207. DOI: 10.1038/nplants.2016.207.

MMM Finland (2020) Forest Reproductive Material and Forest Tree Breeding. Available at: https://mmm.fi/en/forests/forestry/forest-reproductive-material-and-forest-tree-breeding (accessed 3 November 2020).

Mohamed, R., Wang, C.-T., Ma, C., Shevchenko, O., Dye, S.J. *et al.* (2010) *Populus CEN/TFL1* regulates first onset of flowering, axillary meristem identity and dormancy release in *Populus*. *The Plant Journal* 62(4), 674–688. DOI: 10.1111/j.1365-313X.2010.04185.x.

Monson, R.K., Jones, R.T., Rosenstiel, T.N. and Schnitzler, J.P. (2013) Why only some plants emit iso-prene. *Plant, Cell & Environment* 36(3), 503–516. DOI: 10.1111/pce.12015.

Naito, Y., Yoshimura, J., Morishita, S. and Ui-Tei, K. (2009) siDirect 2.0: updated software for designing functional siRNA with reduced seed-dependent off-target effect. *BMC Bioinformatics* 10(1), 392. DOI: 10.1186/1471-2105-10-392.

Napoli, C., Lemieux, C. and Jorgensen, R. (1990) Introduction of a chimeric chalcone synthase gene into petunia results in reversible co-suppression of homologous genes in *trans*. *The Plant Cell* 2(4), 279–289. DOI: 10.2307/3869076.

Obudulu, O., Mähler, N., Skotare, T., Bygdell, J., Abreu, I.N. *et al.* (2018) A multi-omics approach reveals function of secretory Carrier-Associated membrane proteins in wood formation of *Populus* trees. *BMC Genomics* 19(1), 11. DOI: 10.1186/s12864-017-4411-1.

Payn, T., Carnus, J.-M., Freer-Smith, P., Kimberley, M., Kollert, W. *et al.* (2015) Changes in planted for-ests and future global implications. *Forest Ecology and Management* 352, 57–67. DOI: 10.1016/j. foreco.2015.06.021.

Qiu, X.-W., Wu, X.-Q., Huang, L. and Ye, J.-R. (2016) Influence of *Bxpel1* gene silencing by dsRNA interfer-ence on the development and pathogenicity of the pine wood nematode, *Bursaphelenchus xylophi-lus*. *International Journal of Molecular Sciences* 17(1), 125. DOI: 10.3390/ijms17010125.

Rennie, E.A. and Scheller, H.V. (2014) Xylan biosynthesis. *Current Opinion in Biotechnology* 26, 100–107. DOI: 10.1016/j.copbio.2013.11.013.

Ricciardi, A. and Simberloff, D. (2009) Assisted colonization: good intentions and dubious risk assess-ment. *Trends in Ecology & Evolution* 24(9), 476–477. DOI: 10.1016/j.tree.2009.05.005.

Rodrigues, T.B., Rieske, L.K., J. Duan, J., Mogilicherla, K. and Palli, S.R. (2017) Development of RNAi method for screening candidate genes to control emerald ash borer, *Agrilus planipennis*. *Scientific Reports* 7(1), 7379. DOI: 10.1038/s41598-017-07605-x.

Rosvall, O. and Mullin, T.J. (2013) Introduction to breeding strategies and evaluation of alternatives. In: Mullin, T.J. and Lee, S. (eds) *Best Practice for Tree Breeding in Europe*. Skogforsk, Uppsala, pp. 7–27.

Ruotsalainen, S. (2014) Increased forest production through forest tree breeding. *Scandinavian Journal of Forest Research* 29(4), 333–344. DOI: 10.1080/02827581.2014.926100.

Séguin, A., Lachance, D., Déjardin, A., Leplé, J. and Pilate, G. (2014) Scientific research related to geneti-cally modified trees. In: Fenning, T. (ed.) *Challenges and Opportunities for the World's Forests in the 21st Century*. Springer, Dordrecht, pp. 525–548.

Sharkey, T.D., Wiberley, A.E. and Donohue, A.R. (2008) Isoprene emission from plants: why and how. *Annals of Botany* 101(1), 5–18. DOI: 10.1093/aob/mcm240.

Smets, G. and Rüdelsheim, P. (2018) Global trends and developments in confined field trials with geneti-cally modified plants. *Collection of Biosafety Reviews* 10, 53–77.

Song, X.-S., Gu, K.-X., Duan, X.-X., Xiao, X.-M., Hou, Y.-P. *et al.* (2018) Secondary amplification of siRNA machinery limits the application of spray-induced gene silencing. *Molecular Plant Pathology* 19(12), 2543–2560. DOI: 10.1111/mpp.12728.

Ste-Marie, C., A. Nelson, E., Dabros, A. and Bonneau, M.-E. (2011) Assisted migration: introduction to a multifaceted concept. *The Forestry Chronicle* 87(06), 724–730. DOI: 10.5558/tfc2011-089.

Strauss, S.H., Costanza, A. and Séguin, A. (2015) Genetically engineered trees: paralysis from good inten-tions. *Science* 349(6250), 794–795. DOI: 10.1126/science.aab0493.

Teichmann, T. and Muhr, M. (2015) Shaping plant architecture. *Frontiers in Plant Science* 6, 233. DOI: 10.3389/fpls.2015.00233.

Tian, X., Xie, J., Zhao, Y., Lu, H., Liu, S. *et al.* (2013) Sense-, antisense- and RNAi-*4CL1* regulate solu-ble phenolic acids, cell wall components and growth in transgenic *Populus tomentosa* Carr. *Plant Physiology and Biochemistry* 65, 111–119. DOI: 10.1016/j.plaphy.2013.01.010.

Tylewicz, S., Tsuji, H., Miskolczi, P., Petterle, A., Azeez, A. *et al.* (2015) Dual role of tree florigen ac-tivation complex component *FD* in photoperiodic growth control and adaptive response path-ways. *Proceedings of the National Academy of Sciences* 112(10), 3140–3145. DOI: 10.1073/pnas.1423440112.

US Forest Service (2020) Climate change and assisted migration. Available at: https://www.fs.usda.gov/rmrs/tools/climate-change%C2%A0and-assisted-migration (accessed 3 November 2020).

Vogel, E., Santos, D., Mingels, L., Verdonckt, T.-W. and Broeck, J.V. (2019) RNA interference in in-sects: protecting beneficials and controlling pests. *Frontiers in Physiology* 9, 1912. DOI: 10.3389/fphys.2018.01912.

Walter, C., Fladung, M. and Boerjan, W. (2010) The 20-year environmental safety record of GM trees. *Nature Biotechnology* 28(7), 656–658. DOI: 10.1038/nbt0710-656.

Wang, M. and Jin, H. (2017) Spray-induced gene silencing: a powerful innovative strategy for crop protec-tion. *Trends in Microbiology* 25(1), 4–6. DOI: 10.1016/j.tim.2016.11.011.

Wassenegger, M., Heimes, S., Riedel, L. and Sänger, H.L. (1994) RNA-directed de novo methylation of genomic sequences in plants. *Cell* 76(3), 567–576. DOI: 10.1016/0092-8674(94)90119-8.

Watson, J.E.M., Evans, T., Venter, O., Williams, B., Tulloch, A. *et al.* (2018) The exceptional value of intact forest ecosystems. *Nature Ecology & Evolution* 2(4), 599–610. DOI: 10.1038/s41559-018-0490-x.

Whiteman, A. (2014) Global trends and outlook for forest resources. In: Fenning, T. (ed.) *Challenges and Opportunities for the World's Forests in the 21st Century*. Springer, Dordrecht, pp. 163–211.

Winder, R., Nelson, E. and Beardmore, T. (2011) Ecological implications for assisted migration in Canadian forests. *The Forestry Chronicle* 87(06), 731–744. DOI: 10.5558/tfc2011-090.

Yan, X., Liu, J., Kim, H., Liu, B., Huang, X. *et al.* (2019) CAD1 and CCR2 protein complex formation in monolignol biosynthesis in *Populus trichocarpa*. *New Phytologist* 222(1), 244–260. DOI: 10.1111/nph.15505.

Yang, Z., Ebright, Y.W., Yu, B. and Chen, X. (2006) HEN1 recognizes 21-24 nt small RNA duplexes and deposits a methyl group onto the 2′ OH of the 3′ terminal nucleotide. *Nucleic Acids Research* 34(2), 667–675. DOI: 10.1093/nar/gkj474.

Yu, B., Yang, Z., Li, J., Minakhina, S. and Yang, M. (2005) Methylation as a crucial step in plant microRNA biogenesis. *Science* 307(5711), 932–935. DOI: 10.1126/science.1107130.

Yu, D., Wildhagen, H., Tylewicz, S., Miskolczi, P.C., Bhalerao, R.P. *et al.* (2019) Abscisic acid signalling me-diates biomass trade-off and allocation in poplar. *New Phytologist* 223(3), 1192–1203. DOI: 10.1111/nph.15878.

Zang, D., Wang, C., Ji, X. and Wang, Y. (2015) *Tamarix hispida* zinc finger protein ThZFP1 participates in salt and osmotic stress tolerance by increasing proline content and SOD and POD activities. *Plant Science* 235, 111–121. DOI: 10.1016/j.plantsci.2015.02.016.

Zang, D., Wang, L., Zhang, Y., Zhao, H. and Wang, Y. (2017) ThDof1.4 and ThZFP1 constitute a tran-scriptional regulatory cascade involved in salt or osmotic stress in *Tamarix hispida*. *Plant Molecular Biology* 94(4-5), 495–507. DOI: 10.1007/s11103-017-0620-x.

Zawaski, C., Kadmiel, M., Ma, C., Gai, Y., Jiang, X. *et al.* (2011) *SHORT INTERNODES*-like genes reg-ulate shoot growth and xylem proliferation in *Populus*. *New Phytologist* 191(3), 678–691. DOI: 10.1111/j.1469-8137.2011.03742.x.

Zhang, W., Yang, G., Mu, D., Li, H., Zang, D. *et al.* (2016) An *Ethylene-responsive Factor BpERF11* nega-tively modulates salt and osmotic tolerance in *Betula platyphylla*. *Scientific Reports* 6(1), 23085. DOI: 10.1038/srep23085.

Zhao, C., Alvarez Gonzales, M.A., Poland, T.M. and Mittapalli, O. (2015) Core RNAi machinery and gene knockdown in the emerald ash borer (*Agrilus planipennis*). *Journal of Insect Physiology* 72, 70–78. DOI: 10.1016/j.jinsphys.2014.12.002.

Zhou, Z., Wang, M.-J., Zhao, S.-T., Hu, J.-J. and Lu, M.-Z. (2010) Changes in freezing tolerance in hybrid poplar caused by up- and down-regulation of *PtFAD2* gene expression. *Transgenic Research* 19(4), 647–654. DOI: 10.1007/s11248-009-9349-x.

Zhou, X., Ren, S., Lu, M., Zhao, S., Chen, Z. *et al.* (2018) Preliminary study of cell wall structure and its mechanical properties of C3H and HCT RNAi transgenic poplar sapling. *Scientific Reports* 8(1), 1–10. DOI: 10.1038/s41598-018-28675-5.

8 Host-induced Gene Silencing and Spray-induced Gene Silencing for Crop Protection Against Viruses

Angela Ricci[1], Silvia Sabbadini[1], Laura Miozzi[2], Bruno Mezzetti[1] and Emanuela Noris[2]*

[1]Department of Agricultural, Food and Environmental Sciences, Università Politecnica delle Marche, Ancona, Italy; [2]Institute for Sustainable Plant Protection, National Research Council of Italy, Torino, Italy

Abstract

Since the beginning of agriculture, plant virus diseases have been a strong challenge for farming. Following its discovery at the very beginning of the 1990s, the RNA interference (RNAi) mechanism has been widely studied and exploited as an integrative tool to obtain resistance to viruses in several plant species, with high target-sequence specificity. In this chapter, we describe and review the major aspects of host-induced gene silencing (HIGS), as one of the possible plant defence methods, using genetic engineering techniques. In particular, we focus our attention on the use of RNAi-based gene constructs to introduce stable resistance in host plants against viral diseases, by triggering post-transcriptional gene silencing (PTGS). Recently, spray-induced gene silencing (SIGS), consisting of the topical application of small RNA molecules to plants, has been explored as an alternative tool to the stable integration of RNAi-based gene constructs in plants. SIGS has great and innovative potential for crop defence against different plant pathogens and pests and is expected to raise less public and political concern, as it does not alter the genetic structure of the plant.

8.1 Introduction

Plant viruses represent a major threat to global agriculture. Viruses have been found in all cultivated plant species and a wide range of wild species (see the Tenth Report of International Committee on Taxonomy of Viruses (ICTV): Lefkowitz et al., 2017). Viruses are infectious particles containing a nucleic acid core of RNA or DNA, surrounded by a protective shell made of one or more coat proteins. They are considered as obligate parasites, as they exploit the host cell machinery for their replication in living cells. In particular, during the infection process, a plant virus penetrates the plant cell through wounds made by, for example, arthropod pests or during agricultural practices (e.g. badly executed pruning); progressively, it colonizes the surrounding cells/tissues and spreads through the whole plant via the phloem. Agricultural practices such as crop rotation, precocious detection and prompt eradications of infected entities, use

*Corresponding author: emanuela.noris@ipsp.cnr.it

of virus-resistant varieties, virus-free certified plants, or chemical prophylaxis against insect vectors can help to contain viral infections (Hull, 2014).

Taking into consideration the serious economic damage caused worldwide by viral diseases, researches have been committed to introduce genetic resistances against viruses in plants. One of the most promising approaches relied on genetic engineering techniques, an integrative strategy to traditional breeding methods for obtaining virus resistance in several crop species.

One of the key studies in this research area, published in the mid-1980s, is commonly referred to as the pathogen-derived resistance (PDR) strategy. The idea behind this concept comprises the ability of plant cells, transformed with specific gene sequences derived from the pathogen, to interfere with the replication or the infection of the pathogen itself (Sanford and Johnston, 1985). For plant viruses, the proof of concept of PDR was reported by Abel *et al.* (1986). In this study, tobacco explants transformed with *Agrobacterium tumefaciens* carrying the coat protein (CP) gene of tobacco mosaic virus (TMV) showed a reduction of virus symptoms when inoculated with TMV. Despite several reports of overexpression of CP or other virus coding sequences, such as replicases, proteinases and movement proteins, the molecular pathways behind this induced resistance were not always clarified (Prins *et al.*, 2008). Later studies revealed that PDR was not always linked to a deregulated synthesis of the corresponding viral proteins, or to the overexpression of dysfunctional viral proteins. Correlations between PDR events and RNA-dependent degradation mechanisms were detected in most cases. This phenomenon was later described as post-transcriptional gene silencing (PTGS) (Lindbo *et al.*, 1993).

8.2 RNA Interference and Virus Resistance

Between the end of the 1980s and the beginning of the 1990s, two different groups conducting studies on the regulation of gene expression in petunia observed that the overexpression of a foreign sequence homologous to an endogenous plant gene led to specific degradation of both sequences, terming this phenomenon 'coordinated suppression' (co-suppression) (Napoli *et al.*, 1990; van der Krol *et al.*, 1990). Two years later, a non-translatable CP gene sequence of tobacco etch virus (TEV) was introduced into tobacco plants (Lindbo and Dougherty, 1992). Some of the transgenic lines expressing the TEV-CP gene transcripts developed feeble symptoms when inoculated with TEV, while some of them were symptomless. Surprisingly, the latter presented low steady-state levels of transgenic mRNA, despite highly active expression. This demonstrated the existence of a cellular-based, sequence-specific, post-transcriptional RNA degradation system induced by the transgenic mRNA, targeting both the transgene transcript and the homologous virus mRNA for degradation. This was therefore the first described PTGS-based example of virus resistance. Starting from these observations, it has been understood that in plant cells the RNA-mediated virus resistance based on PTGS is part of a natural and complex process now universally known as RNA silencing or RNA interference (RNAi) (Baulcombe, 2004).

The activating molecule of the RNAi machinery is represented by double-stranded RNA (dsRNA) precursors (Voinnet, 2008); in the cytoplasm, Dicer-like enzymes identify and specifically cut these dsRNA molecules into small RNAs (sRNAs) composed of 21–24 nt (Hamilton and Baulcombe, 1999; Bernstein *et al.*, 2001; Elbashir *et al.*, 2001; Baulcombe, 2004). The sRNAs sense strand, recruited by the RNA-induced silencing complex (RISC) with the help of Argonaute proteins, is used to scan the cytoplasm in order to find and degrade homologous mRNAs or compromise their translation, thus modulating gene expression (Tijsterman *et al.*, 2002; Denli and Hannon, 2003; Ghildiyal and Zamore, 2009).

Protection against viruses and modulation of endogenous gene expression are the two main fields of activity of RNAi in plants (Vazquez *et al.*, 2010). As a gene expression regulator, RNAi functions also in insects (Kennerdell and Carthew, 1998), fungi (Romano and Macino, 1992), animals (Fire *et al.*, 1998) and mammals (Maillard *et al.*, 2019). Moreover, the characteristics of the RNAi mechanism have been exploited to silence invading viral sequences in order to prevent and/or reduce their accumulation in plants. Two main biotech strategies based on the RNAi system have been exploited for crop defence against viruses, known as HIGS and the more recently studied SIGS method. HIGS depends on the induction of the plant RNAi biological system and is obtained by stable expression of dsRNAs specific for a target virus. As reviewed by Khalid *et al.* (2017), the activation of PTGS against viruses can depend on the characteristics of the gene constructs introduced in the plant to produce dsRNAs. In this chapter, we offer an excursus concerning different hairpin RNA (hpRNA)-based gene construct features and applications, which are definitively considered one of the most powerful tools to induce stable genetic resistance in crops against viruses. On the other side, SIGS, the more recent strategy based on RNAi, relies on the exogenous application of dsRNA molecules that are homologous to the target viral sequences to trigger the natural RNAi-based defence mechanism towards plant viruses. In this chapter, we discuss the major achievements in producing dsRNA molecules on a large scale, using biofactories, and their topical application to plants. Moreover, we discuss the problems and benefits related to the efficacy and stability of SIGS, compared with HIGS, in particular for field conditions.

8.3 Host-induced Gene Silencing (HIGS) Strategy Against Viruses: hpRNA Silencing Approaches

An elegant study published in Nature by Fire *et al.* (1998) showed that in *Caenorhabditis elegans* RNAi was induced by dsRNA molecules and that these molecules were more efficient in inducing silenced phenotypes compared with single-stranded RNA molecules. At the same time, another study demonstrated increased silencing efficiency obtained by the co-expression in the host cell of sense and antisense sequences, compared with their separated expression (Waterhouse *et al.*, 1998).

Later, the expression of dsRNAs was achieved in plants mainly by introducing hpRNA gene constructs, and these were also designed to induce PTGS against viruses. These gene constructs normally include short inverted sequences homologous to vital viral genes, usually split by a non-coding sequence, such as an intron, all under the control of specific promoters and terminators (Lemgo *et al.*, 2013).

Such a construct strategy was described by Smith *et al.* (2000), who reported the increase of the silencing effect when an intron-based sequence was inserted as a junction between the sense and antisense arms of the hpRNA construct, leading to almost 100% of independently transformed tobacco lines showing silencing against potato virus Y (PVY). It has been supposed that the intron removal throughout splicing may simplify the folding of the hairpin structure and its transit from the nucleus to the cytoplasm (Wesley *et al.*, 2001). As suggested by molecular analysis carried out on transgenic tomato plants expressing intron hpRNA-derived sRNAs and resistant to tomato yellow leaf curl virus (TYLCV), it seems that few unspliced hairpin molecules are processed by DCL 3 into 24 nt sRNAs in the nucleus and used as phloem-mobile silencing inducers. On the contrary, spliced hairpin molecules are processed in the cytoplasm by DCL 4 and DCL 2 into 21 nt and 22 nt sRNAs, respectively, and used as cell-autonomous silencing inducers of the target viral sequence (Fuentes *et al.*, 2006, 2016; Pooggin, 2017).

Concerning the choice of the target viral genome sequence selected to build the short inverted repeats of the hpRNA construct, various aspects have to be considered. All viral genes chosen as RNAi targets for crop defence encode essential proteins necessary for the survival and the replication of the virus in the host, such as coat protein, nuclear capsid protein, replicase and replication-associated proteins (Khalid *et al.*, 2017). Sequences of different lengths have been chosen and inserted into a wide range of plant species (Cillo and Palukaitis, 2014). In general, essential viral genome portions from 300 up to 800 nt are preferred as target regions (Simón-Mateo and García, 2011), but much smaller sequences (from 23 up to 60 nt) have also been successfully used to induce virus resistance (Thomas *et al.*, 2001). The idea behind such preference in terms of sequence length is connected with the concept that hpRNA-mediated silencing occurs when the homologous region between the hp-derived transcripts and the target viral sequence covers more than 100 nt (Pang *et al.*, 1997; Jan *et al.*, 2000).

The 35S cauliflower mosaic virus (CaMV), the first plant promoter identified almost 40 years ago (Covey *et al.*, 1981), is the most broadly exploited promoter sequence in plant biotechnology, also in the case of the hpRNA constructs design, as it causes constitutively high levels of gene expression in a large variety of plant tissues, despite being derived from a pathogenic virus.

Since the dawn of plant biotechnology, tobacco has been widely exploited as a model plant system, mainly to validate the functionality of new gene constructs due to the ease of genetic transformation and virus infection. Since the end of the 1990s, many achievements and failures in terms of RNA and DNA virus defence via hpRNA have been reported, both in model plants and in crops, including several examples where 100% of resistance to the target virus was achieved (reviewed by Khalid *et al.*, 2017). Different hp-gene constructs against several viruses have been evaluated in the model species *Nicotiana tabacum* or *N. benthamiana*, and complete resistance was achieved in 12 cases (ten in *N. benthamiana* and two in *N. tabacum*, respectively). For example, the production of transgenic *N. benthamiana* plants resistant to citrus tristeza virus (CTV) expressing an hp-gene construct targeting P23+3′UTR sequences led to the application of the same approach in citrus (Batuman *et al.*, 2006). However, following transformation via *Agrobacterium* of the citrus 'Alemow' to enable insertion of a hairpin construct (p23UI), potentially capable of inducing CTV resistance via PTGS, none of the transgenic citrus plants exhibited resistance. This example shows that a result achieved in a model plant may not be directly reproduced in a target crop, possibly since specific host factors participate in the infection process. To partially explain this outcome, it was supposed that a virus could be more virulent in its own natural host than in a different experimental host. To integrate the Khalid *et al.* (2017) review, it has to be mentioned that the RNAi mechanism was exploited against plum pox virus (PPV) for the first time by Pandolfini *et al.* (2003) who designed and introduced an hp-gene construct against PPV in the model species *N. benthamiana*. In this study, a 197 bp-long sequence of the PPV strain D genome was chosen to design the *ihprolC-PP197* gene construct, placed as two inverted repeats separated by a non-coding sequence under the control of the phloem-specific *rol C* promoter. When the *ihprolC-PP197* gene construct was employed to transform *N. benthamiana* plants, systemic PPV resistance was obtained. Systemic viral infections are common in fruit trees; thus a comparable construct could be developed to achieve PPV resistance also in *Prunus* spp. (Ilardi and Tavazza, 2015).

The RNAi-based strategy was shown to work also against viruses with a DNA genome, as reviewed in Pooggin (2017). One of the most intriguing examples of hpRNA constructs active against the geminivirus TYLCV consisted of the expression of an hpRNA construct targeting the viral replicase C1 gene (Fuentes *et al.*, 2006). When transgenic lines expressing this construct were tested in field conditions, a long-lasting resistance was demonstrated; moreover, the authors highlighted the possibility that this strategy could induce off-target effects and modify the transcriptome of the transgenic lines, as determined by deep-sequencing approaches (Fuentes *et al.*, 2016; Pooggin, 2017).

8.4 Transgrafting as a Tool to Develop Genetic Resistance Against Viruses in Crops

In worldwide farming, grafting is a very common procedure that basically consists of connecting a portion of a plant (i.e. scion) to another plant (i.e. rootstock), through the junction of their vascular systems. Essentially in a grafted plant system, the rootstock absorbs nutrients from the soil that move to the scion, while the scion synthesizes carbohydrates through photosynthesis that are translocated to the rootstock. The phloem of a grafted organism is where the traffic of plant growth factors, soluble organic compounds, nucleic acids and proteins takes place, creating a dynamic link between rootstock and scion that should lead to an improved growth and yield of the grafted plant (Aloni *et al.*, 2010; Dinant and Suárez-López, 2012; Guelette *et al.*, 2012; Ham *et al.*, 2014). Plant grafting is mostly used for vegetative propagation, to induce resistance against pathogens, to alter plant vigour and increase endurance to abiotic stresses (Gonçalves *et al.*, 2006; Kubota *et al.*, 2008; Aloni *et al.*, 2010; Koepke and Dhingra, 2013).

As explained by Pyott and Molnar (2015), a non-cell autonomous gene silencing signal is 'one whose action extends beyond the cell producing the signal'. In the late 1990s, the transmission of a silencing signal in the form of dsRNA molecules over long distances was demonstrated by two key studies applied on *N. benthamiana* plants (Palauqui *et al.*, 1997; Voinnet and Baulcombe, 1997). In particular, Voinnet and Baulcombe (1997) induced the stable expression of GFP-encoding sequence in *N. benthamiana* plants and, through an optimized *Agrobacterium* infiltration protocol, a temporary GFP silencing was induced in the older leaves of the same treated plants. Probably thanks to the translocation of the silencing molecules, a GFP silenced phenotype was detected also in the upper leaves. In the same year, using a grafting procedure, Palauqui *et al.* (1997) joined wild-type tobacco scions onto transgenic stocks expressing nitrate/nitrite reductase (*Nia/Nii*) transgene. Chlorosis and reduced amounts of *Nia/Nii* mRNA in the scions suggested a movement of *Nia/Nii* silencing signals from the transformed stock to the wild-type scion.

The nature of the systemic RNA silencing signal has been an enigma for researchers. At the beginning, it was supposed that the travelling of long dsRNA precursors should take place in the phloem to achieve systemic silencing (Mallory *et al.*, 2001, 2003), but later reports suggested that systemic RNA silencing depends almost exclusively on sRNAs as mobile molecules (Chiou *et al.*, 2006; Buhtz *et al.*, 2008; Martin *et al.*, 2009; Molnar *et al.*, 2010; Melnyk *et al.*, 2011; Zhang *et al.*, 2014).

The ability of the silencing molecules to move along the plant vascular system can be exploited in a transgrafting system (Song *et al.*, 2015). In this case, transgrafting used as a method to spread sRNAs through the plant and to switch off the replication of a target virus could represent an alternative and promising approach to protect woody plant species against viral diseases. The goal of the hpRNAi transgrafting system would be to obtain a cultivar whose tissues and organs, including pollen and fruits, remain untransformed but which is resistant to one or more target viruses thanks to the translocation of sRNAs from an RNAi transgenic grafted rootstock (Lemgo *et al.*, 2013). This approach is particularly suitable for fruit trees species, which are usually propagated vegetatively and not through seeds. For example, peach (*Prunus persica* L. Batsch), grapevine (*Vitis vinifera*) (Bouquet and Hevin, 1978) and sweet cherry (*Prunus avium*) (Akçay *et al.*, 2008) plants are propagated by grafting to retain the same parental traits in terms of quality and vigour of fruit. Since 1998, transgenic rootstocks have been exploited in grafting systems for woody fruit-bearing plants (reviewed by Song *et al.*, 2015). Two promising examples of virus resistance in non-transgenic scions grafted on transgenic rootstocks were achieved in grapevine (Vigne *et al.*, 2004) and sweet cherry plants (Song *et al.*, 2013; Zhao and Song, 2014); in the latter studies, resistance against prunus necrotic ringspot virus (PNRSV) relies on RNAi mechanism, activated by an hpRNAi-based gene construct integrated in the grafted transgenic rootstock.

Although the HIGS approach applied to transgrafting systems shows several advantages, especially for inducing plant virus resistance, its use is currently hindered by different issues, especially by the need to generate transgenic plants. Furthermore, this process presents several bottlenecks both from a technical point of view and for regulatory and social aspects. In fact, certain crop species are hard to regenerate *in vitro* and/or difficult to transform genetically (Sabbadini *et al.*, 2019). Moreover, the inserted transgenes can be unstable in the host genome, or their expression can be silenced or suppressed in the offspring, making transformation ineffective. In addition, the generation and characterization of transgenic lines can be time consuming for some cultivated crops, making the evaluation of the effective lines unaffordable (Altpeter *et al.*, 2016). To reduce or overcome public concerns and bypass technical difficulties to obtain stable and efficient transgenic lines, the exogenous delivery of RNAi effective molecules (sometimes termed SIGS) has been proposed as an appealing alternative for plant disease control. In this case, the plant host genome is not modified, multi-target strategies are feasible and the products of this strategy can be obtained in a relatively shorter time.

8.5 Spray-induced Gene Silencing (SIGS) Strategy Against Viruses

The first report of the successful use of exogenously applied dsRNAs against plant viruses was that of Tenllado and Díaz-Ruiz (2001). In this pioneering work, RNA-mediated virus resistance was triggered by dsRNA molecules against three different viruses, all with a positive single-stranded RNA genome, such as pepper mild mottle virus (PMMoV), tobacco etch virus (TEV) and alfalfa mosaic virus (AMV). When these viruses were mechanically inoculated on *N. benthamiana* leaves with *in vitro* transcribed dsRNA fragments targeting the PMMoV replicase, the TEV helper component (HcPro) or the AMV RNA3 (fragments of 997 bp, 1483 bp and 1124 bp, respectively), a local antiviral response was elicited, in a dose-dependent manner. However, the authors stated that a certain length of dsRNA was required to reach resistance (Tenllado and Díaz-Ruiz, 2001). Since then, this strategy has been applied on many different plant species targeting different viruses, as reviewed in Dalakouras *et al.* (2020). This work reviews the use of different kinds of formulations of dsRNAs that were delivered on maize plants against sugarcane mosaic virus SCMV) CP (Gan *et al.*, 2010) and on pea against pea seed-borne mosaic virus (PSBMV) CP (Šafářová, D *et al.*, 2014), as well as on the orchid *Brassolaeliocattleya hybrida* against cymbidium mosaic virus (CymMV) CP (Lau *et al.*, 2014). Other constructs were tested on tobacco, targeting the TMV p126 replicase (Konakalla *et al.*, 2016), on cucurbits, targeting zucchini yellow mosaic virus (ZYMV) HcPro (Kaldis *et al.*, 2018), on *N. benthamiana*, targeting a 2611 bp region of the replicase and MP of TMV (Niehl *et al.*, 2018) and on papaya tree against papaya ringspot virus CP (Shen *et al.*, 2014).

For a broad application of dsRNAs in greenhouses and fields, efficient and economically acceptable methods for their large-scale production and purification are required. The initial systems adopted to obtain dsRNAs relied on the *in vitro* enzymatic synthesis of two complementary ssRNA strands, followed by physical annealing (Tenllado and Díaz-Ruiz, 2001; Carbonell *et al.*, 2008). One of the most frequently used enzymes for ssRNA synthesis is the DNA-dependent RNA polymerases (DdRPs) of the bacteriophage T7. For plant virus control, specific target sequences are transcribed by DdRPs from cDNA templates extracted from plants infected by the target virus, using specific primers that carry the T7 promoter at their 5′-end; alternatively, the *in vitro* transcription by

DdRP can occur starting from plasmids carrying the target viral sequences cloned between two T7 promoters (Konakalla *et al.*, 2016). Different kits are commercially available for this purpose, such as the MEGAscript® RNAi Kit (Life Technologies), Replicator™ RNAi Kit (Finnzymes) or T7 RiboMAX™ Express system (Promega, USA). The production of dsRNA molecules specifically targeting a selected pathogen region can be followed, optionally, by digestion with Dicer-like (DCL) enzymes, obtaining a heterogeneous mix of short interfering RNAs (siRNAs), of 18–25 nt in length when the ShortCut® RNase III (NEB, Ipswich, Massachusetts) kit is used or of 25–27 nt when the PowerCut Dicer (Thermo Scientific) kit is employed; the siRNA mixture can be further subjected to cleaning with the mirVana™ miRNA Isolation Kit (Life Technologies, Carlsbad, California) (Koch *et al.*, 2016; Wang *et al.*, 2016). In a more vigorous *in vitro* system, the ssRNA synthesis performed by the T7 RNA polymerase was coupled to a *de novo* primer-independent initiation, using the highly processive RNA-dependent RNA polymerase (RdRP) enzyme of bacteriophage φ6 (Makeyev and Bamford, 2000), a dsRNA virus infecting *Pseudpmonas syringae* cells (Aalto *et al.*, 2007).

To overcome the high costs linked to the *in vitro* dsRNA synthesis, *in vivo* approaches using bacterial cells have been developed, both in *Escherichia coli* (Tenllado *et al.*, 2003; Yin *et al.*, 2009) and in *P. syringae* cells (Aalto *et al.*, 2007; Niehl *et al.*, 2018). In the *E. coli* system, a stably replicating plasmid carrying the target viral sequence cloned within two T7 promoters is introduced into bacteria; following chemical induction of the T7 DdRP gene, which is expressed by a gene cloned in a DE3 prophage or in an additional plasmid, the target sequences are transcribed in both directions; then, the newly generated ssRNA molecules anneal, yielding the desired dsRNAs. Their degradation is inhibited using RNase-III deficient strains, such as *E. coli* HT115 (DE3) or M-JM109lacY, the latter having also a knockout *LacY* permease gene. This easily scaled-up process is reported to yield about 4 µg dsRNA/ml of bacterial culture (Tenllado *et al.*, 2004).

In a pioneering work, Aalto *et al.* (2007) described an *in vivo* dsRNA production system in *P. syringae* that had been engineered in order to express the RdRP of bacteriophage φ6. This system was further improved using the stable carrier cell line amplifying RNA by means of the phage φ6 RdRP (Sun *et al.*, 2004), finally leading to *P. syringae* cells transformed with different plasmids that individually express the viral target sequences, the T7 RdRP, and the φ6 RdRP. The dsRNA amplification takes place within the φ6 polymerase complexes that also provide a protected environment from bacterial RNases (Voloudakis *et al.*, 2015; Niehl *et al.*, 2018). These bacterial dsRNA production systems can be scaled up, allowing cost-effective large-scale production of long dsRNA molecules targeting pathogen genes or genomes, suitable for application in crop protection (Niehl *et al.*, 2018).

However, most studies reporting the delivery of dsRNAs produced *in vitro* or *in vivo* showed that the protective antiviral effect lasts for only a few days, indicating insufficient stability or efficacy of these molecules for practical use. As frequent treatments with dsRNAs would be necessary to protect plants from virus infection, especially for long-lasting crops cultivated in open fields, establishing methods ameliorating the delivery of dsRNA and their stabilization has become a major challenge. dsRNA formulations based on biocompatible and safe materials are currently being evaluated (Pérez-de-Luque, 2017; Vurro *et al.*, 2019); these include packaging of dsRNAs into virus particles or in virion-like particles (VLPs) (reviewed in Zotti *et al.*, 2018 and Dalakouras *et al.*, 2020). Implementation of dsRNA formulations has been achieved by a biotech company with the 'Apse RNA Containers' (ARCs) system (available at www.rnagri.com, accessed 17 March 2020). Here, *E. coli* cells express naturally occurring proteins, such as the CPs from bacteriophage MS2 that can self-assemble and form VLPs. The same cells also contain another plasmid carrying the target RNA precursor signal sequence, linked to a packaging site. During *E. coli* growth, VLPs made of MS2 CP subunits will encapsidate the target RNA molecules.

From another perspective, an elegant breakthrough of the obstacles related to dsRNA delivery relies on the use of non-toxic, degradable, layered double hydroxide (LDH) clay nanosheets of 80–300 nm (BioClay) that bind to dsRNAs and protect them from degradation (Mitter *et al.*, 2017). These BioClay nanostructures are not only resistant to plant watering but also allow gradual release of dsRNAs to the plant cell, leading to more successful inhibition of the propagation of cucumber mosaic virus (Mitter *et al.*, 2017) and bean common mosaic virus in *N. benthamiana* and cowpea (*Vigna unguiculata*) plants, respectively (Worrall *et al.*, 2019).

Other recently developed delivery strategies include direct trunk injection, as in the commercially available Arborjet strategy (available at https://arborjet.com, accessed 17 March 2020) described in Zotti *et al.* (2018) and Dalakouras *et al.* (2018), but their efficacy against viruses affecting woody plants remains to be evaluated. Another delivery strategy, which seems to be appropriate for inducing virus resistance in plants, consists of a high-pressure spraying method inducing a symplastic RNA delivery of the effective dsRNA molecules (Dalakouras *et al.*, 2020). Indeed, this technique, first described by Dalakouras *et al.* (2016), can trigger both local and systemic silencing and the production of secondary siRNAs, especially when 22 nt molecules are sprayed on the plant tissues.

8.6 Biosafety Considerations

Although one of the major problems hindering a widespread use of the HIGS approach includes the cumbersome regulatory procedures to get governmental approval of transgenic plants, the authors would like to highlight 24 examples where all the bureaucratic processes reached a fruitful outcome, described in detail by Khalid *et al.* (2017). Among them was a successful case of intron hpRNA-based transgenic common bean plants resistant to bean golden mosaic virus (BGMV) accepted for commercialization in Brazil (Bonfim *et al.*, 2007), which exhibit durable resistance in open fields, with unaltered agronomic characteristics and nutritional value (Aragão *et al.*, 2013; Carvalho *et al.*, 2015). Examples of virus-resistant fruit tree species approved for commercial release and generated by HIGS technique include the papaya ringspot virus (PRSV)-resistant papaya (Fitch *et al.*, 1992) and the PPV-resistant plum (Scorza *et al.*, 2001).

Although RNAi-based transgenic plants produce only dsRNA molecules complementary to the target pathogen transcripts, without the synthesis of any new protein, possible off-target effects need to be evaluated. These can be caused by dsRNA molecules' complementarity with unintended sequences in the GM plant or in non-target species (Mlotshwa *et al.*, 2008; Auer and Frederick, 2009; Frizzi and Huang, 2010).

Regarding the use of transgrafting to obtain RNA-based virus-resistant rootstocks in arboriculture, it is expected that this technology would cause less public concern and that the risk assessment would be limited to the transgenic rootstock, as the scion, fruits and pollen maintain their genetic inheritance. These aspects could encourage, in principle, a simplified approach for their application in agriculture (Lemgo *et al.*, 2013; Petrick *et al.*, 2013).

From the biosafety side, the most relevant feature of SIGS relies on the fact that the exogenous application of dsRNA does not involve any modification of the plant genome. Moreover, these substances act by means of their specific nucleotide sequence, have higher specificity and a reduced tendency to induce pathogen resistance if managed appropriately. Importantly, and, contrary to chemical pesticides, dsRNAs are biocompatible and biodegradable compounds, ubiquitously occurring in natural conditions inside and outside organisms (Niehl *et al.*, 2018). Based on the expert panel of the Toxicology Forum at its 40th Annual Summer Meeting held in 2015, local delivery of dsRNAs is considered safe for human consumption, as RNA molecules are present in all kinds of food and exogenous RNAs are considered free of residues potentially toxic for the plant, food or the environment (Sherman *et al.*, 2015). Nonetheless, to increase the activity and safety of these molecules, careful design and predictions by bioinformatics tools are necessary on a case-by-case basis, in order to avoid off- and non-target effects on related or non-related organisms with available genomics information (Zotti *et al.*, 2018).

For the policy relevance of this topic, consensus views on dsRNA-based products have not yet been reached and official legislations governing their use are not yet available in Europe (Gathmann, 2019). Nonetheless,

the European scientific community is currently assessing a regulatory framework for such products, as attested by the Organisation for Co-operation and Economic Development (OECD) Conference on RNAi-based pesticides held in April 2019. It is noteworthy that the safety and legislation issues for such products are generating heated debates in many countries. For example, in New Zealand, the official Decision of the Environmental Protection Authority considered that exogenous application of dsRNAs was technically out of the area of interest of the legislations on new organisms, and any environmental risk assessments of such products was unnecessary (EPA, 2018); however, this statement generated an active debate with negative reactions in the scientific community (Heinemann, 2019).

8.7 Conclusions and Future Prospects

In summary, we have presented the major characteristics of HIGS and SIGS strategies so far developed to inhibit the infection and spread of plant viruses. As the majority of plant viruses are transmitted by insects, the reader is also invited to refer to the specific chapters concerning the use of such strategies addressed against insect vectors.

The hpRNA-mediated HIGS strategy is suitable for targeting one or more specific viruses by the integration of one or more copies of the transgene in the plant genome (Stoutjesdijk *et al.*, 2002). During the past 40 years, the *Agrobacterium tumefaciens* T-DNA-mediated gene insertion method has been deeply understood and it is routinely used to transform several plant species, also via the HIGS approach. However, some crops are recalcitrant to *Agrobacterium*-mediated transformation, for which alternative transformation strategies may be attempted, such as electroporation, microinjection, or particle bombardment. Despite the fact that the HIGS strategy is known to be a durable approach for virus control in agriculture, these plants still suffer from low public acceptance and strict rules for their commercialization and/or release into the environment, especially in the European Union.

The exogenous delivery of dsRNAs to trigger the RNAi mechanism against viruses in plants seems to be a reality for the future of plant disease control, considering that these RNAi effective molecules do not fall under the Directive 2001/18/EC (12 March 2001) of the European Commission or the US regulations, since the plant genome is not modified. In the expectation that regulations of small natural molecules for disease control would include these products as biopesticides, researchers are working to stabilize the formulation of dsRNA molecules suitable for field-scale applications at affordable costs.

References

Aalto, A.P., Sarin, L.P., van Dijk, A.A., Saarma, M., Poranen, M.M. *et al.* (2007) Large-scale production of dsRNA and siRNA pools for RNA interference utilizing bacteriophage phi6 RNA-dependent RNA polymerase. *RNA* 13(3), 422–429. DOI: 10.1261/rna.348307.

Abel, P., Nelson, R., De, B., Hoffmann, N., Rogers, S. *et al.* (1986) Delay of disease development in transgenic plants that express the tobacco mosaic virus coat protein gene. *Science* 232(4751), 738–743. DOI: 10.1126/science.3457472.

Akçay, M.E., Fidanci, A. and Burak, M. (2008) Growth and yield of some sweet cherry cultivars grafted on 'Gisela® 5' rootstock. In: Eris, A., Lang, G.A., Gulen, H. and Ipek, A. (eds) *Proceedings of the V International Cherry Symposium, Bursa, Turkey.* Acta Horticulturae article 795_38, pp. 277–282.

Aloni, B., Cohen, R., Karni, L., Aktas, H. and Edelstein, M. (2010) Hormonal signaling in rootstock–scion interactions. *Scientia Horticulturae* 127(2), 119–126. DOI: 10.1016/j.scienta.2010.09.003.

Altpeter, F., Springer, N.M., Bartley, L.E., Blechl, A., Brutnell, T.P. *et al.* (2016) Advancing crop transformation in the era of genome editing. *The Plant Cell* 28, 1510–1520. DOI: 10.1105/tpc.16.00196.

Aragão, F.J.L., Nogueira, E.O.P.L., Tinoco, M.L.P. and Faria, J.C. (2013) Molecular characterization of the first commercial transgenic common bean immune to the bean golden mosaic virus. *Journal of Biotechnology* 166(1-2), 42–50. DOI: 10.1016/j.jbiotec.2013.04.009.

Auer, C. and Frederick, R. (2009) Crop improvement using small RNAs: applications and predictive ecological risk assessments. *Trends in Biotechnology* 27(11), 644–651. DOI: 10.1016/j.tibtech.2009.08.005.

Batuman, O., Mawassi, M. and Bar-Joseph, M. (2006) Transgenes consisting of a dsRNA of an RNAi suppressor plus the 3′ UTR provide resistance to Citrus tristeza virus sequences in *Nicotiana benthamiana* but not in citrus. *Virus Genes* 33, 319–327.

Baulcombe, D. (2004) RNA silencing in plants. *Nature* 431(7006), 356–363. DOI: 10.1038/nature02874.

Bernstein, E., Caudy, A.A., Hammond, S.M. and Hannon, G.J. (2001) Role for a bidentate ribonuclease in the initiation step of RNA interference. *Nature* 409(6818), 363–366. DOI: 10.1038/35053110.

Bonfim, K., Faria, J.C., Nogueira, E.O.P.L., Mendes, E.A. and Aragão, F.J.L. (2007) RNAi-mediated resistance to Bean golden mosaic virus in genetically engineered common bean (*Phaseolus vulgaris*). *Molecular Plant-Microbe Interactions* 20(6), 717–726. DOI: 10.1094/MPMI-20-6-0717.

Bouquet, A. and Hevin, M. (1978) Green grafting between muscadine grape (*Vitis rotundifolia* Michx.) and bunch grapes (*Euvitis* spp.) as a tool for physiological and pathological investigations. *Vitis* 17, 134–138.

Buhtz, A., Springer, F., Chappell, L., Baulcombe, D.C. and Kehr, J. (2008) Identification and characterization of small RNAs from the phloem of *Brassica napus*. *The Plant Journal* 53(5), 739–749. DOI: 10.1111/j.1365-313X.2007.03368.x.

Carbonell, A., Martínez de Alba, Ángel-Emilio., Flores, R. and Gago, S. (2008) Double-stranded RNA interferes in a sequence-specific manner with the infection of representative members of the two viroid families. *Virology* 371(1), 44–53. DOI: 10.1016/j.virol.2007.09.031.

Carvalho, J.L.V., de Oliveira Santos, J., Conte, C., Pacheco, S., Nogueira, E.O.P.L. *et al.* (2015) Comparative analysis of nutritional compositions of transgenic RNAi-mediated virus-resistant bean (event EMB-PV051-1) with its non-transgenic counterpart. *Transgenic Research* 24(5), 813–819. DOI: 10.1007/s11248-015-9877-5.

Chiou, T.-J., Aung, K., Lin, S.-I., Wu, C.-C., Chiang, S.-F. *et al.* (2006) Regulation of phosphate homeostasis by microRNA in *Arabidopsis*. *The Plant Cell* 18(2), 412–421. DOI: 10.1105/tpc.105.038943.

Cillo, F. and Palukaitis, P. (2014) Transgenic resistance. *Advances in Virus Research* 90, 35–146.

Covey, S.N., Lomonossoff, G.P. and Hull, R. (1981) Characterisation of cauliflower mosaic virus DNA sequences which encode major polyadenylated transcripts. *Nucleic Acids Research* 9(24), 6735–6748. DOI: 10.1093/nar/9.24.6735.

Dalakouras, A., Wassenegger, M., McMillan, J.N., Cardoza, V., Maegele, I. *et al.* (2016) Induction of silencing in plants by high-pressure spraying of in vitro-synthesized small RNAs. *Frontiers in Plant Science* 7(99), 1327. DOI: 10.3389/fpls.2016.01327.

Dalakouras, A., Jarausch, W., Buchholz, G., Bassler, A., Braun, M. *et al.* (2018) Delivery of hairpin RNAs and small RNAs into woody and herbaceous plants by trunk injection and petiole absorption. *Frontiers in Plant Science* 9, 1253. DOI: 10.3389/fpls.2018.01253.

Dalakouras, A., Wassenegger, M., Dadami, E., Ganopoulos, I., Pappas, M. *et al.* (2020) GMO-free RNAi: exogenous application of RNA molecules in plants. *Plant Physiology* 182, 38–50.

Denli, A.M. and Hannon, G.J. (2003) RNAi: an ever-growing puzzle. *Trends in Biochemical Sciences* 28(4), 196–201. DOI: 10.1016/S0968-0004(03)00058-6.

Dinant, S. and Suárez-López, P. (2012) *Multitude of long-distance signal molecules acting via phloem. Biocommunication of Plants*. Springer, Berlin, Heidelberg, pp. 89–121.

Elbashir, S.M., Lendeckel, W. and Tuschl, T. (2001) RNA interference is mediated by 21- and 22-nucleotide RNAs. *Genes & Development* 15(2), 188–200. DOI: 10.1101/gad.862301.

EPA (2018) Decision (to determine whether eukaryotic cell lines that have been treated with externally applied double-stranded RNA molecules for the purpose of inducing a transient small interfering RNA (siRNA) response are new organisma for the purposes of the Hazardous Substances and New organisms Act 1996). APP203395. Environmental Protection Authority, New Zealand. Available at: www.epa.govt.nz/assets/FileAPI/hsno-ar/APP203395/APP203395-Decision-FINAL-pdf (accessed 18 November 2020).

Fire, A., Xu, S., Montgomery, M.K., Kostas, S.A., Driver, S.E. *et al.* (1998) Potent and specific genetic interference by double-stranded RNA in *Caenorhabditis elegans*. *Nature* 391(6669), 806–811. DOI: 10.1038/35888.

Fitch, M.M.M., Manshardt, R.M., Gonsalves, D., Slightom, J. and Sanford, J.C. (1992) Virus resistant papaya plants derived from tissues bombarded with the coat protein gene of papaya ringspot virus. *Bio/Technology* 10, 1466.

Frizzi, A. and Huang, S. (2010) Tapping RNA silencing pathways for plant biotechnology. *Plant Biotechnology Journal* 8(6), 655–677. DOI: 10.1111/j.1467-7652.2010.00505.x.

Fuentes, A., Ramos, P.L., Fiallo, E., Callard, D., Sánchez, Y. *et al.* (2006) Intron-hairpin RNA derived from replication associated protein C1 gene confers immunity to tomato yellow leaf curl virus infection in transgenic tomato plants. *Transgenic Research* 15(3), 291–304. DOI: 10.1007/s11248-005-5238-0.

Fuentes, A., Carlos, N., Ruiz, Y., Callard, D., Sánchez, Y. *et al.* (2016) Field trial and molecular characterization of RNAi-transgenic tomato plants that exhibit resistance to tomato yellow leaf curl geminivirus. *Molecular Plant-Microbe Interactions* 29(3), 197–209. DOI: 10.1094/MPMI-08-15-0181-R.

Gan, D., Zhang, J., Jiang, H., Jiang, T., Zhu, S. *et al.* (2010) Bacterially expressed dsRNA protects maize against SCMV infection. *Plant Cell Reports* 29(11), 1261–1268. DOI: 10.1007/s00299-010-0911-z.

Gathmann, A. (2019) *The European perspective on regulatory aspects and experiences with dsRNA-based products*. Presentation at OECD Conference on RNAi based pesticides. 10–12 April 2019, Paris. Available at: www.oecd.org/chemicalsafety/pesticides-biocides/conference-on-rnai-based-pesticides.htm (accessed 18 March 2020).

Ghildiyal, M. and Zamore, P.D. (2009) Small silencing RNAs: an expanding universe. *Nature Reviews Genetics* 10(2), 94–108. DOI: 10.1038/nrg2504.

Gonçalves, B., Moutinho-Pereira, J., Santos, A., Silva, A.P., Bacelar, E. *et al.* (2006) Scion-rootstock interaction affects the physiology and fruit quality of sweet cherry. *Tree Physiology* 26(1), 93–104. DOI: 10.1093/treephys/26.1.93.

Guelette, B.S., Benning, U.F. and Hoffmann-Benning, S. (2012) Identification of lipids and lipid-binding proteins in phloem exudates from *Arabidopsis thaliana*. *Journal of Experimental Botany* 63(10), 3603–3616. DOI: 10.1093/jxb/ers028.

Ham, B.-K., Li, G., Jia, W., Leary, J.A. and Lucas, W.J. (2014) Systemic delivery of siRNA in pumpkin by a plant PHLOEM SMALL RNA-BINDING PROTEIN 1-ribonucleoprotein complex. *The Plant Journal* 80(4), 683–694. DOI: 10.1111/tpj.12662.

Hamilton, A.J. and Baulcombe, D.C. (1999) A species of small antisense RNA in posttranscriptional gene silencing in plants. *Science* 286(5441), 950–952. DOI: 10.1126/science.286.5441.950.

Heinemann, J.A. (2019) Should dsRNA treatments applied in outdoor environments be regulated? *Environment International* 132, 104856. DOI: 10.1016/j.envint.2019.05.050.

Hull, R. (2014) *Matthews' Plant Virology*, 5th edn. Elsevier Academic Press, Cambridge, Massachusetts.

Ilardi, V. and Tavazza, M. (2015) Biotechnological strategies and tools for plum pox virus resistance: trans-, intra-, cis-genesis, and beyond. *Frontiers in Plant Science* 6(379). DOI: 10.3389/fpls.2015.00379.

Jan, F.-J., Fagoaga, C., Pang, S.-Z. and Gonsalves, D. (2000) A minimum length of N gene sequence in transgenic plants is required for RNA-mediated tospovirus resistance. *Journal of General Virology* 81(1), 235–242. DOI: 10.1099/0022-1317-81-1-235.

Kaldis, A., Berbati, M., Melita, O., Reppa, C., Holeva, M. *et al.* (2018) Exogenously applied dsRNA molecules deriving from the Zucchini yellow mosaic virus (ZYMV) genome move systemically and protect cucurbits against ZYMV. *Molecular Plant Pathology* 19(4), 883–895. DOI: 10.1111/mpp.12572.

Kennerdell, J.R. and Carthew, R.W. (1998) Use of dsRNA-mediated genetic interference to demonstrate that *frizzled* and frizzled 2 act in the wingless pathway. *Cell* 95(7), 1017–1026. DOI: 10.1016/S0092-8674(00)81725-0.

Khalid, A., Zhang, Q., Yasir, M. and Li, F. (2017) Small RNA based genetic engineering for plant viral resistance: application in crop protection. *Frontiers in Microbiology* 8(e60829), 43. DOI: 10.3389/fmicb.2017.00043.

Koch, A., Biedenkopf, D., Furch, A., Weber, L., Rossbach, O. *et al.* (2016) An RNAi-based control of *Fusarium graminearum* infections through spraying of long dsRNAs involves a plant passage and is controlled by the fungal silencing machinery. *PLoS Pathogens* 12(10), e1005901. DOI: 10.1371/journal.ppat.1005901.

Koepke, T. and Dhingra, A. (2013) Rootstock scion somatogenetic interactions in perennial composite plants. *Plant Cell Reports* 32(9), 1321–1337. DOI: 10.1007/s00299-013-1471-9.

Konakalla, N.C., Kaldis, A., Berbati, M., Masarapu, H. and Voloudakis, A.E. (2016) Exogenous application of double-stranded RNA molecules from TMV p126 and CP genes confers resistance against TMV in tobacco. *Planta* 244(4), 961–969. DOI: 10.1007/s00425-016-2567-6.

Kubota, C., McClure, M.A., Kokalis-Burelle, N., Bausher, M.G. and Rosskopf, E.N. (2008) Vegetable grafting: history, use, and current technology status in North America. *Horticultural Science* 43, 1664–1669.

Lau, S.E., Mazumdar, P., Hee, T.W., Song, A.L.A., Othman, R.Y. *et al.* (2014) Crude extracts of bacterially-expressed dsRNA protect orchid plants against Cymbidium mosaic virus during transplantation from *in vitro* culture. *The Journal of Horticultural Science and Biotechnlogy* 89(5), 569–576. DOI: 10.1080/14620316.2014.11513122.

Lefkowitz, E.J., Dempsey, D.M., Hendrickson, R.C., Orton, R.J., Siddell, S.G. *et al.* (2017) Virus taxonomy: the database of the International Committee on Taxonomy of Viruses (ICTV). *Nucleic Acids Research* 46(D1), D708–D717. DOI: 10.1093/nar/gkx932.

Lemgo, G.N.Y., Sabbadini, S., Pandolfini, T. and Mezzetti, B. (2013) Biosafety considerations of RNAi-mediated virus resistance in fruit-tree cultivars and in rootstock. *Transgenic Research* 22(6), 1073–1088. DOI: 10.1007/s11248-013-9728-1.

Lindbo, J.A. and Dougherty, W.G. (1992) Untranslatable transcripts of the tobacco etch virus coat protein gene sequence can interfere with tobacco etch virus replication in transgenic plants and protoplasts. *Virology* 189(2), 725–733. DOI: 10.1016/0042-6822(92)90595-G.

Lindbo, J.A., Silva-Rosales, L. and Dougherty, W.G. (1993) Pathogen derived resistance to potyviruses: working, but why? *Seminars in Virology* 4(6), 369–379. DOI: 10.1006/smvy.1993.1036.

Maillard, P.V., van der Veen, A.G., Poirier, E.Z. Reis e Sousa, C. and Sousa, C. (2019) Slicing and dicing viruses: antiviral RNA interference in mammals. *The EMBO Journal* 38(8), e100941. DOI: 10.15252/embj.2018100941.

Makeyev, E.V. and Bamford, D.H. (2000) Replicase activity of purified recombinant protein P2 of double-stranded RNA bacteriophage φ6. *The EMBO Journal* 19(1), 124–133. DOI: 10.1093/emboj/19.1.124.

Mallory, A.C., Ely, L., Smith, T.H., Marathe, R., Anandalakshmi, R. *et al.* (2001) HC-Pro suppression of transgene silencing eliminates the small RNAs but not transgene methylation or the mobile signal. *The Plant Cell* 13(3), 571–583. DOI: 10.1105/tpc.13.3.571.

Mallory, A.C., Mlotshwa, S., Bowman, L.H. Vance, V.B., Ely, L. (2003) The capacity of transgenic tobacco to send a systemic RNA silencing signal depends on the nature of the inducing transgene locus. *The Plant Journal* 35(1), 82–92. DOI: 10.1046/j.1365-313X.2003.01785.x.

Martin, A., Adam, H., Díaz-Mendoza, M., Zurczak, M., González-Schain, N.D. *et al.* (2009) Graft-transmissible induction of potato tuberization by the microRNA miR172. *Development* 136(17), 2873–2881. DOI: 10.1242/dev.031658.

Melnyk, C. W., Molnar, A., Bassett, A. and Baulcombe, D. C. (2011) Mobile 24 nt small RNAs direct transcriptional gene silencing in the root meristems of *Arabidopsis thaliana*. *Current Biology* 21(19), 1678–1683. DOI: 10.1016/j.cub.2011.08.065.

Mitter, N., Worrall, E.A., Robinson, K.E., Li, P., Jain, R.G. *et al.* (2017) Clay nanosheets for topical delivery of RNAi for sustained protection against plant viruses. *Nature Plants* 3(2), 16207. DOI: 10.1038/nplants.2016.207.

Mlotshwa, S., Pruss, G.J. and Vance, V. (2008) Small RNAs in viral infection and host defense. *Trends in Plant Science* 13(7), 375–382. DOI: 10.1016/j.tplants.2008.04.009.

Molnar, A., Melnyk, C.W., Bassett, A., Hardcastle, T.J., Dunn, R. *et al.* (2010) Small silencing RNAs in plants are mobile and direct epigenetic modification in recipient cells. *Science* 328(5980), 872–875. DOI: 10.1126/science.1187959.

Napoli, C., Lemieux, C. and Jorgensen, R. (1990) Introduction of a chimeric chalcone synthase gene into petunia results in reversible co-suppression of homologous genes in trans. *The Plant Cell* 2(4), 279–289. DOI: 10.2307/3869076.

Niehl, A., Soininen, M., Poranen, M.M. and Heinlein, M. (2018) Synthetic biology approach for plant protection using ds RNA. *Plant Biotechnology Journal* 16(9), 1679–1687. DOI: 10.1111/pbi.12904.

Palauqui, J.-C., Elmayan, T., Pollien, J.M. and Vaucheret, H. (1997) Systemic acquired silencing: transgene-specific post-transcriptional silencing is transmitted by grafting from silenced stocks to non-silenced scions. *The EMBO Journal* 16(15), 4738–4745. DOI: 10.1093/emboj/16.15.4738.

Pandolfini, T., Molesini, B., Avesani, L., Spena, A. and Polverari, A. (2003) Expression of self-complementary hairpin RNA under the control of the rolC promoter confers systemic disease resistance to plum pox virus without preventing local infection. *BMC Biotechnology* 3(1), 7. DOI: 10.1186/1472-6750-3-7.

Pang, S.-Z., Jan, F.-J. and Gonsalves, D. (1997) Nontarget DNA sequences reduce the transgene length necessary for RNA-mediated tospovirus resistance in transgenic plants. *Proceedings of the National Academy of Sciences* 94(15), 8261–8266. DOI: 10.1073/pnas.94.15.8261.

Pérez-de-Luque, A. (2017) Interaction of nanomaterials with plants: what do we need for real applications in agriculture? *Frontiers in Environmental Science* 5, 12. DOI: 10.3389/fenvs.2017.00012.

Petrick, J.S., Brower-Toland, B., Jackson, A.L. and Kier, L.D. (2013) Safety assessment of food and feed from biotechnology-derived crops employing RNA-mediated gene regulation to achieve desired traits: a scientific review. *Regulatory Toxicology and Pharmacology* 66(2), 167–176. DOI: 10.1016/j.yrtph.2013.03.008.

Pooggin, M.M. (2017) RNAi-mediated resistance to viruses: a critical assessment of methodologies. *Current Opinion in Virology* 26, 28–35. DOI: 10.1016/j.coviro.2017.07.010.

Prins, M., Laimer, M., Noris, E., Schubert, J., Wassenegger, M. *et al.* (2008) Strategies for antiviral resistance in transgenic plants. *Molecular Plant Pathology* 9, 73–83.

Pyott, D.E. and Molnar, A. (2015) Going mobile: non-cell-autonomous small RNAs shape the genetic landscape of plants. *Plant Biotechnology Journal* 13(3), 306–318. DOI: 10.1111/pbi.12353.

Romano, N. and Macino, G. (1992) Quelling: transient inactivation of gene expression in *Neurospora crassa* by transformation with homologous sequences. *Molecular Microbiology* 6(22), 3343–3353. DOI: 10.1111/j.1365-2958.1992.tb02202.x.

Sabbadini, S., Ricci, A., Limera, C., Baldoni, D., Capriotti, L. *et al.* (2019) Factors affecting the regeneration, via organogenesis, and the selection of transgenic calli in the peach rootstock Hansen 536 (*Prunus persica* × *Prunus amygdalus*) to express an RNAi construct against PPV virus. *Plants* 8(6), 178. DOI: 10.3390/plants8060178.

Šafářová, D., Brázda, P. and Navrátil, M. (2014) Effect of artificial dsRNA on infection of pea plants by pea seed-borne mosaic virus. *Czech Journal of Genetics and Plant Breeding* 50(No. 2), 105–108. DOI: 10.17221/120/2013-CJGPB.

Sanford, J.C. and Johnston, S.A. (1985) The concept of parasite-derived resistance – deriving resistance genes from the parasite's own genome. *Journal of Theoretical Biology* 113(2), 395–405. DOI: 10.1016/S0022-5193(85)80234-4.

Scorza, R., Callahan, A., Levy, L., Damsteegt, V., Webb, K. *et al.* (2001) Post-transcriptional gene silencing in plum pox virus resistant transgenic European plum containing the plum pox potyvirus coat protein gene. *Transgenic Research* 10(3), 201–209. DOI: 10.1023/A:1016644823203.

Shen, W., Tuo, D., Yan, P., Li, X. and Zhou, P. (2014) Detection of papaya leaf distortion mosaic virus by reverse-transcription loop-mediated isothermal amplification. *Journal of Virological Methods* 195, 174–179. DOI: 10.1016/j.jviromet.2013.09.011.

Sherman, J.H., Munyikwa, T., Chan, S.Y., Petrick, J.S., Witwer, K.W. *et al.* (2015) RNAi technologies in agricultural biotechnology: the toxicology forum 40th annual summer meeting. *Regulatory Toxicology and Pharmacology* 73(2), 671–680. DOI: 10.1016/j.yrtph.2015.09.001.

Simón-Mateo, C. and García, J.A. (2011) Antiviral strategies in plants based on RNA silencing. *Biochimica et Biophysica Acta (BBA) - Gene Regulatory Mechanisms* 1809(11-12), 722–731. DOI: 10.1016/j.bbagrm.2011.05.011.

Smith, N.A., Singh, S.P., Wang, M.B., Stoutjesdijk, P.A., Green, A.G. *et al.* (2000) Gene expression: total silencing by intron-spliced hairpin RNAs. *Nature* 407, 319.

Song, G.Q., Sink, K.C., Walworth, A.E., Cook, M.A., Allison, R.F. *et al.* (2013) Engineering cherry rootstocks with resistance to Prunus necrotic ring spot virus through RNAi-mediated silencing. *Plant Biotechnology Journal* 11(6), 702–708. DOI: 10.1111/pbi.12060.

Song, G.-Q., Walworth, A.E. and Loescher, W.H. (2015) Grafting of genetically engineered plants. *Journal of the American Society for Horticultural Science* 140(3), 203–213. DOI: 10.21273/JASHS.140.3.203.

Stoutjesdijk, P.A., Singh, S.P., Liu, Q., Hurlstone, C.J., Waterhouse, P.A. *et al.* (2002) hpRNA-mediated targeting of the *Arabidopsis* FAD2 gene gives highly efficient and stable silencing. *Plant Physiology* 129(4), 1723–1731. DOI: 10.1104/pp.006353.

Sun, Y., Qiao, X. and Mindich, L. (2004) Construction of carrier state viruses with partial genomes of the segmented dsRNA bacteriophages. *Virology* 319(2), 274–279. DOI: 10.1016/j.virol.2003.10.022.

Tenllado, F. and Díaz-Ruíz, J.R. (2001) Double-stranded RNA-mediated interference with plant virus infection. *Journal of Virology* 75(24), 12288–12297. DOI: 10.1128/JVI.75.24.12288-12297.2001.

Tenllado, F., Martínez-García, B., Vargas, M. and Díaz-Ruíz, J. (2003) Crude extracts of bacterially expressed dsRNA can be used to protect plants against virus infections. *BMC Biotechnology* 3(1), 3. DOI: 10.1186/1472-6750-3-3.

Tenllado, F., Llave, C. and Díaz-Ruíz, J.R. (2004) RNA interference as a new biotechnological tool for the control of virus diseases in plants. *Virus Research* 102(1), 85–96. DOI: 10.1016/j.virusres.2004.01.019.

Thomas, C.L., Jones, L., Baulcombe, D.C. and Maule, A.J. (2001) Size constraints for targeting post-transcriptional gene silencing and for RNA-directed methylation in *Nicotiana benthamiana* using a potato virus X vector. *The Plant Journal* 25(4), 417–425. DOI: 10.1046/j.1365-313x.2001.00976.x.

Tijsterman, M., Ketting, R.F. and Plasterk, R.H.A. (2002) The genetics of RNA silencing. *Annual Review of Genetics* 36(1), 489–519. DOI: 10.1146/annurev.genet.36.043002.091619.

van der Krol, A.R., Mur, L.A., Beld, M., Mol, J.N. and Stuitje, A.R. (1990) Flavonoid genes in petunia: addition of a limited number of gene copies may lead to a suppression of gene expression. *The Plant Cell* 2, 291–299.

Vazquez, F., Legrand, S. and Windels, D. (2010) The biosynthetic pathways and biological scopes of plant small RNAs. *Trends in Plant Science* 15(6), 337–345. DOI: 10.1016/j.tplants.2010.04.001.

Vigne, E., Komar, V. and Fuchs, M. (2004) Field safety assessment of recombination in transgenic grapevines expressing the coat protein gene of grapevine fanleaf virus. *Transgenic Research* 13(2), 165–179. DOI: 10.1023/B:TRAG.0000026075.79097.c9.

Voinnet, O. (2008) Post-transcriptional RNA silencing in plant–microbe interactions: a touch of robustness and versatility. *Current Opinion in Plant Biology* 11(4), 464–470. DOI: 10.1016/j.pbi.2008.04.006.

Voinnet, O. and Baulcombe, D.C. (1997) Systemic signalling in gene silencing. *Nature* 389(6651), 553. DOI: 10.1038/39215.

Voloudakis, A.E., Holeva, M.C., Sarin, L.P., Bamford, D.H. and Vargas, M. (2015) Efficient double-stranded RNA production methods for utilization in plant virus control. In: Uyeda, I. and Masuta, C. (eds) *Plant Virology Protocols: New Approaches to Detect Viruses and Host Responses*, 3rd edn. Humana Press, New York, pp. 255–274.

Vurro, M., Miguel-Rojas, C. and Pérez-de-Luque, A. (2019) Safe nanotechnologies for increasing the effectiveness of environmentally friendly natural agrochemicals. *Pest Management Science* 75(9), 2403–2412. DOI: 10.1002/ps.5348.

Wang, M., Weiberg, A., Lin, F.-M., Thomma, B.P.H.J., Huang, H.-D. *et al.* (2016) Bidirectional cross-kingdom RNAi and fungal uptake of external RNAs confer plant protection. *Nature Plants* 2(10), 16151. DOI: 10.1038/nplants.2016.151.

Waterhouse, P.M., Graham, M.W. and Wang, M.-B. (1998) Virus resistance and gene silencing in plants can be induced by simultaneous expression of sense and antisense RNA. *Proceedings of the National Academy of Sciences* 95(23), 13959–13964. DOI: 10.1073/pnas.95.23.13959.

Wesley, S.V., Helliwell, C.A., Smith, N.A., Wang, M., Rouse, D.T. *et al.* (2001) Construct design for effi- cient, effective and high-throughput gene silencing in plants. *The Plant Journal* 27(6), 581–590. DOI: 10.1046/j.1365-313X.2001.01105.x.

Worrall, E.A., Bravo-Cazar, A., Nilon, A.T., Fletcher, S.J., Robinson, K.E. *et al.* (2019) Exogenous applica- tion of RNAi-inducing double-stranded RNA inhibits aphid-mediated transmission of a plant virus. *Frontiers in Plant Science* 10, 265. DOI: 10.3389/fpls.2019.00265.

Yin, G., Sun, Z., Liu, N., Zhang, L., Song, Y. *et al.* (2009) Production of double-stranded RNA for interfer- ence with TMV infection utilizing a bacterial prokaryotic expression system. *Applied Microbiology and Biotechnology* 84(2), 323–333. DOI: 10.1007/s00253-009-1967-y.

Zhang, W., Kollwig, G., Stecyk, E., Apelt, F., Dirks, R. *et al.* (2014) Graft-transmissible movement of inverted-repeat-induced siRNA signals into flowers. *The Plant Journal* 80(1), 106–121. DOI: 10.1111/ tpj.12622.

Zhao, D. and Song, G.-Q. (2014) Rootstock-to-scion transfer of transgene-derived small interfering RNAs and their effect on virus resistance in nontransgenic sweet cherry. *Plant Biotechnology Journal* 12(9), 1319–1328. DOI: 10.1111/pbi.12243.

Zotti, M., dos Santos, E.A., Cagliari, D., Christiaens, O., Taning, C.N.T. *et al.* (2018) RNA interference tech- nology in crop protection against arthropod pests, pathogens and nematodes. *Pest Management Science* 74(6), 1239–1250. DOI: 10.1002/ps.4813.

9 Small Talk and Large Impact: the Importance of Small RNA Molecules in the Fight Against Plant Diseases

Zhen Liao, Kristian Persson Hodén and Christina Dixelius*
Department of Plant Biology, Swedish University of Agricultural Sciences, Uppsala, Sweden

Abstract

This short and general chapter summarizes how plants and pathogens communicate using not only proteins for recognition and signal transduction or other metabolites but also RNA molecules where small RNAs with sizes between 21 to 40 nt are most important. These small RNAs can move between plants and a range of interacting pathogenic organisms in both directions, that is, a 'cross-kingdom' communication process. The first reports on RNA-based communications between plants and plant pathogenic fungi appeared about 10 years ago. Since that time, we have learnt much about sRNA biology in plants and their function in different parasitic organisms. However, many questions on the processes involved remain unanswered. Such information is crucial in order to sustain high crop production. Besides giving a brief background, we highlight the interactions between the potato late blight pathogen and its plant host potato.

9.1 Introduction

In the early 1990s, a significant breakthrough was achieved by the cloning of several resistance (*R*) genes against different pathogenic organisms in tomato, tobacco and, not least, in *Arabidopsis*. The latter is the first plant species with a sequenced genome and for which numerous genetic and molecular tools and information are now available (TAIR, 2020). This *R*-gene work elucidated, for example, the involvement of conserved protein domains. Two main groups of *R* genes are distinguished based on different N-terminal domains: (i) those with a coil–coil (CC) sequence; and (ii) those that share sequence similarity with the *Drosophila* Toll and human interleukin-1 receptor (TIR) domain. These domains can be combined with nucleotide binding sites (NBS) and leucine-rich repeats (LRR). Together the different domains function as cell surface or intracellular receptors, but confer also loss of susceptibility (Kourelis and van der Hoorn, 2018). In parallel, information on regulatory RNA was derived from studies on viruses demonstrating the importance of transcript regulation on RNA levels (Lindbo *et al.*, 1993; Hamilton and Baulcombe, 1999; Baulcombe, 2004). Non-coding RNAs have emerged since then as important and ubiquitous components of eukaryotic transcriptional and post-transcriptional regulatory processes. These

---

*Corresponding author: Christina.Dixelius@slu.se

© CAB International 2021. *RNAi for Plant Improvement and Protection*
(eds B. Mezzetti *et al.*)
DOI: 10.1079/9781789248890.0009

molecules are entities that include both long and short small RNAs (sRNAs) (Ghildiyal and Zamore, 2009) that are commonly present in sizes ranging from 16 to 40 nt in length. Many classes of sRNAs are described today with different biogenesis, functions and targets (Axtell, 2013). One key function is RNA-mediated gene silencing. In plants, sRNAs are classified into two major categories: microRNAs or miRNAs (21–24 nt long); and short interfering RNAs (siRNAs), of which several sub-classes are known. Among these are 21 nt phased siRNAs (phasiRNAs) and the plant-specific transacting

tasiRNAs that are important regulatory factors during plant defence (Deng *et al.*, 2018).

Cross-kingdom sRNA transport refers to sRNAs capable of moving between two taxonomically unrelated organisms (Figs 9.1 and 9.2). As mentioned earlier, sRNA-mediated immunity in plants was first studied with respect to virus infections. Viruses have also evolved a counter-defence strategy by inhibiting the plant's sRNA-mediated antiviral response. This so-called arms race based on pathogen–plant interactions is a natural evolutionary system to defeat the mechanisms that lead to

Fig. 9.1. All species are taxonomically organized in different distinct groups or kingdoms. Four kingdoms (plants, animals, fungi, protists) in eukaryotes were originally found. Today molecular-based analysis has further divided them into several new taxonomic categories. In this context, communication between plants and fungi is an example of a cross-kingdom event.

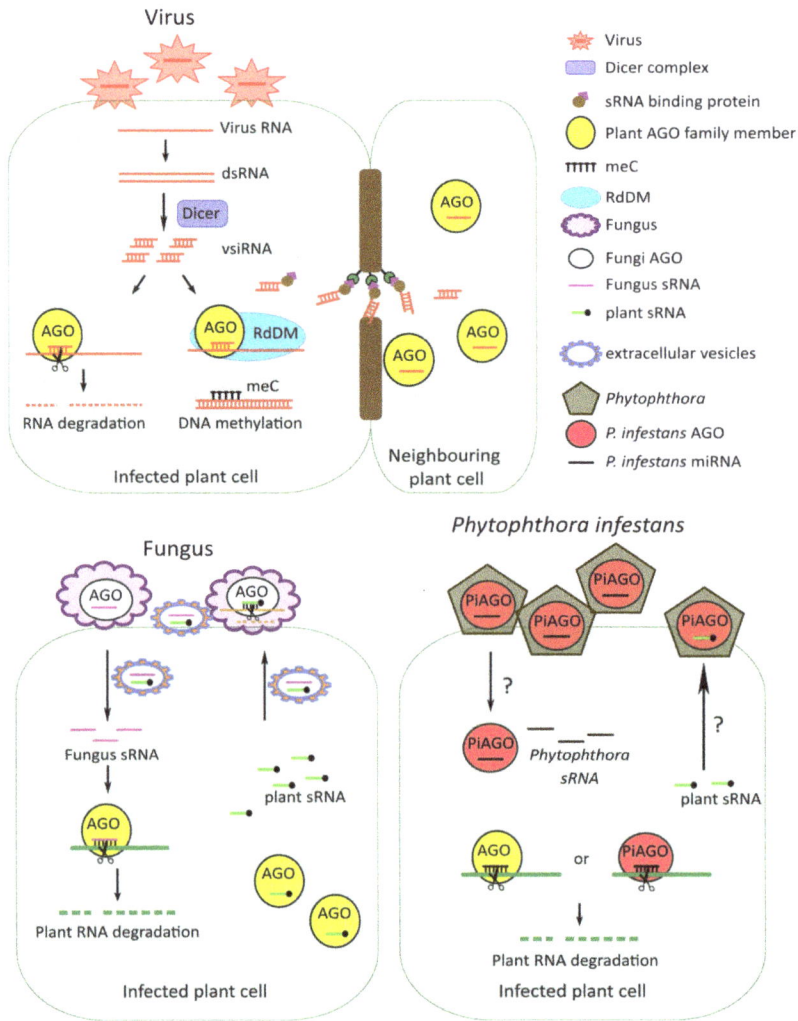

Fig. 9.2. Schematic illustration of 'cross-kingdom' exchange of sRNAs between plants and different intruders. Different main components in the diverse sRNA pathways are listed top right. Argonaute (AGO) has a central role in RNA silencing processes together with Dicer and RNA-directed RNA polymerase, or RNA-directed DNA methylation (RdDM). In plants, DNA is methylated via the cytosine base (meC). There are different classes of sRNAs where the miRNAs are one of the most well-studied. Several protein complexes take part in the different processes. There remains a lot to learn about the transport mechanisms between different organisms (indicated by question marks). The top left panel shows the simplified mechanism of RNA silencing-based antiviral immunity. The plant small RNA binding protein 1 works as a cargo, delivering viral-derived sRNAs to the neighbouring plant cells and therefore amplifying the antiviral immunity. The bottom panel shows sRNA trafficking between plants and two types of eukaryotic pathogens. Extracellular vesicles have been shown to mediate the sRNA movement between plants and fungi. Our current work provided proof-of-concept that potato and *P. infestans* exchange sRNAs during infection (Hu *et al.*, 2020) and vice versa (Jahan *et al.*, 2015). A *P. infestans* miRNA guides either potato AGO or *P. infestans* AGO protein to cleave the target mRNA in the host, promoting infection. Yet, the molecular mechanism of intracellular trafficking remains elusive. (Drawings modified after Zhu *et al.* (2019), Hudzik *et al.* (2020) and Yan *et al.* (2020). © C. Dixelius.)

reduced survival of the organisms. The most common system that is used for making transgenic plants or fungi, the *Agrobacterium*-based system, is per se a typical cross-kingdom system where genes of the soil bacteria can move into plant genomes under natural conditions, resulting in tumour formation (Nester, 2015).

Bacteria species have evolved different secretion systems. The type III secretion system is used by *Pseudomonas syringae* to inject effector molecules into the plant cell, thereby assisting in processes of suppressing plant immunity responses to promote infection and disease. Much understanding of plant immunity responses is based on *P. syringae* and *Arabidopsis* interactions (Xin and He, 2013). In this plant–pathogen system, auxin receptors were shown to be targets of miR393 but this interaction becomes repressed upon treatment with the pathogen-associated molecular pattern (PAMP) effector flagellin of *P. syringae* (Navarro *et al.*, 2006). This work became the starting point of several functional studies of plant miRNAs under various stress conditions. At present, we know that plant miRNAs, tasiRNAs and phasiRNAs target *R*-gene regulation upon stresses. Thus, there is some sort of self-regulation of its own *R* genes resulting in a delicate balance between growth and defence responses. Upon pathogen infection, phasiRNAs and 22 nt miRNAs are downregulated and a subsequent activation of defence responses occurs. Here, some miRNA families seem to be more involved than others, for example miR484 (Yang *et al.*, 2013, 2015).

9.2 Interactions Between Potato Late Blight and its Host Potato Plant

The Solanaceae plant family includes many crops that are grown in almost all countries (Fig. 9.3). One important example is the potato (*Solanum tuberosum*), being the third food crop in the world in terms of human consumption, exceeding 300 million tonnes in annual production (CIP, 2020). To meet the global needs of potato food products and starch, China and India have now advanced ahead of Europe and the USA in acreage and production. Africa, particularly sub-Saharan Africa, has also experienced an increased interest and cultivation of potato (FAO, 2020). Potato tubers are 'easy' to put in soil for multiplication; however, tuber production is sensitive to drought and flooding in the fields. Potato plants suffer from many diseases, of which the potato late blight caused by *Phytophthora infestans* can rapidly destroy the green parts and the tubers both at field levels and in storage (Birch *et al.*, 2012). To prevent infection, plants are commonly protected in the field by recurring chemical sprayings. Estimates of annual costs due to treatments and yield losses worldwide associated with *P. infestans* vary between years but can exceed €10 billion. The problem of generating durable resistance and/or efficient agrochemicals is related to how the genome of this filamentous oomycete is organized. *P. infestans* has a large genome, which encodes close to 1000 genes that could facilitate plant infection. These genes are located in genome regions rich in transposable elements (Haas *et al.*, 2009). Together, these features accelerate genetic changes and adaptation to the surrounding environment imposed by use of new cultivars, chemicals and other new cultivation practices. Introduction of resistance traits often present in related wild *Solanum* species is possible by sexual crossings but commonly takes considerable time, due to the need to remove unwanted DNA introduced into the recipient potato genome along with the desired genes (Vleeshouwers *et al.*, 2011). Thus, the toolbox to combat *P. infestans* needs to be constantly refilled and refined, preferably with different defence components to counter loss of resistance function.

In contrast to plants, *P. infestans* encodes few core components for functional RNA interference pathways: two Dicer-like enzymes, five Argonautes and one RNA-directed RNA polymerase (Vetukuri *et al.*, 2011). After intensive search only one miRNA has been found, compared with plants that could encode hundreds of miRNAs (Fahlgren *et al.*, 2013). There are no specific membrane RNA transporters in *P. infestans* like those found in the nematode *Caenorhabditis elegans*. Neither are such transporters present in plants. In plants, details on mobility of sRNAs, including intercellular, extracellular and long-range mobility, mainly derive from studies on viruses (Yan *et al.*, 2020).

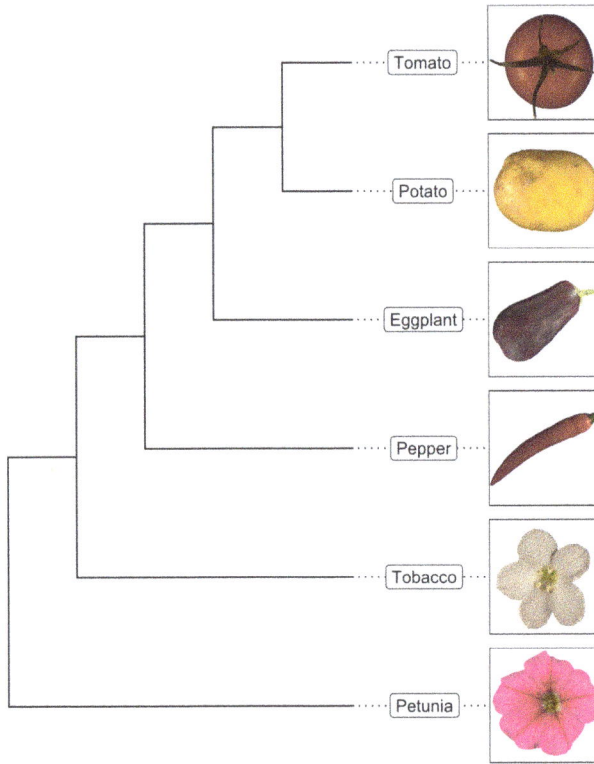

Fig. 9.3. Eggplant, pepper, tomato, potato and petunia are examples of agricultural and horticultural crop species in the Solanaceae family. Potato and tomato are closely related. For more details, see Bombarely *et al.* (2016). (© C. Dixelius.)

Movements of soluble compounds in the plant transport system have been elucidated from studies concerning plant virus infection (Hipper *et al.*, 2013), because viruses use the plant vascular system to spread and colonize the host. It has long been speculated that vesicles in the cellular secretion system can contain sRNAs and be part of the mobility system. To demonstrate their function is a problem, due to their low cellular numbers and because they are therefore difficult to detect. However, this obstacle was recently overcome by the demonstration that *Arabidopsis* sRNAs, protected in extracellular vesicles, could target the fungal virulence genes of *Botrytis cinerea* (Cai *et al.*, 2018). The ability of sRNAs to move from plants to different

pathogens, including *P. infestans*, had been demonstrated earlier via host-induced gene silencing (HIGS) experiments (Jahan *et al.*, 2015) without explicitly explaining the mechanism of movement from one to the other (Fig. 9.2). HIGS requires knowledge on important virulence genes or species-specific active sites in the metabolism to be targeted by sRNAs produced in the properly designed transgenic plant. Interestingly, the single miRNA from *P. infestans* is demonstrated to target a membrane protein localized in the tonoplast of potato, resulting in enhanced susceptibility (Hu *et al.*, 2020). Extensive targeting of potato and pathogen-derived sRNAs to a large number of mRNAs was observed, including 206 sequences coding for *R* genes in the potato genome. Whether genome editing of target sites of these sRNAs in the *R* genes would generate a release of new functional genes against *P. infestans* remains to be shown, as well as the transport mechanism of sRNAs between potato and *P. infestans*.

Direct application of RNA molecules is also under development (Koch *et al.*, 2016). This so-called spray-induced gene silencing (SIGS) has the advantage of not being a genetically modified organism (GMO) strategy but requires upscaling of sRNA quantities for use at field levels, together with overcoming other limiting factors (Song *et al.*, 2018). Several attempts to test this approach against *P. infestans* are ongoing. Besides organism specificity, the designed molecules need to pass national legislations and such regulatory framework is presently not adapted to RNA molecules. Here, studies on environmental RNAi effects will most likely be asked for.

A new form of site-directed nuclease-related class has been developed that originally built on a bacterial defence mechanism against virus phages called clustered regularly interspaced short palindromic repeats associated system number 9 (CRISPR-Cas9). Genome editing approaches, not least by the CRISPR-Cas9 system which is based on RNA biology, may offer new possibilities to induce specific mutations, which gives hope for future crop improvements (Chen *et al.*, 2019) (see also Chapter 6, this volume). The emphasis on resistance breeding during the past century has focused on identification and transfer of resistance genes from related or wild species to crops. However, *R*-genes are rapidly overcome when frequencies of genetic recombination are high, not least in an organism with efficient spore spreading. New technologies such as resistance gene enrichment sequencing (RenSeq) and in combination with association genetics (AgRenSeq) can now facilitate *R* gene identification in complex crop genomes (Jupé *et al.*, 2013) as well as in wild relatives (Arora *et al.*, 2019). We envisage the need to combine genome-wide or genome-selection technologies along with genome-editing approaches to expand the genetic potentials to control late blight disease in potato.

References

Arora, S., Steuernagel, B., Gaurav, K., Chandramohan, S., Long, Y. *et al.* (2019) Resistance gene cloning from a wild crop relative by sequencing capture and association genetics. *Nature Biotechnology* 37(2), 139–143. DOI: 10.1038/s41587-018-0007-9.

Axtell, M.J. (2013) Classification and comparison of small RNAs from plants. *Annual Review of Plant Biology* 64, 137–159. DOI: 10.1146/annurev-arplant-050312-120043.

Baulcombe, D. (2004) RNA silencing in plants. *Nature* 431(7006), 356–363. DOI: 10.1038/nature02874.

Birch, P.R.J., Bryan, G., Fenton, B., Gilroy, E.M., Hein, I. *et al.* (2012) Crops that feed the world 8: Potato: are the trends of increased global production sustainable? *Food Security* 4, 477–508. DOI: 10.1007/s12571-012-0220-1.

Bombarely, A., Moser, M., Amrad, A., Bao, M., Bapaume, L. *et al.* (2016) Insight into the evolution of the Solanaceae from the parental genomes of *Petunia hybrida*. *Nature Plants* 2(6), 16074. DOI: 10.1038/nplants.2016.74.

Cai, Q., Qiao, L., Wang, M., He, B., Lin, F.-M. *et al.* (2018) Plants send small RNAs in extracellular vesicles to fungal pathogen to silence virulence genes. *Science* 360(6393), 1126–1129. DOI: 10.1126/science.aar4142.

Chen, K., Wang, Y., Zhang, R., Zhang, H. and Gao, C. (2019) CRISPR/Cas genome editing and precision plant breeding in agriculture. *Annual Review of Plant Biology* 70, 667–697. DOI: 10.1146/annurev-arplant-050718-100049.

CIP (2020) Potato facts and figures. International Potato Center, Lima, Peru. Available at: www.cip.org (accessed 24 July 2020).

Deng, P., Muhammad, S., Cao, M. and Wu, L. (2018) Biogenesis and regulatory hierachy of phased small interfering RNAs in plants. *Plant Biotechnology Journal* 16(5), 965–975. DOI: 10.1111/pbi.12882.

Fahlgren, N., Bollmann, S.R., Kasschau, K.D., Cuperus, J.T., Press, C.M. *et al*. (2013) *Phytophthora* have distinct endogenous small RNA populations that include short interfering and microRNAs. *PLoS ONE* 8(10), e77181. DOI: 10.1371/journal.pone.0077181.

FAO (2020) Potatoes. Available at: www.fao.org/faostat (accessed 24 July 2020).

Ghildiyal, M. and Zamore, P.D. (2009) Small silencing RNAs: an expanding universe. *Nature Review Genetics* 10(2), 94–108. DOI: 10.1038/nrg2504.

Haas, B.J., Kamoun, S., Zody, M.C., Jiang, R.H.Y., Handsaker, R.E. *et al*. (2009) Genome sequence and analysis of the Irish potato famine pathogen *Phytophthora infestans*. *Nature* 461(7262), 393–398. DOI: 10.1038/nature08358.

Hamilton, A.J. and Baulcombe, D.C. (1999) A species of small antisense RNA in posttranscriptional gene silencing in plants. *Science* 286(5441), 950–952. DOI: 10.1126/science.286.5441.950.

Hipper, C., Brault, V., Ziegler-Graff, V. and Revers, F. (2013) Viral and cellular factors involved in phloem transport of plant viruses. *Frontiers in Plant Science* 4, 154. DOI: 10.3389/fpls.2013.00154.

Hu, X., Persson Hodén, K., Liao, Z., Dölfors, F., Åsman, A. *et al*. (2020) *Phytophthora infestans* Ago1-bound miRNA promotes potato late blight disease. *(Preprint in BioRχiv, January 2020.).*

Hudzik, C., Hou, Y., Ma, W. and Axtell, M.J. (2020) Exchange of small regulatory RNAs between plants and their pests. *Plant Physiology* 182(1), 51–62. DOI: 10.1104/pp.19.00931.

Jahan, S.N., Åsman, A.K.M., Corcoran, P., Fogelqvist, J., Vetukuri, R.R. *et al*. (2015) Plant-mediated gene silencing restricts growth of the potato late blight pathogen *Phytophthora infestans*. *Journal of Experimental Botany* 66(9), 2785–2794. DOI: 10.1093/jxb/erv094.

Jupé, F., Witek, K., Verweij, W., Sliwka, J., Pritchard, L. *et al*. (2013) Resistance gene enrichment sequencing (RenSeq) enables reannotation of the NB-LRR gene family from sequenced plant genomes and rapid mapping of resistance loci in segregating populations. *The Plant Journal* 76(3), 530–544. DOI: 10.1111/tpj.12307.

Koch, A., Biedenkopf, D., Furch, A., Weber, L., Rossbach, O. *et al*. (2016) An RNAi-based control of *Fusarium graminearum* infections through spraying of long dsRNAs involves a plant passage and is controlled by the fungal silencing machinery. *PLoS Pathogen* 12(10), e1005901. DOI: 10.1371/journal.ppat.1005901.

Kourelis, J. and van der Hoorn, R.A.L. (2018) Defended to the nines: 25 years of resistance gene cloning identifies nine mechanisms for R protein function. *The Plant Cell* 30(2), 285–299. DOI: 10.1105/tpc.17.00579.

Lindbo, J.A., Silva-Rosales, L., Proebsting, W.M. and Dougherty, W.G. (1993) Induction of a highly specific antiviral state in transgenic plants: implications for regulation of gene expression and virus resistance. *The Plant Cell* 5(12), 1749–1759. DOI: 10.2307/3869691.

Navarro, L., Dunoyer, P., Jay, F., Arnold, B., Dharmasiri, N., Estelle, M. *et al*. (2006) A plant miRNA contributes to antibacterial resistance by repressing auxin signaling. *Science* 312(5772), 436–439. DOI: 10.1126/science.1126088.

Nester, E.W. (2015) *Agrobacterium*: nature's genetic engineer. *Frontiers in Plant Science* 5, 730. DOI: 10.3389/fpls.2014.00730.

Song, X.-S., Gu, K.-X., Duan, X.-X., Xiao, X.-M., Hou, Y.-P. *et al*. (2018) Secondary amplification of siRNA machinery limits the application of spray-induced gene silencing. *Molecular Plant Pathology* 19(12), 2543–2560. DOI: 10.1111/mpp.12728.

TAIR (2020) The Arabidopsis information resource. Available at: www.arabidopsis.org (accessed 24 July 2020).

Vetukuri, R.R., Avrova, A.O., Grenville-Briggs, L.J., Van West, P., Söderbom, F. *et al*. (2011) Evidence for involvement of Dicer-like, Argonaute and histone deacetylase proteins in gene silencing in *Phytophthora infestans*. *Molecular Plant Pathology* 12(8), 772–785. DOI: 10.1111/j.1364-3703.2011.00710.x.

Vleeshouwers, V.G.A.A., Raffaele, S., Vossen, J.H., Champouret, N., Oliva, R. *et al*. (2011) Understanding and exploiting late blight resistance in the age of effectors. *Annual Reviews of Phytopathology* 49, 507–531. DOI: 10.1146/annurev-phyto-072910-095326.

Xin, X.-F. and He, S.Y. (2013) *Pseudomonas syringae* pv. *tomato* DC3000: A model pathogen for probing disease susceptibility and hormone signaling in plants. *Annual Reviews of Phytopathology* 51(1), 473–498. DOI: 10.1146/annurev-phyto-082712-102321.

Yan, Y., Ham, B.-K., Chong, Y.H., Yeh, S.-D. and Lucas, W.J. (2020) A plant small RNA-binding protein 1 family mediates cell-to-cell trafficking of RNAi signals. *Molecular Plant* 13(2), 321–335. DOI: 10.1016/j.molp.2019.12.001.

Yang, L., Jue, D., Li, W., Zhang, R., Chen, M. *et al.* (2013) Identification of miRNA from eggplant (*Solanum melongena* L.) by small RNA deep sequencing and their response to *Verticillium dahliae* infection. *PLoS ONE* 8, e72840. DOI: 10.1371/journal.pone.0072840.

Yang, L., Mu, X., Liu, C., Cai, J., Shi, K, Zhu, W. *et al.* (2015) Overexpression of potato miR482e enhanced plant sensitivity to *Verticillium dahliae* infection. *Journal of Integrative Plant Biology* 57(12), 1078–1088. DOI: 10.1111/jipb.12348.

Zhu, C., Liu, T., Chang, Y.-N. and Duan, C.-G. (2019) Small RNA functions as a trafficking effector in plant immunity. *International Journal of Molecular Sciences* 20(11), 2816. DOI: 10.3390/ijms20112816.

10 The Stability of dsRNA During External Applications – an Overview

Ivelin Pantchev[1,2]*, Goritsa Rakleova[2] and Atanas Atanassov[2]

[1]*Department of Biochemistry, Sofia University, Sofia, Bulgaria; [2]Joint Genomic Center Ltd, Sofia, Bulgaria*

Abstract

The research community is deeply convinced that RNA is unstable in the environment. Its roots rise from numerous failed attempts to isolate functional cellular RNA molecules. Further support had originated from the fast turnover of RNA in the cells. The situation changed recently with the discovery that externally applied dsRNA can produce targeted gene silencing in plant-feeding insects. First results have demonstrated that external dsRNA can successfully pass the insect gastrointestinal tract and reach its final destination within the body cells. This was somewhat unexpected and sparked new interest in RNA stability in the environment and its fate in the insect organism. In this brief review we make an attempt to summarize current knowledge and to propose a model of how dsRNA can perform its function under these settings.

10.1 Introduction

Since the initial discovery of the phenomenon (Ecker and Davis, 1986; Napoli *et al.*, 1990) and its detailed investigation (Fire *et al.*, 1998), RNA interference (RNAi) technology has gained much attention not only for fundamental research but also for practical applications. During the following decades most components of the interference apparatus were described along with regulatory mechanisms, as well as the constantly widening field of application. Recently, another aspect of RNA interference has gradually focused scientific interest: double-stranded RNA (dsRNA) stability either *in vivo* or *in vitro*.

There were at least two reasons determining this interest. The first was related to better understanding of the regulation of RNA interference pathways in the cell. A second reason emerged upon the discovery that dsRNA might be used as a therapeutic agent in medicine or as a plant protection agent in agriculture. In this case dsRNA is released into the environment in one form or another, which raises biosafety concerns. Initially, the only sources of artificial dsRNA in the environment were transgenic plants. Since dsRNA expression levels were not very high, the risk to the environment was estimated as insignificant (see below). The situation changed radically when a novel application of RNAi appeared as an externally applied insecticide. In this case the amounts of dsRNA directly applied to plants and soil might be significantly higher than those provided by genetically modified (GM) plants. As a result, new concerns about dsRNA biosafety were raised which, in turn, renewed the interest in RNA persistence in the environment. RNA stability in the environment also became a topic of interest for

---

*Corresponding author: ipanchev@abv.bg

© CAB International 2021. *RNAi for Plant Improvement and Protection*
(eds B. Mezzetti *et al.*)
DOI: 10.1079/9781789248890.0010

more practical reasons: how, applied externally as insecticide, dsRNA could survive harsh conditions on leaf surfaces and within insect gastrointestinal tracts during feeding; and how to reach the target cells.

After decades of laboratory work, RNA is generally recognized as a degradation-prone molecule both *in vivo* and *in vitro*. Special precautions are a mandatory part of almost all protocols and manuals dealing with all types of RNA studied. There is a good foundation for both chemical and biochemical reasons. RNA possesses an additional 2′-OH group that makes the molecule more reactively competent than DNA, especially under alkaline conditions. In addition, organisms produce a number of RNA-degrading enzymes, both intracellular and secreted – some of them with very high stability in the environment.

RNA plays numerous roles in the cell, such as a temporary mediator of gene expression (messenger RNA) (mRNA), a structural component of translational apparatus (ribosomal RNA (rRNA), transfer RNA (tRNA)) and regulatory functions (RNAi), to name a few. Since it is not a long-term carrier of genetic information, mRNA is characterized with very high turnover rates, with typical half-life of 30 s in bacteria to 30 min in eukaryotes. rRNA and tRNA appeared to be less prone to degradation (as compared with mRNA) due to either forming nucleoprotein complexes (rRNA) or extensive covalent modifications (tRNA).

The fate of short RNAs involved in interference in the cell is more complicated and has received attention during the past decade. The dsRNA precursor might be bound by specific RNA binding proteins and, eventually, targeted for degradation (Heo *et al.*, 2008) as part of regulatory mechanisms. Once loaded in the RNA-induced silencing complex (RISC), microRNA (miRNA) is relatively stable with a half-life well over that of mRNA (hours or even days).

Further degradation of small RNAs depends on several 5′-to-3′ and 3′-to-5′ miRNA-degrading enzymes. A small RNA degrading nuclease (SDN1) was identified and cloned from *Arabidopsis*. SDN1 uses a 3′-to-5′ exonucleolytic mechanism, yielding a final degradation product of 8–9 nt. SDN1 can degrade single-stranded RNA in the range of 17–27 nt with comparable efficiency, but not premiRNAs, longer RNAs, double-stranded RNA or single-stranded DNA (Ramachandran and Chen, 2008; Wang *et al.*, 2018). Uridylation by terminal

nucleotydil transferase of processed miRNA was also suggested as a degradation-targeting signal (see Ruegger and Grosshans, 2012). Recently, another mechanism for specific degradation of particular miRNA triggered by its target mRNA or another miRNA was identified (Ghini *et al.*, 2018), which is believed to form a complicated network regulating miRNA activity in the cell (Nicassio, 2019).

Bearing in mind the complicated and extensive metabolism of RNA in the cell, the discovery that externally applied dsRNA precursors might efficiently act as an insecticide during feeding was somewhat unexpected from the very early stages (Fire *et al.*, 1998). Since then the mechanisms of the phenomenon have received sufficient attention for both practical and theoretical reasons (Huvenne and Smagghe, 2010). Despite the extensive research, our knowledge about the mechanisms underlying the insecticidal effect of externally applied dsRNA are still fragmented. In this review we attempt to propose a model of how dsRNA acts as an insecticide.

The key questions are how dsRNA reaches its target destination and what are the major factors influencing its stability.

10.2 Challenges to dsRNA Stability in the Environment

The first major stopover of topically applied dsRNA is the leaf surface, where major determinants of its fate are environmental conditions. These can generally be divided into abiotic and biotic factors by their nature. Since little specific research has targeted the effects of these factors on the leaf surface, some hints can be taken from both *in vitro* experiments and data available for two main environmental compartments: water and soil.

RNA in water solution under controlled physiological conditions *in vitro* is a relatively stable molecule with a half-life rate of several months. Its main degradation pathway is via internal phosphoester transfer reaction, promoted by specific base catalysis (Li and Breaker, 1999). The presence of buffer compounds and, especially, Mg^{2+} ions at pH 7 and above can significantly facilitate RNA hydrolysis to half-life times in the range of minutes. The catalytic effect of Mg^{2+} can be reduced by the presence of chelating agents (AbouHaidar and Ivanov, 1999).

Ultraviolet (UV) irradiation is another factor that compromises RNA stability. Exposure to UV can lead to photochemical modifications, crosslinking, and oxidative damage of the molecule (Singer, 1971). Interestingly, the presence of Mg^{2+} can reduce UV damage. Also, single-stranded RNA is more prone to UV damage than dsRNA.

These data suggest that UV irradiation and chemical microenvironment (i.e. pH and presented metal ions) can be considered as the main abiotic factors leading to RNA degradation in the environment (Albright et al., 2017).

The recently opened possibilities for using RNAi technologies in agriculture have sparked new interest in RNA stability, especially for regulatory reasons about its biosafety. One of the key questions was how long dsRNA (either externally applied or produced by transgenic plants) persists in the environment. For example, recombinant Bacillus thuringiensis (Bt) toxins produced by transgenic plants show a half-life in the range of days to weeks (Icoz and Stotzky, 2008). Also, it was demonstrated that these proteins do not accumulate in soil (Sims and Ream, 1997), which was one of the arguments for their biosafety.

Irrespective of its origin, RNA shares the same two main receiving compartments in the environment: soil and water. Early experiments on dsDNA stability in soil revealed a half-life of under 2 h (Greaves and Wilson, 1970; Keown et al., 2004). Recent experiments with ^{32}P-labelled dsRNA applied to 'active' soil samples demonstrated similar half-life times (Parker et al., 2019). The authors identified two main degrading factors: bacterial uptake and extracellular RNases. On the other hand, quantitative evaluation of dsRNA persistence in water reservoirs by qPCR revealed a half-life of approximately 3 days (Fischer et al., 2017). The apparent discrepancy might reflect differences in bacteria as well as extracellular RNase abundance in these two environmental compartments. Also, biotic degradation in the environment appeared to dominate over abiotic factors, especially in soils (Dubelman et al., 2014).

It cannot be wrong to assume that the same factors, determining dsRNA stability in soil and water, also play a role in dsRNA persistence on leaf surfaces.

Leaf surface can be considered as an arid zone with extensive solar irradiation, inconsistent temperature variations and low organic content. Cuticle surface wax renders it water-repellent and does not allow significant water accumulation. Together, these factors lead to changing microenvironments and do not allow extensive bacterial growth. The reduced biotic degradation results in dsRNA half-life of 36 h (Bachman, 2019) and persistence for up to 3 days with sufficient activity.

10.3   Challenges to dsRNA Stability During Insect Feeding

The next major event is the transit of dsRNA in the insect gastrointestinal tract. Here, the main degrading factors are chemical composition (i.e. pH, ions, compounds), secreted RNases and gut microflora.

RNases comprise a large family of enzymes that play different functions in the cells and organisms. They differ by structure, activity, specificity, localization and environmental stability, to name a few. These differences are observed not only among taxonomic groups but also among the enzymes encoded by a particular genome and reflect the many functions that RNases play in the cells. Insect species are no exception and also demonstrate significant differences in RNase (and, more importantly, dsRNase) composition (Singh et al., 2017; Peng et al., 2018, Peng et al., 2020). The involvement of insect dsRNases in RNA interference efficiency through feeding application was demonstrated by knockout of dsRNase genes. Two dsRNase genes named dsRNase1 and dsRNase2 were identified in Queensland fruit fly, Bactrocera tryoni. Their knockout demonstrated significant improvement of the insecticidal effect of externally applied dsRNA (Tayler et al., 2019). These data can lead to the suggestion to consider dsRNase genes as co-targets in complex RNAi insecticide formulations.

Secreted RNases are the main degrading factor that dsRNA encounters during insect feeding. The very first contact occurs in the upper gastrointestinal tract (Lomate and Bonning, 2016; Song et al., 2017). Experiments have revealed that naked dsRNA suffered extensive degradation within 5 min when incubated with saliva of the southern green stink bug, Nezara viridula (Lomate and Bonning, 2016).

In the midgut, dsRNA encounters additional challenges like changes in pH, ionic content and organic compounds that can increase degradation either directly or by destabilizing dsRNA structure, making it more susceptible to dsRNases. However, dsRNase activity along the gastrointestinal tract appeared to have species-specific variances. In *B. tryoni*, dsRNases appeared to be the most important factor determining dsRNA degradation in the midgut (Tayler *et al.*, 2019). On the other hand, in *N. viridula*, dsRNase activity in the midgut is negligible compared with the saliva (Lomate and Bonning, 2016). These seemingly discrepant results suggest that a good knowledge of the biochemistry of the targeted insect is a prerequisite to achieve maximal insecticide activity by RNAi approach.

Since the dsRNase source in these experiments was not clear (insect, bacterial, or both), gut microflora can also be considered as an important degrading factor in the midgut. Although direct evidence is not yet available, experiments on RNA persistence in soil (Parker *et al.*, 2019) might offer a glimpse of its significance.

10.4   Reaching Inside Cells

What is next for the dsRNA molecules that remained intact during their passage through the insect's gastrointestinal tract? In order to express their activity, dsRNA must enter the epithelium cells and, eventually, reach the haemolymph (Garbutt *et al.*, 2013).

First of all, dsRNA must enter the gut epithelium cells. This seems to be carried out by clathrin or caveolin-mediated endocytosis of molecules, adsorbed to the cell surface (Denecke *et al.*, 2018).

There are at least two possible entry mechanisms. The first one might involve formation of complexes between dsRNA and dsRNA-binding proteins in a non-specific manner. Proteins, bearing dsRNA-binding motifs, apparently exist in both prokaryotes and eukaryotes. It might be expected that some proteins might be presented in the midgut, where they form complexes with dsRNA, which might adsorb to the epithelial cell surface and enter via endocytosis.

It can easily be assumed that such adsorption is non-specific but there might be some indication of other more specific mechanisms. Researchers have identified cell membrane-associated DNA protein in human HeLa cells (Siess *et al.*, 2000). Further, an RNA/DNA-binding protein has been demonstrated to relocate to the cell membrane (Ren *et al.*, 2014). Recently, quite interesting data were published that Argonaute proteins can be secreted from the cells (Weaver and Patton, 2020). Together, these results suggest the possibility that dsRNA can be actively imported into the cells via some specific pathway (e.g. receptor-mediated endocytosis).

However, all these data were obtained on human cell lines. Not much data is available for insect (ds)RNA-binding proteins, exposed to cell surfaces. In *Caenorhabditis elegans*, two membrane proteins SID-1 and SID-2 were identified, which are responsible for RNAi uptake and spreading in an endocytosis-independent manner. In insects, SID-2 has no homologues but SID-1 is conserved among almost all species except Diptera. There is no direct evidence for dsRNA binding by SID-1, which makes any conclusions about its role too preliminary (Denecke *et al.*, 2018).

Several possible pathways of dsRNA entry have been suggested (Vélez and Fishilevich, 2018). One proposed pathway might depend on SID-1-like proteins. Another pathway might depend on endocytosis in several aspects. One is related to cholesterol uptake, while the other is related to formation of clathrin vesicles. In the latest case, involvement of yet unidentified dsRNA-binding proteins is suggested (Vélez and Fishilevich, 2018).

Studies on endocytosis in different insect species revealed differences in dsRNA localization and cytoplasm entry routes, which might explain the observed species-specific differences in RNAi efficiency (Vélez and Fishilevich, 2018). Unfortunately, all available data are controversial, which makes it difficult to identify the exact mechanisms.

Once engulfed, dsRNA must escape from the endocytosis vesicles into the cytoplasm. The efficiency of vesicle escape and subsequent intracellular transport are important for triggering the RNAi path (Shukla *et al.*, 2016). This is really terra incognita, since very limited data are available. One can speculate that escape occurs in a manner similar to one exploited by viruses. Since several groups are reporting that work is in progress, one might expect that the first data will appear soon.

The number of dsRNA molecules that eventually reach into the cytoplasm of epithelial cells might be as low as a few molecules per cell. Here, the only viable way to reach effective levels appears to be through an RNA amplification pathway (Zhang and Ruvkun, 2012). Unfortunately, no RNA-dependent RNA polymerase genes were identified in insects (Gordon and Waterhouse, 2007), which puts RNA amplification mechanism beyond consideration. Therefore, in insects, the RNAi effect seems to rely only on molecules, passing from the gut (Ivashuta *et al.*, 2015).

10.5 How dsRNA Appears to Work as an Insecticide and What Improvements Are Needed

In Fig. 10.1 a model of dsRNA delivery from plant surface to insect body is depicted.

Stage 1 refers to the dsRNA application process and its stability on the leaf surface. Here, the most critical factors are environmental conditions like UV, ions, pH and, to some extent, RNases. Since dsRNA in the environment has a half-life of 2–3 days, formulations are necessary to achieve sufficient efficiency.

Stage 2 reflects dsRNA uptake by insects during feeding. At this stage, the main obstacles are dsRNases of the gastrointestinal tract. Since dsRNase patterns differ, the particular set of secreted enzymes might be the first reason for species-specific differences in RNAi efficiency.

During Stage 3, dsRNA must pass through the gastrointestinal tract and reach the epithelial cells in the midgut. Again, dsRNases, either insect-secreted or of bacterial origin, are the main degrading factor. Proper formulations (e.g. nanoparticles) might significantly increase dsRNA stability and, thus, RNAi efficiency.

At Stage 4, dsRNA must either enter epithelial cells or pass into haemolymph. The molecular bases of these processes are not very well understood in insects (Cooper *et al.*, 2019). Obviously, these are basic natural processes like endocytosis and other trans-barrier and trans-membrane trafficking mechanisms, but their exact nature is unrevealed. While most mechanisms of RNAi pathways appear to be conservative (Yoon *et al.*, 2018), it is unclear how dsRNA might express its activity without RNA amplification process (Vélez and Fishilevich, 2018).

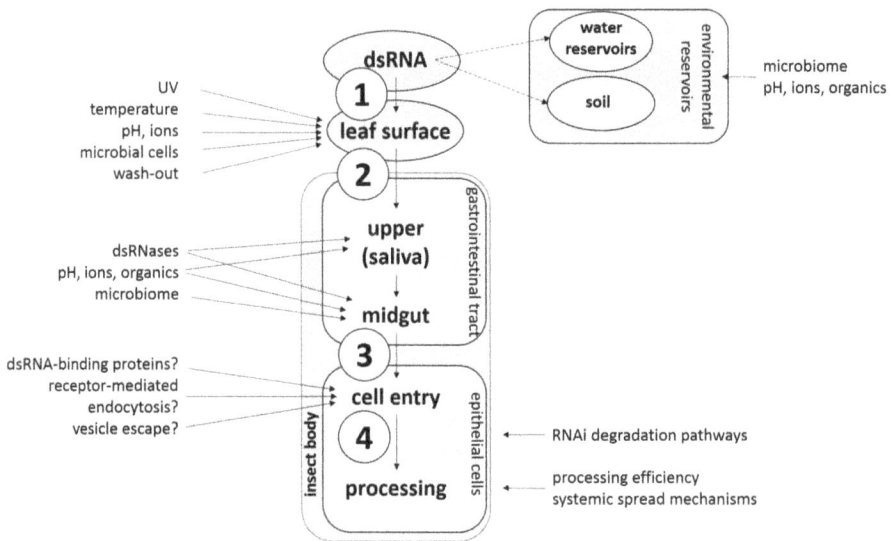

Fig. 10.1. Pathways for dsRNA from plant surface to within insect body.

10.6 Possible Improvements of the RNAi Design

The outlined pathway demonstrates that both biotic and abiotic factors cannot be controlled under field conditions. One possible solution is to design formulations that can improve dsRNA stability for a substantial period of time (for at least 5–7 days). There is an excellent review of delivery systems by Whitten (2019). Briefly, almost all known approaches like chemical condensation, peptide or protein complex formation are providing sufficient increase of RNA stability. Maybe the most promising direction is towards development of protein–RNA complexes with predefined properties. Such complexes have the potential to implement most, if not all, required features for efficient insecticidal effect.

An approach demonstrated by Ghosh et al. (2017) employs a specially formulated diet as an RNA protecting factor. It has been identified that formulations targeted for increased dsRNA stability are an absolute prerequisite and may be the only way to deliver sustainable effect. One could expect that novel solutions, some of them not in the mainstream, will also find their market niche.

Another parameter to be considered for efficient external application is the length of the applied dsRNA molecule. Foliar-applied actin-dsRNA against Colorado potato beetle remained active for 4 weeks under greenhouse conditions but its efficiency depended on the length (San Miguel and Scott, 2016). In similarly designed experiments with *Diabrotica undecimpunctata howardi*, a precursor 27 bases long did not demonstrate toxicity. The efficiency became significant when the length of the precursor was increased to 60 bases and reached a plateau when the length exceeded 70 bases to at least 240 bases (Bolognesi et al., 2012). Another important result was the discovery that a precursor 240 bases long with 100% identity to the target was significantly more efficient than one with the same length but containing the absolute minimum of identical bases (i.e. 21 or 27 base long RNAi target and non-specific carrier). These results clearly demonstrate that using long target-specific precursors is a more effective strategy than short pre-determined analogues of siRNA/miRNA (Wang et al., 2019).

A site cleavage preference during insect dsRNA processing (probably by Dicer) has been described (Guan et al., 2018). The preference appears to be species-specific, thus further explaining differences in RNAi efficiency. Preliminary analysis of such site specificity in the targeted insect might be considered in RNAi design for further efficiency improvement.

10.7 Concluding Remarks

The exact nature of processes that underline dsRNA efficiency as insecticide is largely unclear. Most of them are investigated in great detail, while others are not yet fully revealed even in model organisms. Moreover, it is unclear how these processes interact in order to provide a single pathway of dsRNA from the environment to the insect cells. At the moment, the only possibility is to extrapolate available scientific data in an attempt to generate a hypothetical picture of how dsRNA acts.

References

AbouHaidar, M.G. and Ivanov, I.G. (1999) Non-enzymatic RNA hydrolysis promoted by the combined catalytic activity of buffers and magnesium ions. *Zeitschrift für Naturforschung* 54C(7-8), 542–548. Available at: http://zfn.mpdl.mpg.de/data/Reihe_C/54/ZNC-1999-54c-0542.pdf DOI: 10.1515/znc-1999-7-813.

Albright, V.C., Wong, C.R., Hellmich, R.L. and Coats, J.R. (2017) Dissipation of double-stranded RNA in aquatic microcosms. *Environmental Toxicology and Chemistry* 36(5), 1249–1253. DOI: 10.1002/etc.3648.

Bachman, P. (2019) *Environmental dissipation of dsRNA in soil, aquatic systems and plants*. Presentation at OECD Conference on RNAi based Pesticides, 10–12 April 2019, OECD, Paris. Available at: https://www.oecd.org/chemicalsafety/pesticides-biocides/conference-on-rnai-based-pesticides.htm (accessed 18 March 2020).

Bolognesi, R., Ramaseshadri, P., Anderson, J., Bachman, P., Clinton, W. *et al.* (2012) Characterizing the mechanism of action of double-stranded RNA activity against western corn rootworm (*Diabrotica virgifera vigifera* LeConte). *PLoS ONE* 7(10), e47534. DOI: 10.1371/journal.pone.0047534.

Cooper, A.M., Silver, K., Zhang, J., Park, Y. and Zhu, K.Y. (2019) Molecular mechanisms influencing efficiency of RNA interference in insects. *Pest Management Science* 75(1), 18–28. DOI: 10.1002/ps.5126.

Denecke, S., Swevers, L., Douris, V. and Vontas, J. (2018) How do oral insecticidal compounds cross the insect midgut epithelium? *Insect Biochemistry and Molecular Biology* 103, 22–35. DOI: 10.1016/j.ibmb.2018.10.005.

Dubelman, S., Fischer, J., Zapata, F., Huizinga, K., Jiang, C. *et al.* (2014) Environmental fate of double-stranded RNA in agricultural soils. *PLoS ONE* 9(3), e93155. DOI: 10.1371/journal.pone.0093155.

Ecker, J.R. and Davis, R.W. (1986) Inhibition of gene expression in plant cells by expression of antisense RNA. *Proceedings of the National Academy of Sciences of the United States of America* 83(15), 5372–5376. DOI: 10.1073/pnas.83.15.5372.

Fire, A., Xu, S., Montgomery, M.K., Kostas, S.A., Driver, S.E. *et al.* (1998) Potent and specific genetic interference by double-stranded RNA in *Caenorhabditis elegans*. *Nature* 391(6669), 806–811. DOI: 10.1038/35888.

Fischer, J.R., Zapata, F., Dubelman, S., Mueller, G.M., Uffman, J.P. *et al.* (2017) Aquatic fate of a double-stranded RNA in a sediment–water system following an over-water application. *Environmental Toxicology and Chemistry* 36(3), 727–734. DOI: 10.1002/etc.3585.

Garbutt, J.S., Bellés, X., Richards, E.H. and Reynolds, S.E. (2013) Persistence of double-stranded RNA in insect hemolymph as a potential determiner of RNA interference success: evidence from *Manduca sexta* and *Blattella germanica*. *Journal of Insect Physiology* 59(2), 171–178. DOI: 10.1016/j.jinsphys.2012.05.013.

Ghini, F., Rubolino, C., Climent, M., Simeone, I., Marzi, M.J. *et al.* (2018) Endogenous transcripts control miRNA levels and activity in mammalian cells by target-directed miRNA degradation. *Nature Communications* 9(1), 3119. DOI: 10.1038/s41467-018-05182-9.

Ghosh, S.K.B., Hunter, W.B., Park, A.L. and Gundersen-Rindal, D.E. (2017) Double strand RNA delivery system for plant-sap-feeding insects. *Plos One* 12(2), e0171861, 861. DOI: 10.1371/journal.pone.0171861.

Gordon, K.H.J. and Waterhouse, P.M. (2007) RNAi for insect-proof plants. *Nature Biotechnology* 25(11), 1231–1232. DOI: 10.1038/nbt1107-1231.

Greaves, M.P. and Wilson, M.J. (1970) The degradation of nucleic-acids and montmorillonite-nucleic-acid complexes by soil microorganisms. *Soil Biology and Biochemistry* 2(4), 257–268. DOI: 10.1016/0038-0717(70)90032-5.

Guan, R., Hu, S., Li, H., Shi, Z. and Miao, X. (2018) The *in vivo* dsRNA cleavage has sequence preference in insects. *Frontiers in Physiology* 9, 1768. DOI: 10.3389/fphys.2018.01768.

Heo, I., Joo, C., Cho, J., Ha, M., Han, J. *et al.* (2008) Lin28 mediates the terminal uridylation of let-7 precursor microRNA. *Molecular Cell* 32(2), 276–284. DOI: 10.1016/j.molcel.2008.09.014.

Huvenne, H. and Smagghe, G. (2010) Mechanisms of dsRNA uptake in insects and potential of RNAi for pest control: a review. *Journal of Insect Physiology* 56(3), 227–235. DOI: 10.1016/j.jinsphys.2009.10.004.

Icoz, I. and Stotzky, G. (2008) Fate and effects of insect-resistant Bt crops in soil ecosystems. *Soil Biology and Biochemistry* 40(3), 559–586. DOI: 10.1016/j.soilbio.2007.11.002.

Ivashuta, S., Zhang, Y., Wiggins, B.E., Ramaseshadri, P., Segers, G.C. *et al.* (2015) Environmental RNAi in herbivorous insects. *RNA* 21(5), 840–850. DOI: 10.1261/rna.048116.114.

Keown, H., O'Callaghan, M. and Greenfield, L.G. (2004) Decomposition of nucleic acids in soil. *New Zealand Natural Sciences* 29, 13–19. Available at: http://www.science.canterbury.ac.nz/nzns/issues/vol29-2004/keown.pdf

Li, Y. and Breaker, R.R. (1999) Kinetics of RNA degradation by specific base catalysis of transesterification involving the 2'-hydroxyl group. *Journal of the American Chemical Society* 121(23), 5364–5372. DOI: 10.1021/ja990592p.

Lomate, P.R. and Bonning, B.C. (2016) Distinct properties of proteases and nucleases in the gut, salivary gland and saliva of southern green stink bug, *Nezara viridula*. *Scientific Reports* 6, 27587. DOI: 10.1038/srep27587.

Napoli, C., Lemieux, C. and Jorgensen, R. (1990) Introduction of a chimeric chalcone synthase gene into Petunia results in reversible co-suppression of homologous genes in trans. *The Plant Cell* 2(4), 279–289. DOI: 10.2307/3869076.

Nicassio, F. (2019) Killing miR-softly: new clues to miRNA degradation by RNA targets. *Non-coding RNA Investigation* 3, 5. DOI: 10.21037/ncri.2019.01.03.

Parker, K.M., Barragán Borrero, V., van Leeuwen, D.M., Lever, M.A., Mateescu, B. *et al.* (2019) Environmental fate of RNA interference pesticides: adsorption and degradation of double-stranded RNA molecules in agricultural soils. *Environmental Science & Technology* 53(6), 3027–3036. DOI: 10.1021/acs.est.8b05576.

Peng, Y., Wang, K., Fu, W., Sheng, C. and Han, Z. (2018) Biochemical comparison of dsRNA degrading nucleases in four different insects. *Frontiers in Physiology* 9, 624. DOI: 10.3389/fphys.2018.00624.

Peng, Y., Wang, K., Zhu, G., Han, Q., Chen, J. *et al.* (2020) Identification and characterization of multiple dsRNases from a lepidopteran insect, the tobacco cutworm, *Spodoptera litura* (Lepidoptera: Noctuidae). *Pesticide Biochemistry and Physiology* 162, 86–95. DOI: 10.1016/j.pestbp.2019.09.011.

Ramachandran, V. and Chen, X. (2008) Degradation of microRNAs by a family of exoribonucleases in *Arabidopsis*. *Science* 321(5895), 1490–1492. DOI: 10.1126/science.1163728.

Ren, S., She, M., Li, M., Zhou, Q., Liu, R. *et al.* (2014) The RNA/DNA-binding protein PSF relocates to cell membrane and contributes cells' sensitivity to antitumor drug, doxorubicin. *Cytometry Part A* 85(3), 231–241. DOI: 10.1002/cyto.a.22423.

Ruegger, S. and Grosshans, H. (2012) MicroRNA turnover: when, how, and why. *Trends in Biochemical Sciences* 37(10), 436–446. DOI: 10.1016/j.tibs.2012.07.002.

San Miguel, K. and Scott, J.G. (2016) The next generation of insecticides: dsRNA is stable as a foliar-applied insecticide. *Pest Management Science* 72(4), 801–809. DOI: 10.1002/ps.4056.

Shukla, J.N., Kalsi, M., Sethi, A., Narva, K.E., Fishilevich, E. *et al.* (2016) Reduced stability and intracellular transport of dsRNA contribute to poor RNAi response in lepidopteran insects. *RNA Biology* 13(7), 656–669. DOI: 10.1080/15476286.2016.1191728.

Siess, D.C., Vedder, C.T., Merkens, L.S., Tanaka, T., Freed, A.C. *et al.* (2000) A human gene coding for a membrane-associated nucleic acid-binding protein. *Journal of Biological Chemistry* 275(43), 33655–33662. DOI: 10.1074/jbc.M004461200.

Sims, S.R. and Ream, J.E. (1997) Soil inactivation of the *Bacillus thuringiensis* subsp *kurstaki* CryIIA insecticidal protein within transgenic cotton tissue: laboratory microcosm and field studies. *Journal of Agricultural and Food Chemistry* 45(4), 1502–1505. DOI: 10.1021/jf960647w.

Singer, B. (1971) Chemical modification of viral ribonucleic acid: IX. The effect of ultraviolet irradiation on TMV-RNA and other polynucleotides. *Virology* 45(1), 101–107. DOI: 10.1016/0042-6822(71)90117-6.

Singh, I.K., Singh, S., Mogilicherla, K., Shukla, J.N. and Palli, S.R. (2017) Comparative analysis of double-stranded RNA degradation and processing in insects. *Scientific Reports* 7(1), 17059. DOI: 10.1038/s41598-017-17134-2.

Song, H., Zhang, J., Li, D., Cooper, A.M.W., Silver, K. *et al.* (2017) A double-stranded RNA degrading enzyme reduces the efficiency of oral RNA interference in migratory locust. *Insect Biochemistry and Molecular Biology* 86, 68–80. DOI: 10.1016/j.ibmb.2017.05.008.

Tayler, A., Heschuk, D., Giesbrecht, D., Park, J.Y. and Whyard, S. (2019) Efficiency of RNA interference is improved by knockdown of dsRNA nucleases in tephritid fruit flies. *Open Biology* 9(12), 190198. DOI: 10.1098/rsob.190198.

Vélez, A.M. and Fishilevich, E. (2018) The mysteries of insect RNAi: a focus on dsRNA uptake and transport. *Pesticide Biochemistry and Physiology* 151, 25–31. DOI: 10.1016/j.pestbp.2018.08.005.

Wang, X., Wang, Y., Dou, Y., Chen, L., Wang, J. *et al.* (2018) Degradation of unmethylated miRNA/miRNA*s by a DEDDy-type 3′ to 5′ exoribonuclease Atrimmer 2 in *Arabidopsis*. *Proceedings of the National Academy of Sciences of the United States of America* 115(28), E6659–E6667. DOI: 10.1073/pnas.1721917115.

Wang, K., Peng, Y., Fu, W., Shen, Z. and Han, Z. (2019) Key factors determining variations in RNA interference efficacy mediated by different double-stranded RNA lengths in *Tribolium castaneum*. *Insect Molecular Biology* 28(2), 235–245. DOI: 10.1111/imb.12546.

Weaver, A.M. and Patton, J.G. (2020) Argonautes in extracellular vesicles: artifact or selected cargo? *Cancer Research* 80(3), 379–381. DOI: 10.1158/0008-5472.CAN-19-2782.

Whitten, M.M. (2019) Novel RNAi delivery systems in the control of medical and veterinary pests. *Current Opinion in Insect Science* 34, 1–6. DOI: 10.1016/j.cois.2019.02.001.

Yoon, J.-S., Mogilicherla, K., Gurusamy, D., Chen, X., Chereddy, S.C.R.R. *et al.* (2018) Double-stranded RNA binding protein, Staufen, is required for the initiation of RNAi in coleopteran insects. *Proceedings of the National Academy of Sciences of the United States of America* 115(33), 8334–8339. DOI: 10.1073/pnas.1809381115.

Zhang, C. and Ruvkun, G. (2012) New insights into siRNA amplification and RNAi. *RNA Biology* 9(8), 1045–1049. DOI: 10.4161/rna.21246.

11 Boosting dsRNA Delivery in Plant and Insect Cells with Peptide- and Polymer-based Carriers: Case-based Current Status and Future Perspectives

Kristof De Schutter, Olivier Christiaens, Clauvis Nji Tizi Taning and Guy Smagghe*

Department of Plants and Crops, Ghent University, Ghent, Belgium

Abstract

Since the discovery of this naturally occurring endogenous regulatory and defence mechanism, RNA interference (RNAi) has been exploited as a powerful tool for functional genomic research. In addition, it has evolved as a promising candidate for a sustainable, specific and eco-friendly strategy for pest management and plant improvement. A key element in this technology is the efficient delivery of dsRNAs into the pest or plant tissues. While several examples using transgenic plants expressing the dsRNAs have proved the potential of this technology, non-transgenic approaches are investigated as alternatives, allowing flexibility and circumventing technical limitations of the transgenic approach. However, the efficacy of environmental RNAi is affected by several barriers, such as extracellular degradation of the dsRNA, inefficient internalization of the dsRNA in the cell and low endosomal escape into the cytoplasm, resulting in variable or low RNAi responses. In the medical field, carrier systems are commonly used to enhance RNA delivery and these systems are being rapidly adopted by the agricultural industry.

Using four case studies, this chapter demonstrates the potential of carriers to improve the RNAi response in pest control for aquatic-living mosquito larvae and RNAi-resilient Lepidoptera and to cross the plant cell wall, allowing efficient environmental RNAi in plants.

11.1 Introduction

Plants are crucial for the planet and all organisms living on it. Most essentially for humans, they provide a source of oxygen and food. However, the changing climate poses enormous challenges to the agricultural sector to provide sufficient food for our growing population. Next to the obvious effects on abiotic stress (drought, heat, flooding, etc.), climate change has introduced novel or increased biotic stresses (pests, diseases, etc.) (Peters *et al.*, 2014). To achieve the food demands of our ever-growing world population, agriculture has often practised an unsustainable upscaling of production, leading to reduction of the biodiversity of the terrestrial ecosystems (Pegler *et al.*, 2019). This has included the excessive use of synthetic pesticides

*Corresponding author: Guy.Smagghe@UGent.be

© CAB International 2021. *RNAi for Plant Improvement and Protection*
(eds B. Mezzetti *et al.*)
DOI: 10.1079/9781789248890.0011

to protect crops from biological stresses, which has had a serious detrimental effect on the environment and led to the emergence of resistance to most classes of conventional pesticides. Therefore, there is a dire need for more sustainable and eco-friendly solutions for crop improvement and pest control.

Exploiting RNA interference (RNAi), a natural regulatory and defence mechanism present in most eukaryotic organisms, has emerged as one of the most promising strategies for crop improvement and pest control. This is due to the biodegradability of the active molecule (RNA) and the possibility of designing this natural molecule to be species-selective (Huvenne and Smagghe, 2010; Younis *et al.*, 2014; Cagliari *et al.*, 2019; Taning *et al.*, 2020). In RNAi, the presence of free double-stranded RNA (dsRNA) in the cell triggers and directs the sequence-specific translational repression or degradation of homologous messenger RNA (mRNA) targets, resulting in downregulation or knockdown of protein expression. In the past decade, applications have been developed in the form of genetically modified plants (host-induced gene silencing (HIGS)) expressing specific dsRNAs that silence the expression of essential genes that are required for the survival of the pests (insects, viruses and bacteria), thereby exploiting the RNAi mechanism as a species-selective pest control strategy (Huang *et al.*, 2006; Mansoor *et al.*, 2006; Baum *et al.*, 2007; Mao *et al.*, 2007; Qu *et al.*, 2007). Similarly, the RNAi mechanism can also be exploited as a strategy for crop improvement through the silencing of specific plant genes to provide desired phenotypes or resistance to (a)biotic stress (Li *et al.*, 2009; Younis *et al.*, 2014; Joshi *et al.*, 2018).

Despite the successful development of interesting and promising crop varieties through the HIGS approach, public acceptance of genetically modified crops is very poor (Shew *et al.*, 2017). Moreover, the technical difficulties arising from the lack of established transformation protocols for some cultivated plants, the high cost of production and the long time required from the laboratory to the market have further impeded the adoption of the HIGS approach (Mitter *et al.*, 2017a). These drawbacks have motivated the search for (Scorza *et al.*, 2013) and development of alternative non-GMO (genetically modified organism) strategies for the delivery of dsRNA

molecules. Non-GMO strategies could circumvent the technical limitation of plant transformation and the negative public perception of GMOs and provide an easy-to-use, environmentally friendly and flexible tool to improve plant performance and crop protection (Shew *et al.*, 2017; Cagliari *et al.*, 2018). The non-GMO approach of environmental application of dsRNAs offers an easy design and flexibility to apply relevant dsRNAs when and where needed. This approach has already been shown to offer protection against several pests, such as the Colorado potato beetle (San Miguel and Scott, 2016) and the fungal pathogens *Fusarium* (Koch *et al.*, 2016) and *Botrytis* (Wang *et al.*, 2016). However, a drawback in RNAi-based methods for pest control and plant improvement is the high variability in the RNAi response. Two important factors affecting RNAi efficiency are differences in dsRNA uptake into cells and differences in the stability of the dsRNAs against, for example, dsRNA-degrading enzymes (nucleases).

11.2 Barriers to dsRNA Delivery

Owing to their large size and highly negative charge, dsRNAs cannot easily enter the cells (Whitehead *et al.*, 2009; Scott *et al.*, 2013), making cellular uptake a key factor in explaining the variability in RNAi efficacy in non-GMO applications. Although some core components are known, many questions still remain unanswered concerning the dsRNA uptake pathways (Cappelle *et al.*, 2016; Cooper *et al.*, 2019). In insects, two different uptake mechanisms have been described so far: a pathway mediated by specific dsRNA channels, as also described in nematodes (Winston *et al.*, 2002); and an alternative pathway based on endocytosis-mediated uptake mechanisms (Saleh *et al.*, 2006; Miyata *et al.*, 2014; Cappelle *et al.*, 2016) (Fig. 11.1). A genetic screen in the nematode *Caenorhabditis elegans* identified several genes with a crucial role in the local and systemic RNAi response: the systemic RNA interference deficiency (SID) genes (Winston *et al.*, 2002). In *C. elegans*, the cellular uptake of environmental dsRNA is mediated by the intestinal membrane protein SID-2 (Winston *et al.*, 2007), while the dsRNA-selective dsRNA-gated channel SID-1 is required

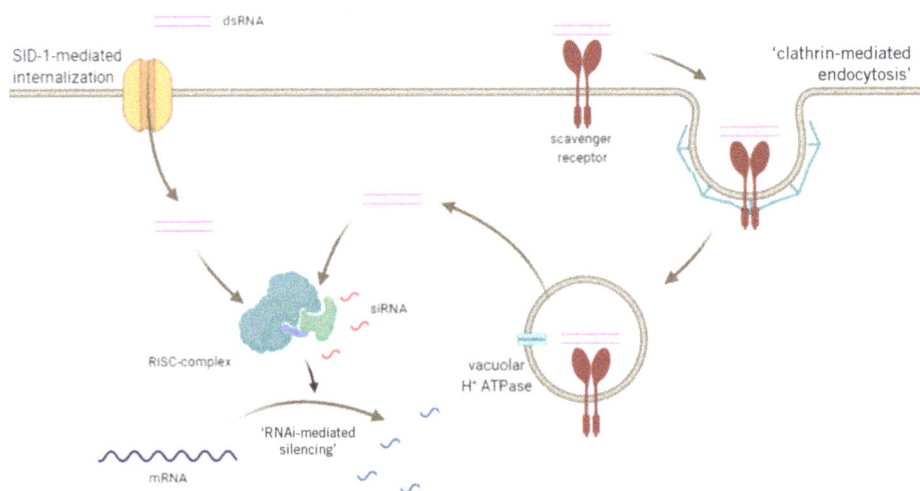

Fig. 11.1. Cellular internalization mechanisms in insects. While SID-1 orthologues are identified in several insect species, they do not play an essential role in dsRNA internalization. In insects, the primary dsRNA uptake mechanism depends on endocytosis. After binding of the dsRNA to membrane-bound scavenger receptors, the complexes are internalized through clathrin-mediated endocytosis. After acidification of the endosomes, the dsRNA is released into the cytoplasm where it is processed by the core RNAi machinery.

for systemic RNAi (Winston *et al.*, 2002). An absence of orthologues of SID-2 in insects suggests that the mediators of dsRNA might be different across metazoa (Cappelle *et al.*, 2016; Vélez and Fishilevich, 2018). Although orthologues of SID-1 are present in some insect species, data suggests that these SID-1 orthologues are not essential for systemic dsRNA uptake (Tomoyasu *et al.*, 2008). In the closely related coleopteran species *Diabrotica virgifera* and *Tribolium castaneum*, the former has two SID-1 orthologues which are both involved in dsRNA uptake (Miyata *et al.*, 2014), while the three SID-1 orthologues in the latter seem not to be necessary, suggesting an alternative uptake mechanism (Tomoyasu *et al.*, 2008). In *Drosophila melanogaster*, no SID-1 orthologues have been identified; however, uptake of dsRNA has been demonstrated by receptor-mediated endocytosis (Saleh *et al.*, 2006; Ulvila *et al.*, 2006). This endocytosis-mediated uptake mechanism makes use of (pattern recognition) scavenger receptors and clathrin-dependent endocytosis (Saleh *et al.*, 2006; Cappelle *et al.*, 2016) (Fig. 11.1). In humans, a clathrin-independent (caveolae) endocytic pathway contributes to the cellular uptake mechanisms

(Kasai *et al.*, 2019), but a similar pathway is not involved in dsRNA uptake in *D. melanogaster* or *T. castaneum* (Saleh *et al.*, 2006; Xiao *et al.*, 2015). In *D. melanogaster* S2 cells, two scavenger receptors, SR-CI and Eater, account for 90% of the dsRNA uptake (Ulvila *et al.*, 2006) (Fig. 11.1). Analysis of the components in this alternative uptake mechanism in *C. elegans* suggested this mechanism might be evolutionarily conserved (Saleh *et al.*, 2006). In plants, the cell wall poses an additional barrier for the internalization of the dsRNAs. While it was shown that exogenously applied RNAs can spread locally and systemically through the plant and induce RNAi-mediated plant pathogen resistance, the understanding of the mechanisms for uptake of extracellular nucleic acids is limited and data are scarce and inconsistent (Bhat and Ryu, 2016; Mermigka *et al.*, 2016; Dubrovina *et al.*, 2019). Similar to the endocytosis-mediated uptake mechanisms present in animals, pattern recognition receptors are shown to be involved, but further research is needed to shed light on the mechanisms of extracellular dsRNA uptake (Dubrovina *et al.*, 2019).

With endocytosis established as the major cellular internalization mechanism in plants and insects, the next barrier is the release of the dsRNA from the endosomes into the cytoplasm, where they are processed by the RNAi machinery (Dicer and RISC) (Saleh *et al.*, 2006) (Fig. 11.2). Endosomal release occurs after acidification of the endosomes. A vacuolar H+-ATPase was suggested to play a role in this endosomal escape (Saleh *et al.*, 2006). However, this escape from the endosomes is not always efficient and impaired endosomal release was demonstrated as a cause of low sensitivity to RNAi (Shukla *et al.*, 2016; Yoon *et al.*, 2017) (Fig. 11.2).

Besides cellular uptake and endosomal release, stability of the dsRNA is also an important factor undermining RNAi efficacy (Fig. 11.2). Despite being considered as an unstable molecule, dsRNA can persist on leaves for up to 20 days in greenhouse conditions (Mitter *et al.*, 2017a, b). Experiments with photochambers and in field conditions showed that UV radiation is not a major contributor to instability of the dsRNA (Bachman *et al.*, 2020). In contrast, wash-off by rain or dew is an important factor in foliar application (Bachman *et al.*, 2020). In an aquatic environment, dsRNA can persist up to 4–7 days (Fischer *et al.*, 2017); however, in soil the dsRNA is only stable up to 24–36 h

(Dubelman *et al.*, 2014). The instability of dsRNA is mainly attributed to the presence of microbial nucleases (Dubelman *et al.*, 2014). Next to the microbial nucleases, damage to the plant (during dsRNA application) can result in the release of nucleases and subsequent degradation of the exogenously applied dsRNA. Especially in insects, extracellular degradation of dsRNA by nucleases in the gut has been identified as a key factor explaining reduced RNAi efficacy (Christiaens *et al.*, 2014, 2016, 2018; Prentice *et al.*, 2017; Guan *et al.*, 2018; Ghodke *et al.*, 2019; Castellanos *et al.*, 2019). Next to the gut nucleases, also the extracts from saliva exhibit nuclease activity that can cause the rapid degradation of the dsRNA (Allen and Walker, 2012; Christiaens *et al.*, 2014). Although the characterization of these nucleases requires further study, several candidates have been identified in the insect gut (Arimatsu *et al.*, 2007; Liu *et al.*, 2012; Wynant *et al.*, 2012; Almeida Garcia *et al.*, 2017; Spit *et al.*, 2017; Prentice *et al.*, 2019).

Increasing RNAi efficacy can be achieved by the use of dsRNA carrier systems. These systems are designed to efficiently deliver their dsRNA cargo into the cells by avoiding RNAi barriers such as an inefficient cellular uptake, a low endosomal release and extracellular

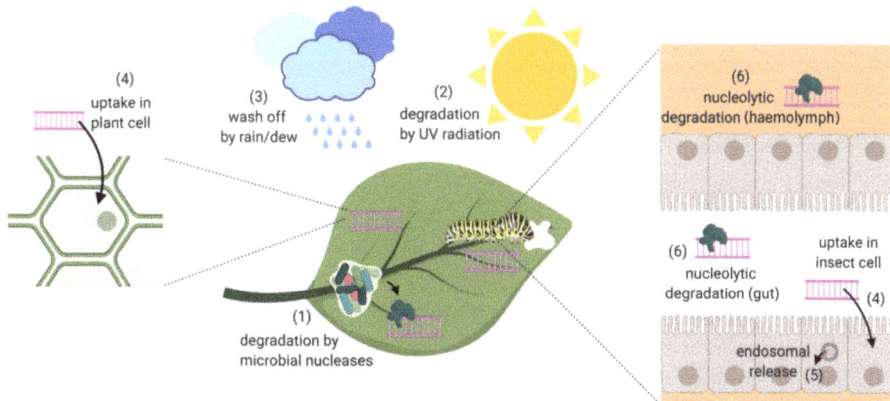

Fig. 11.2. Barriers of environmental RNAi. External and internal barriers can affect the efficiency of the RNAi response. External factors include the degradation of the dsRNA by microbial nucleases (1) and UV radiation (2) and the wash-off of the applied dsRNA by rain or dew into the soil, where it is rapidly degraded by nucleases (3). Internal factors include the inefficient cellular uptake of the dsRNA (4), low endosomal release (5) and the presence of nucleases in the salivary glands, midgut and haemolymph of insects (6).

degradation of the dsRNA (Akinc *et al.*, 2008; Vogel *et al.*, 2019). Complexation of the dsRNA with the carriers increases the environmental stability of the dsRNA molecules, protects them against degradation by nucleases, improves cellular internalization and/or stimulates endosomal release, and this without affecting the ability to silence the target genes. These carriers can be based on naturally occurring or synthetic molecules and may include viral particles, lipids, metals, sugars, peptides, proteins and polymers. The peptide- and polymer-based carriers are the best studied (Vogel *et al.*, 2019; Christiaens *et al.*, 2020a, b). Polymers are macromolecules of variable sizes, composed of many repeating subunits, and can be naturally occurring or synthetically designed. The use of polymeric carrier systems has a long history to enhance RNA delivery in medical applications but applications in the agricultural industry are growing rapidly (Christiaens *et al.*, 2018). Similarly, the use of naturally occurring peptides or proteins to direct the delivery of dsRNA provides a promising prospect and has already been developed in the biomedical and pharmaceutical fields (Milletti, 2012). Peptide-based carriers mainly make use of cell membrane penetrating peptides (CMPPs), which are small polycationic or amphipathic peptides that can facilitate cellular uptake of various molecular cargo, including nucleotides (Wang *et al.*, 2014; Gillet *et al.*, 2017).

While different synthetic and natural carrier systems have been investigated in relation to RNAi efficacy, this chapter presents a selection of four case studies to demonstrate the potential of the carrier systems to overcome specific barriers and improve RNAi efficacy in plants and insects: (1) a natural polymer for the control of aquatic-living mosquito larvae; (2) a synthetic polymer for the protection of dsRNA in the strong alkaline environment of lepidopteran guts; and (3) a polymer- and peptide-based carrier system to improve environmental RNAi in plants by assisting the RNAs to cross the cell wall. In the fourth case, we focus on the design of a peptide-based carrier, showing the potential of adding additional domains to improve its functionality. For a more comprehensive review on the barriers and the use of carrier systems to improve RNAi responses in plants and insects, refer to the recent reviews by Vogel *et al.* (2019) and Christiaens *et al.* (2020a, b).

11.3   Case 1: Delivery of dsRNA in Insects in Aquatic Environments

Blood-feeding mosquitoes serve as vectors for disease-causing agents responsible for the death of more than one million people each year (Zhang *et al.*, 2015) and also act as vectors of infectious diseases that affect animal production (Bartlow *et al.*, 2019). While the direct injection of dsRNA in adult mosquitoes has been shown to effectively trigger RNAi, microinjection delivery is not feasible as an application method for vector control in the field (Zhang *et al.*, 2010, 2015). A viable strategy for the RNAi-based control of mosquitoes would be through the delivery of the interfering RNA with the ingested food at larval stage. However, the aquatic lifestyle of the mosquito larvae poses several technical challenges, such as the instability of the dsRNA and the dispersion of the dsRNA from the food, causing a low dose of dsRNA in the organism and subsequently an inadequate RNAi response (Zhang *et al.*, 2010) (Fig. 11.3). To overcome these challenges, Zhang *et al.* (2010) developed a delivery system based on a natural polymer, chitosan. Chitosan is a non-toxic and biodegradable polymer prepared by deacetylation of chitin, the most abundant natural polymer after cellulose (Dass and Choong, 2008). The chitosan/dsRNA nanoparticles are formed by self-assembly through electrostatic forces between the polycationic chitosan and the negatively charged dsRNA (Zhang *et al.*, 2015). Using the chitosan/dsRNA nanoparticles in feeding experiments with *Anopheles gambiae* and *Aedes aegypti* larvae significantly improved the RNAi efficacy (Zhang *et al.*, 2010, 2015; Mysore *et al.*, 2013; Kumar *et al.*, 2016). The application of these nanoparticles improved the retention of the dsRNA in the food gel, an important element in feeding-based RNAi in aquatic environments; in addition, the nanoparticles significantly stabilized the dsRNA and enhanced delivery into the gut epithelial cells (Zhang *et al.*, 2010) (Fig. 11.3). Although the mechanisms by which the cellular internalization is achieved were not completely elucidated, it is suggested that the nanoparticle carriers may facilitate dsRNA uptake by the endocytosis pathway in the gut (Zhang *et al.*, 2010).

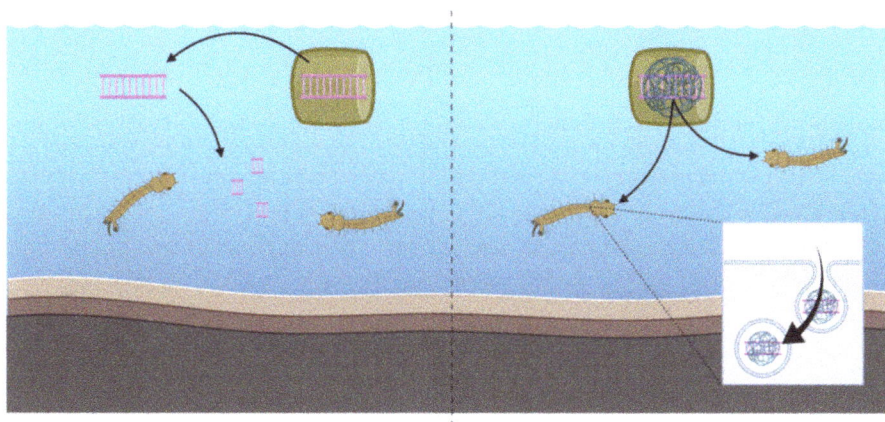

Fig. 11.3. Barriers of RNAi-based pest control in mosquito larvae. Treatment of insects with an aquatic lifestyle is challenging, due to instability of the dsRNA in water and the dispersion of the dsRNA from the food. This causes a low dose of ingested dsRNA and subsequently an inadequate RNAi effect. Complexation of the dsRNA with chitosan leads to improved retention of the dsRNA complex in the food, stabilization of the dsRNA and an enhanced delivery into the gut epithelial cells.

11.4   Case 2: Overcoming the High pH in the Lepidoptera Midgut

Not all carrier systems are appropriate for all applications. This implies that carriers must be optimized or sometimes even tailor-designed to the biology of the target organism in question (Christiaens *et al.*, 2020a). The synthetic character of certain polymer-based carriers allows the design of carrier systems adapted for specific conditions. Lepidoptera can be considered as a worst-case scenario for RNAi-mediated pest control. Due to their slow cellular uptake and the strong dsRNA-degrading capacity of nucleases in the very alkaline (pH > 9) gut environment, these insects are generally very resilient to RNAi, especially upon oral delivery (Terenius *et al.*, 2010; Garbutt *et al.*, 2013; Christiaens and Smagghe, 2014; Christiaens *et al.*, 2018) (Fig. 11.4). To overcome these barriers, a formulation was needed to protect the dsRNA against nucleolytic degradation to allow uptake into the cells and that was stable at high pH (Christiaens *et al.*, 2018). Despite their proven efficiency in Diptera, the natural polymers were found to be unsuitable for use in Lepidoptera as the complexation was not stable in the strong alkaline environment in the gut of most Lepidoptera

(Christiaens *et al.*, 2018). Therefore, a series of nanoparticles based on cationic polymethacrylate derivates were designed to specifically shield the dsRNA from the degrading effects at high pH. The stability of the nanoparticles at high pH was enhanced by the modification of the polymer with protective guanidine side groups. Complexation of the dsRNA with these guanylated polymers resulted in an increased RNAi efficacy in *in vivo* feeding experiments with *Spodoptera exigua* larvae (Christiaens *et al.*, 2018). This increased RNAi efficacy was shown to be due to an improved protection of the dsRNA against nucleolytic degradation, protecting the dsRNA for up to 30 h against *S. exigua* gut juice, and an enhanced cellular uptake of the dsRNA (Christiaens *et al.*, 2018) (Fig. 11.4).

In addition to the protection in the high-pH environment by polymers, uptake in the midgut epithelium is needed for an RNAi response. Parsons *et al.* (2018) suggested that the synthetic polymer mimics cell-penetrating peptides to allow efficient internalization into *Spodoptera frugiperda* midgut epithelial cells. Similarly, synthetically modified cationic polymers have been shown to facilitate dsRNA uptake in feeding experiments with larvae of the Asian corn borer, *Ostrinia furnacalis*, and significantly improve the

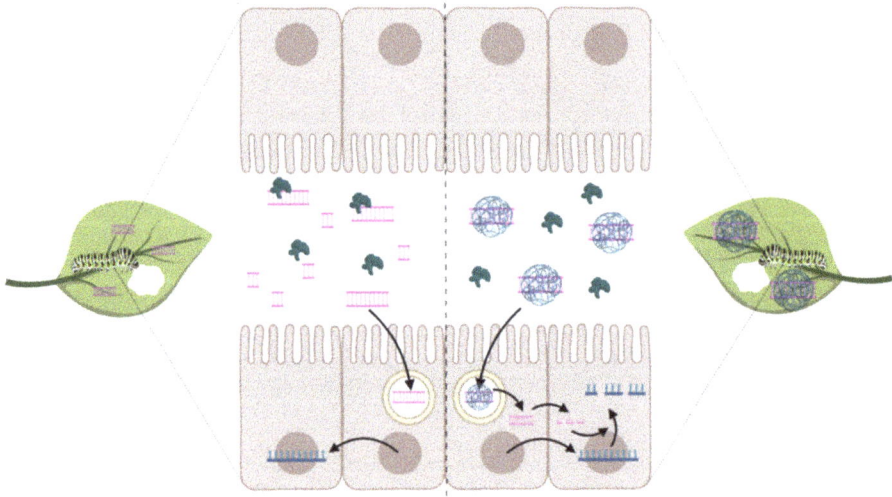

Fig. 11.4. Barriers in Lepidoptera. The strong dsRNA-degrading capacity (nucleases and very alkaline pH) of the gut, the slow cellular uptake and the low endosomal release hampers RNAi in Lepidoptera. Complexation of the dsRNA with a guanylated polymer, designed to remain stable at high pH, protects the dsRNA against degradation in the gut and enhances cellular uptake.

RNAi efficacy in this insect (He *et al.*, 2013). Next to the improved internalization and protection against extracellular endonucleases, polymer-based carrier systems have been suggested to improve RNAi efficacy by promoting endosomal escape. Especially in Lepidoptera, the endosomal release can be inefficient and contribute to the low RNAi response (Shukla *et al.*, 2016; Yoon *et al.*, 2017). According to the protein sponge theory, the buffering capacity of the nanoparticles could lead to osmotic swelling and rupture of the endosomes (Akinc *et al.*, 2005) (Fig. 11.4).

11.5 Case 3: Crossing the Plant Cell Wall

As well as the challenge posed by the difficulties for the dsRNA to cross the cell membrane in animal cells, delivery of interfering RNAs in plant cells is faced with another barrier: the cell wall. In several studies, polymer-based carriers have been shown to deliver plasmid DNA and proteins into intact plant cells (Chang *et al.*, 2013; Hussain *et al.*, 2013; Martin-Ortigosa

et al., 2012; Demirer *et al.*, 2019), suggesting the potential of these systems to deliver interfering RNAs. However, the use of these carriers to improve non-GMO RNAi in plants remains understudied, with only a few papers reporting the delivery of RNAi molecules into the plant cell using nanoparticles (Demirer and Landry, 2017). The study by Mitter *et al.* (2017b) showed delivery of dsRNA into *Nicotiana tabacum* using layered double hydroxide clay nanosheets (BioClay). When loaded with dsRNA, this nanoparticle led to the sustained release of the dsRNA as the BioClay degraded (Mitter *et al.*, 2017b). The slow release allowed the detection of dsRNA for up to 30 days after being sprayed on the plant and led to successful antiviral effects for at least 20 days (Mitter *et al.*, 2017b). In another study, single-walled carbon nanotubes were used to improve the cellular delivery of small interfering RNAs in *Nicotiana benthamiana* plants (Demirer *et al.*, 2019). Infiltration of complexed sense and antisense siRNA leads to efficient uptake of the complexes and subsequent desorption and hybridizing of the complementary siRNA strands activating an RNAi response (Demirer *et al.*, 2019). Similar to the protection against nucleolytic degradation observed in insects, the polymeric carrier protects the siRNA against degradation by RNaseA (Demirer *et al.*, 2019).

In addition to the polymer-based carriers, peptide-based carriers have been shown to be able to deliver dsRNA cargo into plant cells. Peptide-based carrier systems using cell membrane penetrating peptide (CMPP) domains have been successfully used to initiate rapid and efficient RNAi-mediated silencing of exogenous and endogenous genes in leaves of diverse plant species, such as *Arabidopsis thaliana*, *Nicotiana benthamiana*, *Solanum lycopersicum* and poplar (Numata *et al.*, 2014, 2018), *N. tabacum* suspension-cell cultures (Unnamalai *et al.*, 2004; Numata *et al.*, 2018) and rice callus tissue (Numata *et al.*, 2018). While these results showed the potential of these delivery systems, it is likely that the delivery of dsRNA can be improved by altering the lengths and/or amino acid composition of the peptides (Unnamalai *et al.*, 2004). It is suggested that longer polypeptides with many positive charges might form complexes too tight to dissociate inside the cell, leading to a lower RNAi efficacy (Bettinger *et al.*, 2001). The influence of the amino acid composition was shown in a comparative study of 55 CMPP-based carriers, revealing that the cell penetrating efficiency of Lys-containing CMPP-based carriers is relatively higher in plant cells than in animal cells (Numata *et al.*, 2018). In addition, several CMPPs were found to function with specific plants or tissues. The inability to identify one peptide carrier with high cell-penetration efficiency for all plant species and cell types suggests that optimization of the CMPP domain will be essential for each application (Numata *et al.*, 2018).

11.6 Case 4: Modifying carriers to improve functionality, uptake and endosomal release

Optimization of the CMPP-based carrier systems can significantly improve their ability to provoke an RNAi response. Within the CMPPs, the short cationic arginine-rich transactivating transcriptional activator (Tat) peptide of the human immunodeficiency virus 1 (HIV-1) has been specifically studied and engineered to improve its uptake efficiency and endosomal escape (Vivès *et al.*, 1997; Wadia *et al.*, 2004; Gillet *et al.*, 2017) (Fig. 11.5). To improve the oral delivery of dsRNA in the cotton boll weevil, *Anthonomus grandis*, the Tat peptide was enhanced with the inclusion of a haemagglutinin peptide to destabilize the membrane of the endocytic vesicle and promote endosomal escape (Wadia *et al.*, 2004; Erazo-Oliveras *et al.*, 2012; Gillet *et al.*, 2017). Direct conjugation of the cationic CMPP domains to anionic RNAs results in charge neutralization, which renders the carrier system inactive and limits delivery into the cells. In addition, this causes the aggregation/precipitation of the complex and leads to cytotoxicity (Turner *et al.*, 2007; Meade and Dowdy, 2008; Eguchi *et al.*, 2009). To circumvent the charge neutralization, the engineered peptide transduction domain was fused to the dsRNA-binding domain of the human protein kinase R (Eguchi *et al.*, 2009). These modifications allowed a swift internalization of the complexes into the cell through endocytosis, an efficient endosomal escape and protection against nucleolytic degradation in the insect gut, leading to an enhanced RNAi response (Gillet *et al.*, 2017).

11.7 Perspectives

RNAi-mediated pest control and improvement of plant performance have emerged as one of the most promising strategies, combining specificity and sustainability. Although exogenous application of RNA molecules is known to trigger RNAi responses in plants and insects, several barriers impede the use of non-GMO-based RNAi. Among these barriers, stability of the dsRNA and efficiency of the cellular internalization are the major challenges. The conjugation of RNA to different types of carriers is reported to improve the stability of the dsRNA, protect the dsRNA against nucleolytic degradation and facilitate an efficient internalization into the plant or insect cells, resulting in an improved RNAi response. The exploitation of chemical creativity to design carriers with specific properties, and the large biological diversity in which novel interesting proteins to direct dsRNA delivery can be identified, provides us with a wide diversity of

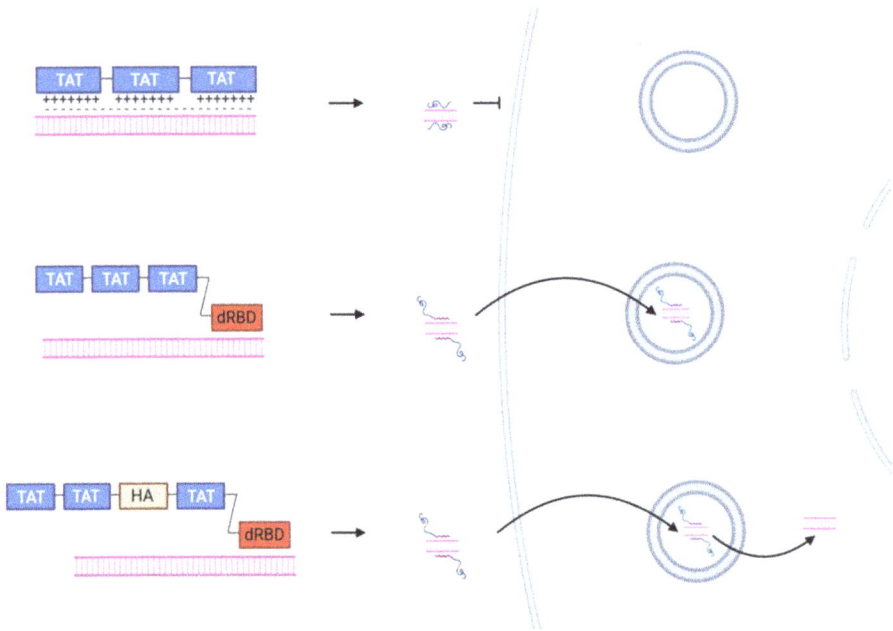

Fig. 11.5. Modification of the TAT-based peptide carrier enhances its ability to provoke an RNAi response. Direct conjugation of the positive-charged TAT domain to the negative-charged dsRNA results in charge neutralization, causing a limited delivery into the cells. Fusion of a dsRNA-binding domain (dRBD) circumvents the charge neutralization and allows efficient cellular internalization. To promote endosomal escape, the carrier was modified with a haemagglutinin domain to destabilize the membrane of the endocytic vesicle.

untested candidates that will allow the discovery of many potentially interesting carriers.

An interesting class of proteins for the design of protein-based carriers is the carbohydrate-binding proteins or lectins, allowing carbohydrate-targeted delivery of dsRNAs. Many lectins are shown to be efficiently internalized by insect cells and can even be transported across the epithelium into the underlying tissues (Powell *et al.*, 1998; Caccia *et al.*, 2012; Shen *et al.*, 2017). In addition, many lectins were found to be stable in a large range of pH and temperatures and are resilient to proteolytic degradation (Chan *et al.*, 2012; Walski *et al.*, 2014), suggesting these proteins could offer protection to the dsRNA. These properties could make lectin-based carrier systems powerful tools for oral delivery of dsRNA. One example is the development of a lectin-based carrier using the mannose-specific *Galanthus nivalis* agglutinin (GNA) (Van Damme

et al., 1987; Shibuya *et al.*, 1988). Previously this lectin was used for the delivery of peptides and proteins in various insect cells through the generation of fusion proteins (Raemaekers *et al.*, 1999; Raemaekers, 2000; Fitches *et al.*, 2002, 2004; Down *et al.*, 2006). Similarly, fusion to a dsRNA-binding domain would enable the cellular delivery of dsRNA (Bogaert *et al.*, 2005; Cao, 2016), although further research is needed to investigate the potential of this dsRNA carrier.

As the functionality of dsRNA carrier systems has been shown, some caution must be taken when working with these carrier systems (Vogel *et al.*, 2019). Many of the carrier systems are inspired by those used in the medical field, which implies that these could be capable of entering mammalian cells as well as arthropod or plant cells (Vogel *et al.*, 2019). These aspects need to be taken into consideration in the development of potential applications.

Several studies have already confirmed the potential of carrier systems to improve RNAi for applications in pest control and plant improvement after environmental application of dsRNA. However, improving our knowledge on the factors affecting dsRNA stability and uptake mechanisms of carriers and dsRNA, in both plants and insects, will allow the generation of improved carrier systems, consequently improving this technology for future applications.

Acknowledgements

This research was funded by the Research Foundation-Flanders (FWO-Vlaanderen, Belgium) and the Special Research Fund from the Ghent University. Olivier Christiaens is a recipient of a postdoctoral scholarship of the Research Foundation-Flanders (FWO). The authors declare no competing financial interests. Figures for this chapter were created with BioRender.com.

References

Akinc, A., Thomas, M., Klibanov, A.M. and Langer, R. (2005) Exploring polyethylenimine-mediated DNA transfection and the proton sponge hypothesis. *The Journal of Gene Medicine* 7(5), 657–663. DOI: 10.1002/jgm.696.

Akinc, A., Zumbuehl, A., Goldberg, M., Leshchiner, E.S., Busini, V. *et al.* (2008) A combinatorial library of lipid-like materials for delivery of RNAi therapeutics. *Nature Biotechnology* 26(5), 561–569. DOI: 10.1038/nbt1402.

Allen, M.L. and Walker, W.B. (2012) Saliva of *Lygus lineolaris* digests double stranded ribonucleic acids. *Journal of Insect Physiology* 58(3), 391–396. DOI: 10.1016/j.jinsphys.2011.12.014.

Almeida Garcia, R., Lima Pepino Macedo, L., Cabral do Nascimento, D., Gillet, F.-X., Moreira-Pinto, C.E. *et al.* (2017) Nucleases as a barrier to gene silencing in the cotton boll weevil, *Anthonomus grandis*. *PLoS ONE* 12, e0189600. DOI: 10.1371/journal.pone.0189600.

Arimatsu, Y., Furuno, T., Sugimura, Y., Togoh, M., Ishihara, R. *et al.* (2007) Purification and properties of double-stranded RNA-degrading nuclease, dsRNase, from the digestive juice of the silkworm, *Bombyx mori*. *Journal of Insect Biotechnology and Sericology* 76, 57–62.

Bachman, P., Fischer, J., Song, Z., Urbanczyk-Wochniak, E. and Bennet, E. (2020) Environmental dissipation of dsRNA and barriers to plant uptake. *Frontiers in Plant Science* 11, 21.

Bartlow, A.W., Manore, C., Xu, C., Kaufeld, K.A., Del Valle, S. *et al.* (2019) Forecasting zoonotic infectious disease response to climate change: mosquito vectors and a changing environment. *Veterinary Sciences* 6(2), 40. DOI: 10.3390/vetsci6020040.

Baum, J.A., Bogaert, T., Clinton, W., Heck, G.R., Feldmann, P. *et al.* (2007) Control of coleopteran insect pests through RNA interference. *Nature Biotechnology* 25(11), 1322–1326. DOI: 10.1038/nbt1359.

Bettinger, T., Carlisle, R.C., Read, M.L., Ogris, M. and Seymour, L.W. (2001) Peptide-mediated RNA delivery: a novel approach for enhanced transfection of primary and post-mitotic cells. *Nucleic Acids Research* 29(18), 3882–3891. DOI: 10.1093/nar/29.18.3882.

Bhat, A. and Ryu, C.-M. (2016) Plant perceptions of extracellular DNA and RNA. *Molecular Plant* 9(7), 956–958. DOI: 10.1016/j.molp.2016.05.014.

Bogaert, T.A.O.E., Zwaal, R., Plaetinck, G., De Kerpel, J.O. and Kaletta, T.J. (2005) Method and constructs for delivering double stranded RNA to pest organisms. *Patent* 20090263364.

Caccia, S., Van Damme, E.J.M., De Vos, W.H. and Smagghe, G. (2012) Mechanism of entomotoxicity of the plant lectin from *Hippeastrum hybrid* (Amaryllis) in *Spodoptera littoralis* larvae. *Journal of Insect Physiology* 58(9), 1177–1183. DOI: 10.1016/j.jinsphys.2012.05.014.

Cagliari, D., Santos, E.A., dos Dias, N., Smagghe, G. and Zotti, M. (2018) Modulating gene expression – abridging the RNAi and CRISPR-Cas9 technologies. In: Singh, A. and Khan, M.W. (eds) *Nontransformative Strategies for RNAi in Crop Protection*. InTech Open, London, pp. 1–18.

Cagliari, D., Dias, N.P., Galdeano, D.M., dos Santos, Ericmar Ávila., Smagghe, G. *et al.* (2019) Management of pest insects and plant diseases by non-transformative RNAi. *Frontiers in Plant Science* 10, 1319. DOI: 10.3389/fpls.2019.01319.

Cao, M. (2016) Limitations of RNA interference as a potential technique for crop protection against insect pests. PhD thesis. Durham University, Durham, UK.

Cappelle, K., de Oliveira, C.F.R., Van Eynde, B., Christiaens, O. and Smagghe, G. (2016) The involvement of clathrin-mediated endocytosis and two Sid-1-like transmembrane proteins in double-stranded RNA uptake in the Colorado potato beetle midgut. *Insect Molecular Biology* 25(3), 315–323. DOI: 10.1111/imb.12222.

Castellanos, N.L., Smagghe, G., Sharma, R., Oliveira, E.E. and Christiaens, O. (2019) Liposome encapsulation and EDTA formulation of dsRNA targeting essential genes increase oral RNAi-caused mortality in the Neotropical stink bug *Euschistus heros*. *Pest Management Science* 75(2), 537–548. DOI: 10.1002/ps.5167.

Chan, Y.S., Wong, J.H., Fang, E.F., Pan, W. and Ng, T.B. (2012) Isolation of a glucosamine binding leguminous lectin with mitogenic activity towards splenocytes and anti-proliferative activity towards tumor cells. *PLoS ONE* 7(6), e38961. DOI: 10.1371/journal.pone.0038961.

Chang, F.P., Kuang, L.Y., Huang, C.A., Jane, W.N., Hung, Y. *et al.* (2013) A simple plant gene delivery system using mesoporous silica nanoparticles as carriers. *Journal of Materials Chemistry B* 1(39), 5279–5287. DOI: 10.1039/c3tb20529k.

Christiaens, O. and Smagghe, G. (2014) The challenge of RNAi-mediated control of hemipterans. *Current Opinion in Insect Science* 6, 15–21. DOI: 10.1016/j.cois.2014.09.012.

Christiaens, O., Swevers, L. and Smagghe, G. (2014) DsRNA degradation in the pea aphid (*Acyrthosiphon pisum*) associated with lack of response in RNAi feeding and injection assay. *Peptides* 53, 307–314. DOI: 10.1016/j.peptides.2013.12.014.

Christiaens, O., Prentice, K., Pertry, I., Ghislain, M., Bailey, A. *et al.* (2016) RNA interference: a promising biopesticide strategy against the African sweetpotato weevil *Cylas brunneus*. *Scientific Reports* 6(1), 38836. DOI: 10.1038/srep38836.

Christiaens, O., Tardajos, M.G., Martinez Reyna, Z.L., Dash, M., Dubruel, P. *et al.* (2018) Increased RNAi efficacy in *Spodoptera exigua* via the formulation of dsRNA with guanylated polymers. *Frontiers in Physiology* 9, 316. DOI: 10.3389/fphys.2018.00316.

Christiaens, O., Petek, M., Smagghe, G. and Taning, C.N.T. (2020a) The use of nanocarriers to improve the efficiency of RNAi-based pesticides in agriculture. In: Fraceto, L., de Castro, S.S., Grillo, R, V.L., Avila, D., Caixeta Oliveira, H., *et al.* (eds) *Nanopesticides: From Research and Development to Mechanisms of Action and Sustainable Use in Agriculture*. Springer International, pp. 49–68.

Christiaens, O., Whyard, S., Vélez, A.M. and Smagghe, G. (2020b) Double-stranded RNA technology to control insect pests: current status and challenges. *Frontiers in Plant Science* 11, 451. DOI: 10.3389/fpls.2020.00451.

Cooper, A.M.W., Silver, K., Zhang, J., Park, Y. and Zhu, K.Y. (2019) Molecular mechanisms influencing efficiency of RNA interference in insects. *Pest Management Science* 75(1), 18–28. DOI: 10.1002/ps.5126.

Dass, C.R. and Choong, P.F.M. (2008) Chitosan-mediated orally delivered nucleic acids: a gutful of gene therapy. *Journal of Drug Targeting* 16(4), 257–261. DOI: 10.1080/10611860801900801.

Demirer, G.S. and Landry, M.P. (2017) Delivering genes to plants. *Chemical Engineering Progress* 113, 40–45.

Demirer, G.S., Zhang, H., Goh, N.S., Chang, R. and Landry, M.P. (2019) Nanotubes effectively deliver siRNA to intact plant cells and protect siRNA against nuclease degradation. *BioRxiv*.

Down, R.E., Fitches, E.C., Wiles, D.P., Corti, P., Bell, H.A. *et al.* (2006) Insecticidal spider venom toxin fused to snowdrop lectin is toxic to the peach-potato aphid, *Myzus persicae* (Hemiptera: Aphididae) and the rice brown planthopper, *Nilaparvata lugens* (Hemiptera: Delphacidae). *Pest Management Science* 62(1), 77–85. DOI: 10.1002/ps.1119.

Dubelman, S., Fischer, J., Zapata, F., Huizinga, K., Jiang, C. *et al.* (2014) Environmental fate of double-stranded RNA in agricultural soils. *PLoS ONE* 9(3), e93155, 42. DOI: 10.1371/journal.pone.0093155.

Dubrovina, A., Aleynova, O., Kalachev, A., Suprun, A., Ogneva, Z. *et al.* (2019) Induction of transgene suppression in plants via external application of synthetic dsRNA. *International Journal of Molecular Sciences* 20(7), 1585. DOI: 10.3390/ijms20071585.

Eguchi, A., Meade, B.R., Chang, Y.-C., Fredrickson, C.T., Willert, K. *et al.* (2009) Efficient siRNA delivery into primary cells by a peptide transduction domain–dsRNA binding domain fusion protein. *Nature Biotechnology* 27(6), 567–571. DOI: 10.1038/nbt.1541.

Erazo-Oliveras, A., Muthukrishnan, N., Baker, R., Wang, T.-Y. and Pellois, J.-P. (2012) Improving the endosomal escape of cell-penetrating peptides and their cargos: strategies and challenges. *Pharmaceuticals* 5(11), 1177–1209. DOI: 10.3390/ph5111177.

Fischer, J.R., Zapata, F., Dubelman, S., Mueller, G.M., Uffman, J.P. *et al.* (2017) Aquatic fate of a double-stranded RNA in a sediment-water system following an over-water application. *Environmental Toxicology and Chemistry* 36(3), 727–734. DOI: 10.1002/etc.3585.

Fitches, E., Audsley, N., Gatehouse, J.A. and Edwards, J.P. (2002) Fusion proteins containing neuropeptides as novel insect control agents: snowdrop lectin delivers fused allatostatin to insect haemolymph following oral ingestion. *Insect Biochemistry and Molecular Biology* 32(12), 1653–1661. DOI: 10.1016/S0965-1748(02)00105-4.

Fitches, E., Edwards, M.G., Mee, C., Grishin, E., Gatehouse, A.M.R. *et al.* (2004) Fusion proteins containing insect-specific toxins as pest control agents: snowdrop lectin delivers fused insecticidal spider venom toxin to insect haemolymph following oral ingestion. *Journal of Insect Physiology* 50(1), 61–71. DOI: 10.1016/j.jinsphys.2003.09.010.

Garbutt, J.S., Bellés, X., Richards, E.H. and Reynolds, S.E. (2013) Persistence of double-stranded RNA in insect hemolymph as a potential determiner of RNA interference success: evidence from *Manduca sexta* and *Blattella germanica*. *Journal of Insect Physiology* 59(2), 171–178. DOI: 10.1016/j.jinsphys.2012.05.013.

Ghodke, A.B., Good, R.T., Golz, J.F., Russell, D.A., Edwards, O. *et al.* (2019) Extracellular endonucleases in the midgut of *Myzus persicae* may limit the efficacy of orally delivered RNAi. *Scientific Reports* 9(1), 11898. DOI: 10.1038/s41598-019-47357-4.

Gillet, F.-X., Garcia, R.A., Macedo, L.L.P., Albuquerque, E.V.S., Silva, M.C.M. *et al.* (2017) Investigating engineered ribonucleoprotein particles to improve oral RNAi delivery in crop insect pests. *Frontiers in Physiology* 8, 256. DOI: 10.3389/fphys.2017.00256.

Guan, R.-B., Li, H.-C., Fan, Y.-J., Hu, S.-R., Christiaens, O. *et al.* (2018) A nuclease specific to lepidopteran insects suppresses RNAi. *Journal of Biological Chemistry* 293(16), 6011–6021. DOI: 10.1074/jbc.RA117.001553.

He, B., Chu, Y., Yin, M., Müllen, K., An, C. *et al.* (2013) Fluorescent nanoparticle delivered dsRNA toward genetic control of insect pests. *Advanced Materials* 25(33), 4580–4584. DOI: 10.1002/adma.201301201.

Huang, G., Allen, R., Davis, E.L., Baum, T.J. and Hussey, R.S. (2006) Engineering broad root-knot resistance in transgenic plants by RNAi silencing of a conserved and essential root-knot nematode parasitism gene. *Proceedings of the National Academy of Sciences of the USA* 103(39), 14302–14306. DOI: 10.1073/pnas.0604698103.

Hussain, H.I., Yi, Z., Rookes, J.E., Kong, L.X. and Cahill, D.M. (2013) Mesoporous silica nanoparticles as a biomolecule delivery vehicle in plants. *Journal of Nanoparticle Research* 15(6), 1676. DOI: 10.1007/s11051-013-1676-4.

Huvenne, H. and Smagghe, G. (2010) Mechanisms of dsRNA uptake in insects and potential of RNAi for pest control: a review. *Journal of Insect Physiology* 56(3), 227–235. DOI: 10.1016/j.jinsphys.2009.10.004.

Joshi, R., Sahoo, K.K., Tripathi, A.K., Kumar, R., Gupta, B.K. *et al.* (2018) Knockdown of an inflorescence meristem-specific cytokinin oxidase - OsCKX2 in rice reduces yield penalty under salinity stress conditions. *Plant Cell & Environment* 41(5), 936–946. DOI: 10.1111/pce.12947.

Kasai, H., Inoue, K., Imamura, K., Yuvienco, C., Montclare, J.K. *et al.* (2019) Efficient siRNA delivery and gene silencing using a lipopolypeptide hybrid vector mediated by a caveolae-mediated and temperature-dependent endocytic pathway. *Journal of Nanobiotechnology* 17(1), 11. DOI: 10.1186/s12951-019-0444-8.

Koch, A., Biedenkopf, D., Furch, A., Weber, L., Rossbach, O. *et al.* (2016) An RNAi-based control of *Fusarium graminearum* infections through spraying of long dsRNAs involves a plant passage and is controlled by the fungal silencing machinery. *PLoS Pathogens* 12(10), e1005901. DOI: 10.1371/journal.ppat.1005901.

Kumar, P.M., Murugan, K., Madhiyazhagan, P., Kovendan, K., Amerasan, D. *et al.* (2016) Biosynthesis, characterization, and acute toxicity of *Berberis tinctoria*-fabricated silver nanoparticles against the Asian tiger mosquito, *Aedes albopictus*, and the mosquito predators *Toxorhynchites splendens* and *Mesocyclops thermocyclopoides*. *Parasitology Research* 115(2), 751–759. DOI: 10.1007/s00436-015-4799-y.

Li, D., Liu, H., Yang, Y.-li., Zhen, P.-ping. and Liang, J.-sheng. (2009) Down-regulated expression of RACK1 gene by RNA interference enhances drought tolerance in rice. *Rice Science* 16(1), 14–20. DOI: 10.1016/S1672-6308(08)60051-7.

Liu, J., Swevers, L., Iatrou, K., Huvenne, H. and Smagghe, G. (2012) *Bombyx mori* DNA/RNA non-specific nuclease: expression of isoforms in insect culture cells, subcellular localization and functional assays. *Journal of Insect Physiology* 58(8), 1166–1176. DOI: 10.1016/j.jinsphys.2012.05.016.

Mansoor, S., Amin, I., Hussain, M., Zafar, Y. and Briddon, R. (2006) Engineering novel traits in plants through RNA interference. *Trends in Plant Science* 11, 559–565. DOI: 10.1016/j.tplants.2006.09.010.

Mao, Y.-B., Cai, W.-J., Wang, J.-W., Hong, G.-J., Tao, X.-Y. *et al.* (2007) Silencing a cotton bollworm P450 monooxygenase gene by plant-mediated RNAi impairs larval tolerance of gossypol. *Nature Biotechnology* 25(11), 1307–1313. DOI: 10.1038/nbt1352.

Martin-Ortigosa, S., Valenstein, J.S., Sun, W., Moeller, L., Fang, N. *et al.* (2012) Parameters affecting the efficient delivery of mesoporous silica nanoparticle materials and gold nanorods into plant tissues by the biolistic method. *Small* 8(3), 413–422. DOI: 10.1002/smll.201101294.

Meade, B.R. and Dowdy, S.F. (2008) Enhancing the cellular uptake of siRNA duplexes following noncovalent packaging with protein transduction domain peptides. *Advances in Drug Delivery Reviews* 60(4-5), 530–536. DOI: 10.1016/j.addr.2007.10.004.

Mermigka, G., Verret, F. and Kalantidis, K. (2016) RNA silencing movement in plants. *Journal of Integrative Plant Biology* 58(4), 328–342. DOI: 10.1111/jipb.12423.

Milletti, F. (2012) Cell-penetrating peptides: classes, origin, and current landscape. *Drug Discovery Today* 17(15-16), 850–860. DOI: 10.1016/j.drudis.2012.03.002.

Mitter, N., Worrall, E.A., Robinson, K.E., Xu, Z.P. and Carroll, B.J. (2017a) Induction of virus resistance by exogenous application of double-stranded RNA. *Current Opinions in Virology* 26, 49–55. DOI: 10.1016/j.coviro.2017.07.009.

Mitter, N., Worrall, E.A., Robinson, K.E., Li, P., Jain, R.G. *et al.* (2017b) Clay nanosheets for topical delivery of RNAi for sustained protection against plant viruses. *Nature Plants* 3(2), 16207. DOI: 10.1038/nplants.2016.207.

Miyata, K., Ramaseshadri, P., Zhang, Y., Segers, G., Bolognesi, R. *et al.* (2014) Establishing an in vivo assay system to identify components involved in environmental RNA interference in the western corn rootworm. *PLoS ONE* 9(7), e101661. DOI: 10.1371/journal.pone.0101661.

Mysore, K., Flannery, E.M., Tomchaney, M., Severson, D.W. and Duman-Scheel, M. (2013) Disruption of *Aedes aegypti* olfactory system development through chitosan/siRNA nanoparticle targeting of semaphorin-1a. *PLoS Neglected Tropical Diseases* 7(5), e2215. DOI: 10.1371/journal.pntd.0002215.

Numata, K., Ohtani, M., Yoshizumi, T., Demura, T. and Kodama, Y. (2014) Local gene silencing in plants via synthetic dsRNA and carrier peptide. *Plant Biotechnology Journal* 12(8), 1027–1034. DOI: 10.1111/pbi.12208.

Numata, K., Horii, Y., Oikawa, K., Miyagi, Y., Demura, T. *et al.* (2018) Library screening of cell-penetrating peptide for BY-2 cells, leaves of *Arabidopsis*, tobacco, tomato, poplar, and rice callus. *Scientific Reports* 8(1), 10966. DOI: 10.1038/s41598-018-29298-6.

Parsons, K.H., Mondal, M.H., McCormick, C.L. and Flynt, A.S. (2018) Guanidinium-functionalized inter-polyelectrolyte complexes enabling RNAi in resistant insect pests. *Biomacromolecules* 19(4), 1111–1117. DOI: 10.1021/acs.biomac.7b01717.

Pegler, J., Oultram, J., Grof, C. and Eamens, A. (2019) Profiling the abiotic stress responsive microRNA landscape of *Arabidopsis thaliana*. *Plants* 8(3), 58. DOI: 10.3390/plants8030058.

Peters, K., Breitsameter, L. and Gerowitt, B. (2014) Impact of climate change on weeds in agriculture: a review. *Agronomy for Sustainable Development* 34(4), 707–721. DOI: 10.1007/s13593-014-0245-2.

Powell, K.S., Spence, J., Bharathi, M., Gatehouse, J.A. and Gatehouse, A.M.R. (1998) Immunohistochemical and developmental studies to elucidate the mechanism of action of the snowdrop lectin on the rice brown planthopper, *Nilaparvata lugens* (Stal). *Journal of Insect Physiology* 44(7-8), 529–539. DOI: 10.1016/S0022-1910(98)00054-7.

Prentice, K., Christiaens, O., Pertry, I., Bailey, A., Niblett, C. *et al.* (2017) RNAi-based gene silencing through dsRNA injection or ingestion against the African sweet potato weevil *Cylas puncticollis* (Coleoptera: Brentidae). *Pest Management Science* 73(1), 44–52. DOI: 10.1002/ps.4337.

Prentice, K., Smagghe, G., Gheysen, G. and Christiaens, O. (2019) Nuclease activity decreases the RNAi response in the sweetpotato weevil *Cylas puncticollis*. *Insect Biochemistry and Molecular Biology* 110, 80–89. DOI: 10.1016/j.ibmb.2019.04.001.

Qu, J., Ye, J. and Fang, R. (2007) Artificial microRNA-mediated virus resistance in plants. *Journal of Virology* 81(12), 6690–6699. DOI: 10.1128/JVI.02457-06.

Raemaekers, R.J.M. (2000) *Expression of functional plant lectins in heterologous systems*. Durham University, Durham, UK.

Raemaekers, R.J.M., de Muro, L., Gatehouse, J.A. and Fordham-Skelton, A.P. (1999) Functional phyto-hemagglutinin (PHA) and *Galanthus nivalis* agglutinin (GNA) expressed in *Pichia pastoris. European Journal of Biochemistry* 265, 394–403.

Saleh, M.-C., van Rij, R.P., Hekele, A., Gillis, A., Foley, E. *et al.* (2006) The endocytic pathway mediates cell entry of dsRNA to induce RNAi silencing. *Nature Cell Biology* 8, 793–802. DOI: 10.1038/ncb1439.

San Miguel, K. and Scott, J.G. (2016) The next generation of insecticides: dsRNA is stable as a foliar-applied insecticide. *Pest Management Science* 72(4), 801–809. DOI: 10.1002/ps.4056.

Scorza, R., Callahan, A., Dardick, C., Ravelonandro, M., Polak, J. *et al.* (2013) Genetic engineering of *Plum pox virus* resistance: 'Honeysweet' plum – from concept to product. *Plant Cell, Tissue and Organ Culture* 115(1), 1–12. DOI: 10.1007/s11240-013-0339-6.

Scott, J.G., Michel, K., Bartholomay, L.C., Siegfried, B.D., Hunter, W.B. *et al.* (2013) Towards the elements of successful insect RNAi. *Journal of Insect Physiology* 59(12), 1212–1221. DOI: 10.1016/j.jinsphys.2013.08.014.

Shen, Y., De Schutter, K., Walski, T., Van Damme, E.J.M. and Smagghe, G. (2017) Toxicity, membrane binding and uptake of the *Sclerotinia sclerotiorum* agglutinin (SSA) in different insect cell lines. *In Vitro Cellular & Developmental Biology – Animal* 53(8), 691–698. DOI: 10.1007/s11626-017-0176-8.

Shew, A.M., Danforth, D.M., Nalley, L.L., Nayga, R.M., Tsiboe, F. *et al.* (2017) New innovations in agricultural biotech: consumer acceptance of topical RNAi in rice production. *Food Control* 81, 189–195. DOI: 10.1016/j.foodcont.2017.05.047.

Shibuya, N., Goldstein, I.J., Van Damme, E.J.M. and Peumans, W.J. (1988) Binding properties of a mannose-specific lectin from the snowdrop (*Galanthus nivalis*) bulb. *Journal of Biological Chemistry* 263, 728–734.

Shukla, J.N., Kalsi, M., Sethi, A., Narva, K.E., Fishilevich, E. *et al.* (2016) Reduced stability and intracellular transport of dsRNA contribute to poor RNAi response in lepidopteran insects. *RNA Biology* 13(7), 656–669. DOI: 10.1080/15476286.2016.1191728.

Spit, J., Philips, A., Wynant, N., Santos, D., Plaetinck, G. *et al.* (2017) Knockdown of nuclease activity in the gut enhances RNAi efficiency in the Colorado potato beetle, *Leptinotarsa decemlineata*, but not in the desert locust, *Schistocerca gregaria. Insect Biochemistry and Molecular Biology* 81, 103–116. DOI: 10.1016/j.ibmb.2017.01.004.

Taning, C.N.T., Arpaia, S., Christiaens, O., Dietz-Pfeilstetter, A., Jones, H. *et al.* (2020) RNA-based biocontrol compounds: current status and perspectives to reach the market. *Pest Managemant Science* 76(3), 841–845. DOI: 10.1002/ps.5686.

Terenius, O., Papanicolaou, A., Garbutt, J.S., Eleftherianos, I., Huvenne, H. *et al.* (2010) RNA interference in Lepidoptera: an overview of successful and unsuccessful studies and implications for experimental design. *Journal of Insect Physiology* 57(2), 231–245. DOI: 10.1016/j.jinsphys.2010.11.006.

Tomoyasu, Y., Miller, S.C., Tomita, S., Schoppmeier, M., Grossmann, D. *et al.* (2008) Exploring systemic RNA interference in insects: a genome-wide survey for RNAi genes in *Tribolium. Genome Biology* 9(1), R10. DOI: 10.1186/gb-2008-9-1-r10.

Turner, J.J., Jones, S., Fabani, M.M., Ivanova, G., Arzumanov, A.A. *et al.* (2007) RNA targeting with peptide conjugates of oligonucleotides, siRNA and PNA. *Blood Cells, Molecules, and Diseases* 38(1), 1–7. DOI: 10.1016/j.bcmd.2006.10.003.

Ulvila, J., Parikka, M., Kleino, A., Sormunen, R., Ezekowitz, R.A. *et al.* (2006) Double-stranded RNA is internalized by scavenger receptor-mediated endocytosis in *Drosophila* S2 cells. *Journal of Biological Chemistry* 281(20), 14370–14375. DOI: 10.1074/jbc.M513868200.

Unnamalai, N., Kang, B.G. and Lee, W.S. (2004) Cationic oligopeptide-mediated delivery of dsRNA for post-transcriptional gene silencing in plant cells. *FEBS Letters* 566(1-3), 307–310. DOI: 10.1016/j.febslet.2004.04.018.

Van Damme, E.J.M., Allen, A.K. and Peumans, W.J. (1987) Isolation and characterization of a lectin with exclusive specificity towards mannose from snowdrop (*Galanthus nivalis*) bulbs. *FEBS Letters* 215(1), 140–144. DOI: 10.1016/0014-5793(87)80129-1.

Vélez, A.M. and Fishilevich, E. (2018) The mysteries of insect RNAi: a focus on dsRNA uptake and transport. *Pesticide Biochemistry and Physiology* 151, 25–31. DOI: 10.1016/j.pestbp.2018.08.005.

Vivès, E., Brodin, P. and Lebleu, B. (1997) Truncated HIV-1 Tat protein basic domain rapidly translocates through the plasma membrane and accumulates in the cell nucleus. *Journal of Biological Chemistry* 272(25), 16010–16017. DOI: 10.1074/jbc.272.25.16010.

Vogel, E., Santos, D., Mingels, L., Verdonckt, T.-W. and Vanden Broeck, J. (2019) Rna interference in insects: protecting beneficials and controlling pests. *Frontiers in physiology* 9, 1912. DOI: 10.3389/fphys.2018.01912.

Wadia, J.S., Stan, R.V. and Dowdy, S.F. (2004) Transducible TAT-HA fusogenic peptide enhances escape of TAT-fusion proteins after lipid raft macropinocytosis. *Nature Medicine* 10(3), 310–315. DOI: 10.1038/nm996.

Walski, T., Van Damme, E.J.M. and Smagghe, G. (2014) Penetration through the peritrophic matrix is a key to lectin toxicity against *Tribolium castaneum*. *Journal of Insect Physiology* 70, 94–101. DOI: 10.1016/j.jinsphys.2014.09.004.

Wang, F., Wang, Y., Zhang, X., Zhang, W., Guo, S. *et al.* (2014) Recent progress of cell-penetrating peptides as new carriers for intracellular cargo delivery. *Journal of Controlled Release* 174, 126–136. DOI: 10.1016/j.jconrel.2013.11.020.

Wang, M., Weiberg, A., Lin, F.-M., Thomma, B.P.H.J., Huang, H.-D. *et al.* (2016) Bidirectional cross-kingdom RNAi and fungal uptake of external RNAs confer plant protection. *Nature Plants* 2(10), 16151. DOI: 10.1038/nplants.2016.151.

Whitehead, K.A., Langer, R. and Anderson, D.G. (2009) Knocking down barriers: advances in siRNA delivery. *Nature Reviews Drug Discovery* 8(2), 129–138. DOI: 10.1038/nrd2742.

Winston, W.M., Molodowitch, C. and Hunter, C.P. (2002) Systemic RNAi in *C. elegans* requires the putative transmembrane protein SID-1. *Science* 295(5564), 2456–2459. DOI: 10.1126/science.1068836.

Winston, W.M., Sutherlin, M., Wright, A.J., Feinberg, E.H. and Hunter, C.P. (2007) *Caenorhabditis elegans* SID-2 is required for environmental RNA interference. *Proceedings of the National Academy of Sciences of the USA* 104(25), 10565–10570. DOI: 10.1073/pnas.0611282104.

Wynant, N., Verlinden, H., Breugelmans, B., Simonet, G. and Vanden Broeck, J. (2012) Tissue-dependence and sensitivity of the systemic RNA interference response in the desert locust, *Schistocerca gregaria*. *Insect Biochemistry and Molecular Biology* 42(12), 911–917. DOI: 10.1016/j.ibmb.2012.09.004.

Xiao, D., Gao, X., Xu, J., Liang, X., Li, Q. *et al.* (2015) Clathrin-dependent endocytosis plays a predominant role in cellular uptake of double-stranded RNA in the red flour beetle. *Insect Biochemistry and Molecular Biology* 60, 68–77. DOI: 10.1016/j.ibmb.2015.03.009.

Yoon, J.-S., Gurusamy, D. and Palli, S.R. (2017) Accumulation of dsRNA in endosomes contributes to inefficient RNA interference in the fall armyworm, *Spodoptera frugiperda*. *Insect Biochemistry and Molecular Biology* 90, 53–60. DOI: 10.1016/j.ibmb.2017.09.011.

Younis, A., Siddique, M.I., Kim, C.-K. and Lim, K.-B. (2014) RNA interference (RNAi) induced gene silencing: a promising approach of hi-tech plant breeding. *International Journal of Biological Sciences* 10, 1150–1158. DOI: 10.7150/ijbs.10452.

Zhang, X., Zhang, J. and Zhu, K.Y. (2010) Chitosan/double-stranded RNA nanoparticle-mediated RNA interference to silence chitin synthase genes through larval feeding in the African malaria mosquito (*Anopheles gambiae*). *Insect Molecular Biology* 19(5), 683–693. DOI: 10.1111/j.1365-2583.2010.01029.x.

Zhang, X., Mysore, K., Flannery, E., Michel, K., Severson, D.W. *et al.* (2015) Chitosan/interfering RNA nanoparticle mediated gene silencing in disease vector mosquito larvae. *Journal of Visualized Experiments* 97, 52523. DOI: 10.3791/52523.

12 Environmental Safety Assessment of Plants Expressing RNAi for Pest Control

Salvatore Arpaia[1]*, Olivier Christiaens[2], Paul Henning Krogh[3], Kimberly M. Parker[4] and Jeremy Sweet[5]

[1]ENEA, Research Centre Trisaia, Rotondella (MT), Italy; [2]Faculty of Bioscience Engineering, Ghent University, Belgium; [3]Department of Bioscience, Aarhus University, Denmark; [4]Department of Energy, Environmental, and Chemical Engineering, Washington University in St Louis, Missouri, USA; [5]Sweet Environmental Consultants, Cambridge, UK

12.1 Introduction

Problem formulation (PF) is normally considered the first part of the environmental risk assessment (ERA) process and involves the identification of the possible hazards associated with a stressor (e.g. genetically modified (GM) RNA interference (RNAi)-expressing plants or RNAi-based pesticides). This initially requires an examination of all existing information to determine which hazards are identified by current scientific literature or experiences with the stressor and similar organisms or products. It also requires an element of brain storming in order to envisage new potential hazards that might arise, particularly considering how the new stressor will be used and managed. The hazards identified in the PF are characterized in order to determine whether they have the potential to cause adverse environmental impacts and the potentially harmful characteristics become the main focus for the risk assessment. The PF also examines information on the potential receiving environments for the new stressor in order to determine what other biota might be exposed and which ecosystem functions might be affected. In addition, the PF identifies

where there is lack of knowledge or experience with a new stressor and/or its receiving environments and therefore what studies are required to determine its environmental impacts. The risk hypotheses developed from the PF are used to hypothesize pathways to risk and to support the design of experimental studies to determine environmental impacts.

Any environmental risk assessment needs to provide quantitative information on two main components of the pathway to risk: exposure and hazard. Each of the two components can be determined based on the evaluation of several factors to estimate the exposure function $f(\exp)$ and the hazard function $f(\text{haz})$.

12.2 Exposure to dsRNA Expressed in Genetically Modified Plants

12.2.1 Environmental exposure and fate of dsRNA, siRNA and miRNA

Environmental risk assessment of RNAi-based pesticides (i.e. double-stranded RNA (dsRNA),

*Corresponding author: salvatore.arpaia@enea.it

© CAB International 2021. *RNAi for Plant Improvement and Protection*
(eds B. Mezzetti *et al.*)
DOI: 10.1079/9781789248890.0012

small interfering RNA (siRNA), microRNA (miRNA)) involves a characterization of the potential ecological effects of exposure to these pesticides combined with a characterization of the anticipated concentrations of RNAi-based pesticides in environmental systems to which organisms will be exposed (Auer and Frederick, 2009; Lundgren and Duan, 2013). Developing estimates of environmental concentrations of RNAi-based pesticides requires specific knowledge on the entry and fate of RNAi-based pesticides in these environmental systems, which primarily are expected to be agricultural soils and adjacent surface water (Parker and Sander, 2017).

The release of RNAi-based pesticides to receiving environments from GM plant tissue differs greatly from the environmental release of sprayable RNAi-based pesticides or conventional synthetic pesticides. Whereas the amount of an exogenously applied pesticide entering the receiving environment depends primarily on its application rate, the amount of an RNAi-based pesticide produced in the tissue of a GM plant is determined by production and processing of the dsRNA within the plant tissue and the route(s) of entry from the plant tissue into the environment.

The amount of RNAi-based pesticides entering receiving environments has not yet been quantified from either spray application or GM crops. In the case of the latter, release rates from certain GM crops for which data are available have been estimated using release rates of Cry proteins from GM plant tissue to receiving environments (Clark *et al.*, 2005) and reported concentrations of both RNAi-based pesticides and Cry proteins in GM plant tissue (US Environmental Protection Agency, 2015). From this available information, release rates of RNAi-based pesticides from GM crops to agricultural soil are estimated to occur at levels of micrograms per hectare (3–4 orders of magnitude lower than Cry protein release rates), resulting in nanogram or lower concentrations of RNAi-based pesticides per gram of soil (Parker and Sander, 2017; Parker *et al.*, 2019).

One validated method that uses quantitative reverse transcription-polymerase chain reaction (RT-qPCR) is able to quantify RNAi-based pesticides at low levels applicable to release rates of RNAi-based pesticides from GM crops

in receiving environments (Zhang *et al.*, 2020). Alternative methods to measure RNAi-based pesticides in receiving environments require them to be present at relatively high concentrations (Fischer *et al.*, 2016) or to be radio-isotopically labelled (Parker *et al.*, 2019) and therefore may be unable to quantify RNAi-based pesticides in the field.

After entry into receiving environments, the fate of RNAi-based pesticides is determined by the relative rates and extents of multiple processes, including abiotic, enzymatic or microbial degradation and adsorption to solid–water interfaces (Parker and Sander, 2017; Parker *et al.*, 2019). A few studies using the aforementioned hybridization assay have reported dissipation of detectable dsRNA pesticides or model dsRNA analogues applied at relatively high concentrations (i.e. µg/ml or µg/g levels) to soil or sediment–water microcosms (Dubelman *et al.*, 2014; Albright *et al.*, 2017; Fischer *et al.*, 2017). These studies estimated half-lives for dsRNA dissipation ranging from hours to days. One study conducted using 32-phosphorus (^{32}P) labelled dsRNA investigated the fate of dsRNA at lower concentrations (ng/g) in soil microcosms (Parker *et al.*, 2019). In addition to enabling experiments to be conducted at lower dsRNA concentrations that are closer to expected concentrations in receiving environments, the use of ^{32}P-labelled dsRNA enabled delineation of specific fate processes, including dsRNA degradation and adsorption to solid–water interfaces (Parker *et al.*, 2019). These experiments revealed that both processes affecting dsRNA fate occur simultaneously in soils and therefore must be further evaluated to determine expected concentrations of RNAi-based pesticides in receiving environments.

Degradation of RNAi-based pesticides may occur by abiotic, enzymatic or microbial pathways. Abiotic degradation pathways include denaturation of dsRNA to single-stranded RNA and acid- or base-catalysed hydrolysis of the ribose–phosphodiester bonds comprising the RNA backbone (Parker and Sander, 2017). Microorganisms in receiving environments may accelerate the degradation of RNAi-based pesticides either through the production of extracellular hydrolases competent towards the pesticides, or through direct uptake and utilization of the pesticides. While reducing the abundance of viable microorganisms through either

X-ray sterilization or filtration of solutions extracted from soils only slightly decreased the degradation of ^{32}P-labelled dsRNA in microcosm experiments, it dramatically reduced the formation of specific ^{32}P-labelled degradation products indicative of microbial utilization (Parker *et al.*, 2019). Together, these results provide a preliminary indication that both extracellular enzyme activity and microorganism viability lead to biological degradation of RNAi-based pesticides in receiving environments.

RNAi-based pesticides are also expected to adsorb to solid–water interfaces on particles in soil or sediment. In soil microcosms, adsorption of ^{32}P-labelled dsRNA to particles was found to be rapid and extensive, particularly in soils with finer texture (Parker *et al.*, 2019). Adsorption of RNAi-based pesticides in environmental media is expected to result primarily from electrostatic interactions between negatively charged phosphodiester groups along the pesticide backbone and positively charged soil constituents (e.g. iron and aluminium (oxyhydr-)oxides), as previously observed for DNA (Cai *et al.*, 2006). Adsorption sites may be limited in abundance, particularly in the presence of competing adsorbates including other nucleic acids, phosphate, and organic acids co-occurring with RNAi-based pesticides in environmental media (Cai *et al.*, 2007). Saturation of adsorption sites may explain the absence of significant adsorption observed in microcosm experiments conducted at high RNAi-based pesticide concentrations (Albright *et al.*, 2017). Once adsorbed to an interface, longer RNAi-based pesticides (i.e. dsRNA molecules) are hypothesized to form train-and-loop structures common among linear polyelectrolytes, resulting in kinetically slow desorption which requires simultaneous detachment of the pesticide from all points of attachment to the interface (Parker and Sander, 2017). Relative to dissolved RNAi-based pesticides, RNAi-based pesticides adsorbed to sediment and soil particles appear to undergo slower degradation (Fischer *et al.*, 2017; Parker *et al.*, 2019); slower degradation of adsorbed nucleic acids relative to dissolved molecules has been widely attributed to protection of the adsorbed molecules from hydrolases (Aardema *et al.*, 1983; Lorenz and Wackernagel, 1987; Romanowski *et al.*, 1991; Paget *et al.*, 1992; Blum *et al.*, 1997; Crecchio and Stotzky, 1998).

Taken together, current results suggest an important role for both degradation and adsorption to solid–water interfaces in determining the fate of RNAi-based pesticides in receiving environments. To constrain estimates of anticipated concentrations of RNAi-based pesticides in environmental systems, the rates and extents of these processes must be determined as a function of physicochemical and biological properties of the soil or other receiving environment (i.e. soil pH, texture, biological activity), as well as the properties of the RNAi-based pesticide (i.e. length, sequence) and the concentration at which it occurs. Furthermore, the use of delivery formulations to increase the stability and/or cellular uptake of the RNAi-based pesticide may impact these processes, for example by reducing degradation rates or inhibiting adsorption to sediment and soil particles. In addition, the link between environmental concentrations and organism exposure must be established by characterizing: (i) the bioavailability of RNAi-based pesticides adsorbed to solid–water interfaces; and (ii) the bioactivity of RNAi-based pesticides after partial degradation in the receiving environment.

12.2.2 Environmental exposure routes from plants and plant products to invertebrates

The principal pathway of exposure from plants to invertebrates involves herbivore organisms feeding on plants that, upon ingestion, introduce a number of compounds that are channelled to the digestive system. Herbivore organisms in any ecosystem, including agro-ecosystems, are numerous and normally linked to a few plant species (oligophagy) as their food source. Herbivores are active both on the aerial parts of the plant (initiating grazing food chains) and in the rhizosphere (detritus food chains). Trophic chains can be rather complex and it is not surprising to find, in many agro-ecosystems, organisms active at the fourth trophic level (e.g. Gillespie and Wratten, 2017).

The three main channels through which environmental exposure for invertebrates to plant-expressed components can occur are air, plants and soil. Exposure through the air is initiated when pollen or seeds are dispersed from the plants into the wider environment and may involve organisms living in sites outside the cropped area. This type of exposure route is expected to implicate mainly herbivores, e.g. pollen feeders like bees, ladybirds, etc. or seed feeders like many beetles. An indirect exposure to other herbivores can occur when wind-dispersed pollen grains dust leaves of wild or cultivated plants where more herbivore organisms could be affected (Pleasants *et al.*, 2001; Perry *et al.*, 2010).

The exposure to plant-expressed compounds through trophic chains can initiate in any moment of the cropping season when a herbivorous species starts feeding on the plant. However, it does not stop at harvest, since plant residues may remain on soil for some time and can be moved incidentally by mechanical operations or naturally dispersed in nearby environments, including water bodies (Palm *et al.*, 1996; Zwahlen *et al.*, 2003; Rosi-Marshall *et al.*, 2007; Tank *et al.*, 2010). Herbivores and higher-order consumers can then become exposed offsite. Finally, exposure through the soil is also expected due to the emission of root exudates and litter to which soil-dwelling organisms at different trophic levels can then be exposed.

In the framework of ERA of genetically modified plants, different types of data need to be collected to provide estimates on the likelihood of exposure through the above-mentioned channels. First and foremost, data on the expression of dsRNA in various plant parts along the cropping season need to be collected. Scientific literature is rather poor in quantitative data referring to dsRNA expression *in planta*, which is normally derived only by comparison with the expression of housekeeping genes. The benchmark study in this respect was conducted by Bachman *et al.* (2016) during the characterization of the MON 87411 maize event that expresses dsRNA targeting the *DvSnf7* gene, which was developed to provide an additional mode of action to confer protection against corn rootworm species. *In planta* studies were also conducted on the same maize event by Ahmad *et al.* (2016).

As stated above, the exposure is not limited to herbivore arthropods but can involve indirectly organisms at higher trophic levels (e.g. predators, parasitoids, hyperparasitoids) along the food chains based on the host plants expressing new compounds. In the specific case of dsRNA

expressed in genetically modified organisms, information on the actual exposure along the food chain is very limited. The most compelling evidence of movement of dsRNA along the food chain comes from tritrophic studies conducted by Garbian *et al.* (2012), who investigated bidirectional transfer of RNAi between the honey bee *Apis mellifera* and its parasite spider mite, *Varroa destructor*. A dsRNA targeting *V. destructor* was supplied to a bee colony which was successively infested with *Varroa* mites. dsRNA was detected in *Varroa* individuals and, over time, the population of the parasite was sensibly reduced, demonstrating that movement of dsRNA along the food chain did not impair its biological activity. These individuals were also able to induce reverse movement when put in contact with a new honey bee colony. This particular example indicates a possible profitable use of dsRNA in the beekeeping sector. However, opposite scenarios could occur if a similar movement of dsRNA would affect predators of insect pests and jeopardize the contribution of natural pest control in the field. While several studies are ongoing to estimate the hazardous characteristics of dsRNA on some natural enemies (see below), studies aimed at identifying their possible indirect exposure in natural conditions are still scarce.

12.2.3   RNAi efficiency and (cellular) uptake of dsRNA in invertebrates

Several steps are necessary before exposure of an organism to the noxious substance present in the environment actually occurs. First of all, the compound needs to enter the target organism to exert its effect, then the substance (e.g. dsRNA) needs to undergo metabolic processes inside the body (e.g. cellular uptake, cleavage to siRNA) for the physiological exposure to occur.

RNAi efficiency is known to be very variable among invertebrate species, especially when dsRNA is taken up through the oral route. In nematodes, *Caenorhabditis elegans* is considered a model species for RNAi studies for different reasons, one of them being a very high sensitivity to RNAi. However, many other nematode species, including animal parasites and even other closely related *Caenorhabditis* soil-living species, show a much less robust RNAi response. In arthropods, coleopteran insect species (beetles) contain some of the most RNAi-sensitive species, while insects belonging to other orders are often recalcitrant (e.g. Lepidoptera) or display a variable efficiency at best (e.g. Diptera, Hemiptera). Many studies have investigated potential factors explaining this variability, which include differences in RNAi core machinery gene repertoire, the stability of dsRNA in the insect body, the efficiency of cellular uptake of dsRNA from the gut lumen, the endosomal release inside the cell and the influence of viruses on the RNAi core machinery. A great amount of research is also conducted to improve RNAi efficiency in these less sensitive species, for example by using different formulations to increase the dsRNA stability and cellular uptake. Here, an overview is given on the variability of RNAi efficiency, focusing mainly on nematodes and arthropods since data on molluscs and annelids are very scarce at this moment.

C. elegans is the model species for RNAi research. This is partly because RNAi was first described in this species (Fire et al., 1998) and because C. elegans had already been a model species for biological, genetic and molecular research for several decades (Kaletta and Hengartner, 2006). Undoubtedly, it is also facilitated by the fact that C. elegans is highly sensitive to dsRNA taken up from the environment. Efficient RNAi gene silencing can be achieved by injecting the worms with dsRNA but also by oral or transdermal uptake (Timmons, 2006). Although unknown during the early years of RNAi, many other nematodes show a much less robust response to dietary uptake of dsRNA. A meta-study looking into the RNAi response and RNAi machinery in a wide range of nematodes, including plant and animal parasites, found that C. elegans is a rather special case, having an expanded RNAi machinery gene repertoire (Dalzell et al., 2011). The study found that many non-Caenorhabditis species possess less than half the number of RNAi-related genes considered to be involved in C. elegans. Particularly, genes that are known to be involved in cellular uptake of dsRNA were found to be absent in many parasitic nematodes. The study also showed that C. elegans has a highly evolved cellular uptake mechanism for dsRNA, involving different pathways and specific channel proteins encoded by sid genes. Uptake from the gut happens via Sid-2-mediated endocytosis

and the dsRNA is then released in the cytoplasm from the internalized vesicles via Sid-1 channel proteins (McEwan et al., 2012). The same Sid-1 is also responsible for cellular export of dsRNA and uptake in cells that are not lining the gut. Therefore, Sid-1 is a major component of the successful systemic RNAi that is observed in C. elegans. Dalzell et al. (2011) found that not all nematodes, particularly parasitic species, possess these genes in their genome.

In arthropods, the cellular uptake pathways have not been completely characterized yet. While most insects do have one or more sid-1-like (sil) genes in their genome, their role in cellular dsRNA uptake is not clear. In all species where the involvement of (clathrin-mediated) endocytosis in cellular dsRNA uptake was investigated, this pathway turned out to be heavily involved. However, silencing of sil genes only affected RNAi efficacy in some of the species. An overview of this was presented in a cellular uptake study in the Colorado potato beetle, which is known to be highly sensitive to RNAi (Cappelle et al., 2016). In dipteran insects, there are clear indications that cellular uptake of dsRNA, at least from the midgut, is inefficient in some species. Studies in Drosophila melanogaster and Drosophila suzukii have shown that liposome encapsulation of dsRNA, aimed to improve cellular uptake, greatly increases RNAi efficacy (Whyard et al., 2009; Taning et al., 2016). Also in some other dipteran insects, such as mosquitoes, studies have shown that delivery formulations are necessary for efficient oral RNAi (Zhang et al., 2010).

A study by Shukla et al. (2016) showed that cytoplasmic release of dsRNA from internalized vesicles also plays a role in some lepidopteran insects. These researchers could see, through confocal microscopy, that fluorescently labelled dsRNA was taken up by the lepidopteran cells into endosomic vesicles but then not released into the cytoplasm. This was further confirmed when no siRNAs could be detected in these lepidopteran cells, indicating that no processing of the long dsRNA happened (Shukla et al., 2016; Yoon et al., 2017).

Another factor explaining the higher RNAi sensitivity in C. elegans and possibly other nematodes compared with arthropods is the presence of an RNA-dependent RNA polymerase-dependent amplification system, whereby secondary siRNAs are created (Sijen *et al.*, 2001). In this way, the silencing signal can be amplified and prolonged. In fact, most of the siRNAs in *C. elegans* are such secondary siRNAs, highlighting the importance of this amplification pathway (Pak and Fire, 2007). In arthropods, the presence of these RNA-dependent RNA polymerases (RdRPs) has only been reported in some ticks and mites (Kurscheid *et al.*, 2009; Grbić *et al.*, 2011). No homologues for this particular RdRP have been identified in insects so far. However, given the sensitivity of some insects to RNAi, it cannot be excluded that other, different amplification systems might be present in these species.

While variable cellular uptake efficiency and the lack of an amplification mechanism clearly play a role in some groups of insects, the most important factor affecting RNAi efficacy in insects might be the stability of dsRNA in the insect body. Many studies have shown that nucleases that are present in haemolymph, saliva and especially the midgut of a wide range of insect species are capable of causing rapid nucleolytic degradation of dsRNA that is taken up in the insect body. Nucleolytic degradation in saliva has been demonstrated in the saliva and haemolymph of Hemiptera and in the midgut of Coleoptera, Orthoptera and Lepidoptera. A study investigating RNAi efficacy and dsRNA persistence in the insect gut of three Coleoptera revealed a clear positive correlation between the two (Prentice *et al.*, 2017). In their *in vitro* gut juice incubation assays, the most sensitive of the three showed a very long persistence (more than 10 h) while the dsRNA in the least sensitive of these three insects was degraded within 30 min. Several studies have discovered several nucleases in the genome of insect species, for example in *Bombyx mori*, *Schistocerca gregaria*, *Locusta migratoria*, *Cylas puncticollis* and *Anthonomus grandis* (Liu *et al.*, 2012; Wynant *et al.*, 2014; Song *et al.*, 2017; Prentice *et al.*, 2017; Almeida Garcia *et al.*, 2017, respectively). Finally, in the RNAi-insensitive lepidopteran *Ostrinia furnacalis*, a nuclease was discovered and characterized which is specific for lepidopteran insects and which negatively affects RNAi efficacy in this species (Guan *et al.*, 2018).

More information on these cellular and physiological barriers in insects can be found in a review by Cooper *et al.* (2019). These types of physiological barriers will prove a challenge for scientists and industry to apply this technology against a wide range of insect species. Advances in dsRNA delivery, for example by using nanoparticles or other delivery vehicles, might help us to overcome these barriers (Joga *et al.*, 2016; Christiaens *et al.*, 2018b). Of course, these new delivery methods will have an impact on the risk assessment of these RNAi products. For example, these delivery vehicles will prolong the persistence in the environment, they might expose the dsRNA to non-target organisms (NTOs) which would otherwise not be exposed and they can also overcome barriers in NTOs that might otherwise prevent dsRNA from being taken up by its cells. These considerations will have to be taken into account during the development and regulation of future RNAi-based products.

While RNAi efficiency and its variability among invertebrates is obviously of great importance for product developers, it can also have implications for risk assessment. In the current pesticide as well as genetically modified organism (GMO) regulatory frameworks, toxicity testing on NTOs is an important stage. Knowledge on the efficiency of dsRNA uptake in invertebrates could guide us in the choice of NTOs to perform these tests on. For example, it could be questioned whether it is useful to test a novel product on a phylogenetically distant species that is known to be insensitive to environmental RNAi, while a more closely related NTO known to be highly sensitive could be chosen. Of course, potential formulations and delivery methods will impact these choices, as they could lead to exposure in species that would not be exposed to naked dsRNA.

12.3 Hazards of dsRNA Expressed in Genetically Modified Plants

12.3.1 Off-target, non-target and unintended effects of RNAi-based GM plants

Due to the mode of action, RNA interference is a potentially very specific means of silencing genes, e.g. in pest insects, mites or pathogens (Mysore *et al.*, 2018; Niu *et al.*, 2018; Zotti *et al.*, 2018). Within the body of a sensitive species, long dsRNA is cleaved by the enzyme Dicer into siRNAs, which are the effective molecules involved in silencing genes that produce RNA with a complementary sequence. siRNAs are about 20–22 nt in length and can therefore be effectively designed to attack specific sequences within genes of interest, usually involving lethal effects on target species or drastically reducing reproductive performance (e.g. Whyard, 2018). However, several possibilities of harm to non-target organisms have been hypothesized (Lundgren and Duan, 2013). These unwanted effects might be related to sequence-dependent mechanisms if the same target sequence is found in non-target organisms. Also, sequence-dependent mechanisms might be the cause of harm to target or non-target organisms if the same sequence targeted by the siRNA is found in other parts of the genome (off-target effects).

A sequence-dependent mechanism was the cause of a silencing effect on the *vATPase A* gene in two ladybird beetles when fed dsRNA designed to target the same gene in the western corn rootworm (WCR), *Diabrotica virgifera virgifera* (Haller *et al.*, 2019). The extent of the silencing and its biological impact were different in the two predatory species, being higher in *Coccinella septempunctata*, in which a significantly reduced survival rate in the bioassays was recorded. In *Adalia bipunctata*, during laboratory bioassays the authors only noted a prolonged developmental time. When the genome of the two species was studied in bioinformatics analyses, there was a difference in the number of 21 nt matches of the dsRNA with the *vATPase A* of *C. septempunctata* (34 matches) and that of *A. bipunctata* (six matches). This indicates that the degree of the negative effective could be attributed to the different presence of target sites in the genome. Further studies including additional species of Coccinellidae (Pan *et al.*, 2020) confirmed that taxonomic similarities are a good proxy to estimate the possibility of non-target effects, since taxonomically related species share a higher percentage of genomes. However, gene silencing has been noted in some cases also on quite distant species, i.e. belonging to completely different insect orders. For example, Chen *et al.* (2015) studied the effects of dsRNA targeting *rpl19* gene from *Bactrocera dorsalis* on a number of non-target species by measuring silencing with RT-PCR. The maximum effect was obtained

on the co-specific *B. minax*, but significant effects were also obtained on the hymenopteran *Diachasmimorpha longicaudata*, which shared 72% sequence homology with *B. dorsalis*. The available studies clearly indicate the necessity of characterizing the possible sensitivity of non-target species to the dsRNA in an early phase of the development of a new RNAi-based product.

For the development of the MON 87411 maize expressing Cry3Bb1, Cry34Ab1/Cry35Ab1 and DvSnf7 dsRNA to induce multiple insect resistance, Bachman *et al.* (2016) characterized the spectrum of insecticidal activity of a 240 nt dsRNA targeting the Snf7 gene in *D. virgifera virgifera*. Insects belonging to ten different families and four different orders were tested in continuous feeding diet bioassays with DvSnf7 dsRNA. The results demonstrated that the spectrum of activity for DvSnf7 was narrow and activity was only evident in a group of beetles within the Galerucinae subfamily of Chrysomelidae, which show > 90% identity with WCR Snf7. A shared sequence length of ≥ 20 nt seemed to be required for efficacy against *D. virgifera virgifera* and all orthologues susceptible for gene silencing by DvSnf7 contained at least three 21 nt matches with the DvSnf7 sequence. However, these sequence identity requirements could be different between insect species, as research has shown that the length of siRNAs which are the result of Dicer-2 processing of long dsRNA is variable (20–22 nt) between species of different orders (Santos *et al.*, 2019). Therefore, further research is needed to elucidate the sequence identity requirements for efficient RNAi. Taning *et al.* (2020) investigated potential effects of feeding dsRNA specific to pollen beetle (*Brassicogethes aeneus*) target genes by the bumblebee *Bombus terrestris*. Besides observing that this dsRNA had no effect on lethal and sublethal endpoints, the authors also investigated expression changes of 24 *B. terrestris* genes, which had the highest sequence identity with the non-target dsRNA. They found no changes in expression for any of these genes, despite siRNA matches of up to 20 nt.

Off-target effects are commonly related to the siRNA sequence and may occur when a partial complementarity of the siRNA to an unintended target within an organism is found (Jackson *et al.*, 2006). It is not uncommon that off-target binding sites exist in several different organisms, given the small sizes of siRNAs and the large genome of even quite simple organisms (Qiu *et al.*, 2005). As shown above, sequence complementarity is needed to trigger off target effects; however, siRNAs containing some mismatches may still effectively trigger silencing (Christiaens *et al.*, 2018a). The most striking example of off-target effects was shown in experiments with honey bees that were fed diet containing dsRNA targeting *gfp*, a gene that does not exist in the bee's genome (Nunes *et al.*, 2013). Although dsGFP is not expected to induce a response in honey bees, the authors reported phenotypical effects on specimens of *Apis mellifera* (i.e. altered pupal pigmentation and larval development). Examples are not limited to insects: Zhou *et al.* (2014) conducted a study on *C. elegans* and showed that nuclear Ago NRDE-3 protein associates with off-target silencing effects following administration of exogenous RNAi.

Unintended effects or RNA interference might sometimes occur, due to non-sequence-dependent mechanisms. A saturation of the RNAi machinery (e.g. on the protein RISC complex) is possible when a large number of dsRNAs enters the body of an organism, with consequent temporary inhibition of cellular use of RNA and compromise of some of its natural functions. However, to our knowledge this mechanism has never been proved in invertebrates.

RNA in invertebrate species is involved in the functioning of the immune system, especially against virus infections. The presence of exogenous RNA is known to trigger this response in mammals, and due to the similarities between the mammalian and arthropod immune system (Lundgren and Jurat-Fuentes, 2012) a possible alteration of the immune system functioning has been hypothesized (Lundgren and Duan, 2013), though rarely experimentally proven in arthropods.

12.3.2 Activity spectrum on soil- and plant-dwelling organisms

Due to the very general mechanism involved in RNA interference, theoretically speaking any gene can be silenced in species of interest (e.g. plant pests or pathogens) with the use of dsRNA.

Highly conserved genes could therefore represent a common target among a high number of species, which are not meant to be affected if exposed non-target organisms. The extensive review by Christiaens *et al.* (2018a) was based on a thorough literature search in July 2016 regarding possible silencing effects on invertebrates due to RNA interference. As of June 2019, no new studies had addressed soil invertebrates as revealed by doing a literature search on Web of Science (WoS) using the search terms of Christiaens *et al.* (2018a), but restricted to the main soil invertebrate taxa in agricultural soils: collembolans, mites, enchytraeids and lumbricids.

The open literature is still void of testing results involving RNAi and soil invertebrates. Ecotoxicological testing of RNAi with soil invertebrates has been reported only twice in the literature, i.e. for DvSnf7 (Bachman *et al.*, 2016) and for *v-ATPase A* dsRNAs (Pan *et al.*, 2016). In these studies, collembolans were exposed to the dsRNA active ingredient through food and the earthworm was exposed through soil. Worst-case scenarios were explored by manipulating the dsRNA similar to application of sprayable RNAi pesticides. None of the studies employed an increasing dose approach enabling an LC (lethal concentration) or EC (effect concentration) estimation, but this is not warranted if range-finding indicates no effects and a high often unrealistic dose is an option in a limit test (OECD Test Guideline 232; see OECD, 2016). So, ten times the expected maximum environmental concentration was tested for DvSnf7. *In planta* exposure was not addressed as recommended by Arpaia *et al.* (2017) and in this case the choice of test species should be litter feeders and litter decomposers. Both Bachman *et al.* (2016) and Pan *et al.* (2016) concluded that there were no effects on the soil invertebrates. However, a typical dilemma of the assessment occurred for the collembolan exposed to *v-ATPase A* dsRNAs: the developmental time was decreased. This would be interpreted as a case of hormesis if it was observed for a chemical, but for an RNAi it was not considered to indicate adversity and the RNAi was deemed harmless. It remains to be elucidated if such an effect is due to unintended effects or a stress response leading to hormesis.

A bioinformatics approach aiming for *in silico* screening of potential risks is possible for the two most tested soil invertebrates, the earthworm *Eisenia fetida* and the collembolan *Folsomia candida*, as their genomes and transcriptomes are available (Faddeeva *et al.*, 2015; Bhambri *et al.*, 2018, respectively). However, available genomic or transcriptomic information for a broad range of soil NTOs is still needed. Currently the database of the US National Center for Biotechnology Information (NCBI) includes the genome of three annelids, 17 collembolans and 30 mites, and the Transcriptome Shotgun Assembly (TSA) database of NCBI includes eight collembolans, 25 mites and two earthworms (available at www.ncbi.nlm.nih.gov, accessed 12 November 2020), but these transcriptomes are not evenly distributed across the taxonomic and functional diversity of soil invertebrates.

For soil invertebrates, we still need candidate dsRNAs with a reproducible effect to include as a positive control. Positive controls are available for chemical testing in OECD Guidelines for the Testing of Chemicals programme (OECD, 1994), but not for exposure through food or soil of dsRNA. Protection of soil ecosystem services has received increasing attention at the European Union (EU) level (Krogh, 2021), but hitherto no assessment protocols are available, and the ERA is stuck with assessment of life history parameters and biodiversity.

The paragraph above on the cellular uptake of dsRNA in invertebrates gives a good overview of the species that might constitute possible effective targets for dsRNA-based pesticides or prolactin-induced proteins (PIPs). Obviously, most of the knowledge in this area comes from insects and nematodes, due to their relevant role as pests in agriculture. Particular attention has been given to the use of dsRNA as a control for noxious organisms that are quite recalcitrant to other forms of pest control (for example, *D. virgifera virgifera* could not be satisfactorily controlled with the use of *Bacillus thuringiensis (Bt)-expressing genetically modified maize)*. This is reflected in the results of the review by Christiaens *et al.* (2018a), who indicated that the great majority of studies had been conducted with the aim of silencing genes in insects (2862 studies), though there was a bias towards the model species *Drosophila melanogaster* (Diptera) which, alone, was the subject of 1243 publications. The current research trend is in line with the existing literature, and emphasis

is being given to several insect pests, such as sap feeders (e.g. Castellanos *et al.*, 2019; Sun *et al.*, 2019; Tian *et al.*, 2019) or invasive species (e.g. Christiaens *et al.*, 2018b; Bento *et al.*, 2019) that are still known to represent difficult pests to manage in many agro-ecosystems in different areas worldwide. Yet very little is being published regarding non-target species not taxonomically related to targets and for which there is not much information about their genome.

12.4 Conclusions and Knowledge Gaps

RNAI-based mechanisms are a promising new means of pest control which could couple high effectiveness, due to completely new modes of action, to an extreme selectivity as a consequence of carefully selected target sequences in the genome of pests. However, even an accurate design of the dsRNA to induce interference does not exclude the possibility of off-target or non-target effects. Bioinformatics can give important support in designing RNA sequences that target genes expressed specifically in target pests. Nevertheless, the limited availability of genomic sequences of arthropod pests, the possibility of gene silencing if mismatches between the target and the siRNA sequences exist, and the likelihood of sequence-independent silencing suggest that laboratory, or higher-tier, bioassays remain fundamental in assessing possible environmental risks for non-target organisms due to the use of dsRNA. Current environmental risk assessment frameworks regarding possible effects on non-target organisms are expected to be effective to estimate RNAi-based products (Arpaia *et al.*, 2020; Papadopoulou *et al.*, 2020), though for some soil invertebrates more realistic exposure scenarios need to be developed. Arpaia *et al.* (2017) indicated some features of the dsRNA mode of action that need also to be considered during environmental risk assessment. For example, since mRNAs are transcribed only when needed by the organism, it is important to specifically consider the environmental conditions under which tests are being conducted. This is true for assessing silencing in both non-target genes and potentially existing off-target sequences.

Moreover, there is uncertainty about the possible modes of action, and consequent effects, in case of off-target silencing. Due to the various modes of action that can be activated while silencing genes with dsRNA, it is very unlikely that one single set of test species will serve as an adequate proxy of non-target species for all products using RNAi technology.

Several knowledge gaps need to be filled in order to have a thorough understanding of the possible environmental impacts of this new means of pest control. It must be noted that most scientific publications describing RNAi-expressing GM plants did not explicitly investigate the potential exposure of invertebrates to dsRNA expressed in such plants. For example, the actual presence of dsRNA in different plant parts over time has only occasionally been reported. As indicated above, only a few soil-dwelling species have been specifically tested for sensitivity to dsRNA, therefore studies encompassing more species, especially those known to be involved in providing ecological services to agriculture, are certainly needed.

The movement of dsRNA along food chains has been studied to only a limited extent. Indirect exposure was demonstrated for *Varroa* mites when feeding on honey bee colonies to which dsRNA was added to the diet (Garbian *et al.*, 2012). In this study, not only was transfer of the nucleic acid ascertained, but it was also found that dsRNA remained biologically active and the transfer could be reversed from mite individuals to new honey bee families. Data on other pests and their natural enemies are urgently needed. Likewise, some recent reports of potential interference of exogenous dsRNA with the immune system in bees (e.g. Niu *et al.*, 2016; Brutscher *et al.*, 2017) need confirmation of the mechanism and its consequences.

Finally, we note that the molecular mechanisms for uptake have mostly been studied in *C. elegans* and therefore we still need to fill relevant gaps for other arthropod systems of relevance.

References

Aardema, B.W., Lorenz, M.G. and Krumbein, W.E. (1983) Protection of sediment-adsorbed transforming DNA against enzymatic inactivation. *Applied and Environmental Microbiology* 46(2), 417–420. DOI: 10.1128/AEM.46.2.417-420.1983.

Ahmad, A., Negri, I., Oliveira, W., Brown, C., Asiimwe, P. *et al.* (2016) Transportable data from non-target arthropod field studies for the environmental risk assessment of genetically modified maize expressing an insecticidal double-stranded RNA. *Transgenic Research* 25(1), 1–17. DOI: 10.1007/s11248-015-9907-3.

Albright, V.C., Wong, C.R., Hellmich, R.L. and Coats, J.R. (2017) Dissipation of double-stranded RNA in aquatic microcosms. *Environmental Toxicology and Chemistry* 36(5), 1249–1253. DOI: 10.1002/etc.3648.

Almeida Garcia, R., Lima Pepino Macedo, L., Cabral do Nascimento, D., Gillet, F.-X., Moreira-Pinto, C.E. *et al.* (2017) Nucleases as a barrier to gene silencing in the cotton boll weevil, *Anthonomus grandis*. *PloS one* 12, e0189600. DOI: 10.1371/journal.pone.0189600.

Arpaia, S., Birch, A.N.E., Kiss, J., van Loon, J.J.A., Messéan, A. *et al.* (2017) Assessing environmental impacts of genetically modified plants on non-target organisms: the relevance of *in planta* studies. *Science of The Total Environment* 583, 123–132. DOI: 10.1016/j.scitotenv.2017.01.039.

Arpaia, S., Christiaens, O., Giddings, K., Jones, H., Mezzetti, B. *et al.* (2020) Biosafety of GM crop plants expressing dsRNA: data requirements and EU regulatory considerations. *Frontiers in Plant Science* 11, 940. DOI: 10.3389/fpls.2020.00940.

Auer, C. and Frederick, R. (2009) Crop improvement using small RNAs: applications and predictive ecological risk assessments. *Trends in Biotechnology* 27(11), 644–651. DOI: 10.1016/j.tibtech.2009.08.005.

Bachman, P.M., Huizinga, K.M., Jensen, P.D., Mueller, G., Tan, J. *et al.* (2016) Ecological risk assessment for DvSnf7 RNA: a plant-incorporated protectant with targeted activity against western corn rootworm. *Regulatory Toxicology and Pharmacology* 81, 77–88. DOI: 10.1016/j.yrtph.2016.08.001.

Bento, F.M., Marques, R.N., Campana, F.B., Demétrio, C.G., Leandro, R.A. *et al.* (2019) Gene silencing by RNAi via oral delivery of dsRNA by bacteria in the South American tomato pinworm, *Tuta absoluta*. *Pest Management Science* 76(1), 287-295. DOI: 10.1002/ps.5513.

Bhambri, A., Dhaunta, N., Patel, S.S., Hardikar, M., Bhatt, A. *et al.* (2018) Large scale changes in the transcriptome of *Eisenia fetida* during regeneration. *PLoS ONE* 13(9), e0204234. DOI: 10.1371/journal.pone.0204234.

Blum, S.A.E., Lorenz, M.G. and Wackernagel, W. (1997) Mechanism of retarded DNA degradation and prokaryotic origin of DNases in nonsterile soils. *Systematic and Applied Microbiology* 20(4), 513–521. DOI: 10.1016/S0723-2020(97)80021-5.

Brutscher, L.M., Daughenbaugh, K.F. and Flenniken, M.L. (2017) Virus and dsRNA-triggered transcriptional responses reveal key components of honey bee antiviral defense. *Scientific Reports* 7(1), 6448. DOI: 10.1038/s41598-017-06623-z.

Cai, P., Huang, Q., Jiang, D., Rong, X. and Liang, W. (2006) Microcalorimetric studies on the adsorption of DNA by soil colloidal particles. *Colloids and Surfaces B: Biointerfaces* 49(1), 49–54. DOI: 10.1016/j.colsurfb.2006.02.011.

Cai, P., Huang, Q., Zhu, J., Jiang, D., Zhou, X. *et al.* (2007) Effects of low-molecular-weight organic ligands and phosphate on DNA adsorption by soil colloids and minerals. *Colloids and Surfaces B: Biointerfaces* 54(1), 53–59. DOI: 10.1016/j.colsurfb.2006.07.013.

Cappelle, K., de Oliveira, C.F.R., Van Eynde, B., Christiaens, O. and Smagghe, G. (2016) The involvement of clathrin-mediated endocytosis and two Sid-1-like transmembrane proteins in double-stranded RNA uptake in the Colorado potato beetle midgut. *Insect Molecular Biology* 25(3), 315–323. DOI: 10.1111/imb.12222.

Castellanos, N.L., Smagghe, G., Sharma, R., Oliveira, E.E. and Christiaens, O. (2019) Liposome encapsulation and EDTA formulation of dsRNA targeting essential genes increase oral RNAi-caused mortality in the Neotropical stink bug *Euschistus heros*. *Pest Management Science* 75(2), 537–548. DOI: 10.1002/ps.5167.

Chen, A., Zheng, W., Zheng, W. and Zhang, H. (2015) The effects of RNA interference targeting *Bactrocera dorsalis* ds-Bdrpl19 on the gene expression of rpl19 in non-target insects. *Ecotoxicology* 24(3), 595–603. DOI: 10.1007/s10646-014-1407-3.

Christiaens, O., Dzhambazova, T., Kostov, K., Arpaia, S., Joga, M.R. *et al.* (2018a) Literature review of baseline information on RNAi to support the environmental risk assessment of RNAi-based GM plants. *EFSA Supporting Publications* 15(5), 1424E. DOI: 10.2903/sp.efsa.2018.EN-1424.

Christiaens, O., Tardajos, M.G., Martinez Reyna, Z.L., Dash, M., Dubruel, P. *et al.* (2018b) Increased RNAi efficacy in *Spodoptera exigua* via the formulation of dsRNA with guanylated polymers. *Frontiers in Physiology* 9, 316. DOI: 10.3389/fphys.2018.00316.

Clark, B.W., Phillips, T.A. and Coats, J.R. (2005) Environmental fate and effects of *Bacillus thuringiensis* (Bt) proteins from transgenic crops: a review. *Journal of Agricultural and Food Chemistry* 53(12), 4643–4653. DOI: 10.1021/jf040442k.

Cooper, A.M., Silver, K., Zhang, J., Park, Y. and Zhu, K.Y. (2019) Molecular mechanisms influencing efficiency of RNA interference in insects. *Pest Management Science* 75(1), 18–28. DOI: 10.1002/ps.5126.

Crecchio, C. and Stotzky, G. (1998) Binding of DNA on humic acids: effect on transformation of *Bacillus subtilis* and resistance to DNase. *Soil Biology and Biochemistry* 30(8-9), 1061–1067. DOI: 10.1016/S0038-0717(97)00248-4.

Dalzell, J.J., McVeigh, P., Warnock, N.D., Mitreva, M., Bird, D.M. *et al.* (2011) RNAi effector diversity in nematodes. *PLoS Neglected Tropical Diseases* 5(6), e1176. DOI: 10.1371/journal.pntd.0001176.

Dubelman, S., Fischer, J., Zapata, F., Huizinga, K., Jiang, C. *et al.* (2014) Environmental fate of double-stranded RNA in agricultural soils. *PLoS ONE* 9(3), e93155. DOI: 10.1371/journal.pone.0093155.

Faddeeva, A., Studer, R.A., Kraaijeveld, K., Sie, D., Ylstra, B. *et al.* (2015) Collembolan transcriptomes highlight molecular evolution of hexapods and provide clues on the adaptation to terrestrial life. *PLoS ONE* 10(6), e0130600. DOI: 10.1371/journal.pone.0130600.

Fire, A., Xu, S., Montgomery, M.K., Kostas, S.A., Driver, S.E. *et al.* (1998) Potent and specific genetic interference by double-stranded RNA in *Caenorhabditis elegans*. *Nature (London)* 391(6669), 806–811. DOI: 10.1038/35888.

Fischer, J.R., Zapata, F., Dubelman, S., Mueller, G.M., Jensen, P.D. *et al.* (2016) Characterizing a novel and sensitive method to measure dsRNA in soil. *Chemosphere* 161, 319–324. DOI: 10.1016/j.chemosphere.2016.07.014.

Fischer, J.R., Zapata, F., Dubelman, S., Mueller, G.M., Uffman, J.P. *et al.* (2017) Aquatic fate of a double-stranded RNA in a sediment-water system following an over-water application. *Environmental Toxicology and Chemistry* 36(3), 727–734. DOI: 10.1002/etc.3585.

Garbian, Y., Maori, E., Kalev, H., Shafir, S. and Sela, I. (2012) Bidirectional transfer of RNAi between honey bee and *Varroa destructor*: Varroa gene silencing reduces Varroa population. *PLoS Pathogens* 8(12), e1003035. DOI: 10.1371/journal.ppat.1003035.

Gillespie, M.A.K. and Wratten, S.D. (2017) The role of ecosystem disservices in pest management. In: Coll, M. and Wajnberg, E. (eds) *Environmental Pest Management: Challenges for Agronomists, Ecologists, Economists and Policymakers*. Wiley, Oxford, UK, pp. 175–194.

Grbić, M., Van Leeuwen, T., Clark, R.M., Rombauts, S., Rouzé, P. *et al.* (2011) The genome of *Tetranychus urticae* reveals herbivorous pest adaptations. *Nature* 479(7374), 487–492. DOI: 10.1038/nature10640.

Guan, R.-B., Li, H.-C., Fan, Y.-J., Hu, S.-R., Christiaens, O. *et al.* (2018) A nuclease specific to lepidopteran insects suppresses RNAi. *Journal of Biological Chemistry* 293(16), 6011–6021. DOI: 10.1074/jbc.RA117.001553.

Haller, S., Widmer, F., Siegfried, B.D., Zhuo, X. and Romeis, J. (2019) Responses of two ladybird beetle species (Coleoptera: Coccinellidae) to dietary RNAi. *Pest Management Science* 75(10), 2652–2662. DOI: 10.1002/ps.5370.

Jackson, A.L., Burchard, J., Schelter, J., Chau, B.N., Cleary, M. *et al.* (2006) Widespread siRNA 'off-target' transcript silencing mediated by seed region sequence complementarity. *RNA* 12(7), 1179–1187. DOI: 10.1261/rna.25706.

Joga, M.R., Zotti, M.J., Smagghe, G. and Christiaens, O. (2016) RNAi efficiency, systemic properties, and novel delivery methods for pest insect control: what we know so far. *Frontiers in Physiology* 7, 553. DOI: 10.3389/fphys.2016.00553.

Kaletta, T. and Hengartner, M.O. (2006) Finding function in novel targets: *C. elegans* as a model organism. *Nature Reviews Drug Discovery* 5, 387–399. DOI: 10.1038/nrd2031.

Krogh, P.H. (2021) Ecological risk assessment for soil invertebrate biodiversity and ecosystem services. In: Chaurasia, A., Hawksworth, D.L. and Pessoa de Miranda, M. (eds) *GMOs: Implications for Biodiversity Conservation and Ecological Processes*. Springer International.

Kurscheid, S., Lew-Tabor, A.E., Rodriguez Valle, M., Bruyeres, A.G., Doogan, V.J. *et al.* (2009) Evidence of a tick RNAi pathway by comparative genomics and reverse genetics screen of targets with known loss-of-function phenotypes in Drosophila. *BMC Molecular Biology* 10, 26. DOI: 10.1186/1471-2199-10-26.

Liu, J., Swevers, L., Iatrou, K., Huvenne, H. and Smagghe, G. (2012) *Bombyx mori* DNA/RNA non-specific nuclease: expression of isoforms in insect culture cells, subcellular localization and functional assays. *Journal of Insect Physiology* 58(8), 1166–1176. DOI: 10.1016/j.jinsphys.2012.05.016.

Lorenz, M.G. and Wackernagel, W. (1987) Adsorption of DNA to sand and variable degradation rates of adsorbed DNA. *Applied and Environmental Microbiology* 53(12), 2948–2952. DOI: 10.1128/AEM.53.12.2948-2952.1987.

Lundgren, J.G. and Duan, J.J. (2013) RNAi-based insecticidal crops: potential effects on nontarget species. *Bioscience* 63(8), 657–665. DOI: 10.1525/bio.2013.63.8.8.

Lundgren, J. and Jurat-Fuentes, J. (2012) Physiology and ecology of host defense against microbial invaders. In: Vega, F.E. and Kaya, H.K. (eds) *Insect Pathology*, 2nd edn. Academic Press, Amsterdam, pp. 461–480.

McEwan, D.L., Weisman, A.S. and Hunter, C.P. (2012) Uptake of extracellular double-stranded RNA by SID-2. *Molecular Cell* 47(5), 746–754. DOI: 10.1016/j.molcel.2012.07.014.

Mysore, K.S., Babitha, K.C. and Ramu, V.S. (2018) RNAi and microRNA technologies to combat plant insect pests. In: Emain, C. (ed.) *The Biology of Plant–Insect Interactions*. CRC Press, Boca Raton, Florida, pp. 150–177.

Niu, J., Smagghe, G., De Coninck, D.I.M., Van Nieuwerburgh, F., Deforce, D. *et al.* (2016) *In vivo* study of Dicer-2-mediated immune response of the small interfering RNA pathway upon systemic infections of virulent and avirulent viruses in *Bombus terrestris*. *Insect Biochemistry and Molecular Biology* 70, 127–137. DOI: 10.1016/j.ibmb.2015.12.006.

Niu, J., Shen, G., Christiaens, O., Smagghe, G., He, L. *et al.* (2018) Beyond insects: current status and achievements of RNA interference in mite pests and future perspectives. *Pest Management Science* 74(12), 2680–2687. DOI: 10.1002/ps.5071.

Nunes, F.M.F., Aleixo, A.C., Barchuk, A.R., Bomtorin, A.D., Grozinger, C.M. *et al.* (2013) Non-target effects of green fluorescent protein (GFP)-derived double-stranded RNA (dsRNA-GFP) used in honey bee RNA interference (RNAi) assays. *Insects* 4(1), 90–103. DOI: 10.3390/insects4010090.

OECD (1994) *OECD Guidelines for the Testing of Chemicals*. Organisation for Economic Cooperation and Development, Paris. Available at: www.oecd-ilibrary.org/environment/oecd-guidelines-for-the-testing-of-chemicals_72d77764-en (accessed 9 November 2020).

OECD (2016) *Test No. 232: Collembolan Reproduction Test in Soil. In: OECD Guidelines for the Testing of Chemicals, Section 2*. OECD Publishing, Paris.

Paget, E., Monrozier, L.J. and Simonet, P. (1992) Adsorption of DNA on clay minerals: protection against DNaseI and influence on gene transfer. *FEMS Microbiology Letters* 97(1-2), 31–39. DOI: 10.1111/j.1574-6968.1992.tb05435.x.

Pak, J. and Fire, A. (2007) Distinct populations of primary and secondary effectors during RNAi in *C. elegans. Science* 315(5809), 241–244. DOI: 10.1126/science.1132839.

Palm, C.J., Seidler, R.J., Schaller, D.L. and Donegan, K.K. (1996) Persistence in soil of transgenic plant produced *Bacillus thuringiensis* var. *kurstaki* δ-endotoxin. *Canadian Journal of Microbiology* 42(12), 1258–1262. DOI: 10.1139/m96-163.

Pan, H., Xu, L., Noland, J.E., Li, H., Siegfried, B.D. *et al.* (2016) Assessment of potential risks of dietary RNAi to a soil micro-arthropod, *Sinella curviseta* brook (Collembola: Entomobryidae). *Frontiers in Plant Science* 7, 1028. DOI: 10.3389/fpls.2016.01028.

Pan, H., Yang, X., Romeis, J., Siegfried, B.D. and Zhou, X. (2020) Dietary RNAi toxicity assay exhibits differential responses to ingested dsRNAs among lady beetles. *Pest Management Science* 76(11), 3606–3614. DOI: 10.1002/ps.5894.

Papadopoulou, N., Devos, Y., Álvarez-Alfageme, F., Lanzoni, A. and Waigmann, E. (2020) Risk assessment considerations for genetically modivied RNAi plants: EFSA's activities and perspective. *Frontiers in Plant Science* 11. DOI: 10.3389/fpls.2020.00445.

Parker, K.M. and Sander, M. (2017) Environmental fate of insecticidal plant-incorporated protectants from genetically modified crops: knowledge gaps and research opportunities. *Environmental Science & Technology* 51(21), 12049–12057. DOI: 10.1021/acs.est.7b03456.

Parker, K.M., Barragán Borrero, V., van Leeuwen, D.M., Lever, M.A., Mateescu, B. *et al.* (2019) Environmental fate of RNA interference pesticides: adsorption and degradation of double-stranded RNA molecules in agricultural soils. *Environmental Science & Technology* 53(6), 3027–3036. DOI: 10.1021/acs.est.8b05576.

Perry, J.N., Devos, Y., Arpaia, S., Bartsch, D., Gathmann, A. *et al.* (2010) The usefulness of a mathematical model of exposure for environmental risk assessment. *Proceedings of the Royal Society B* 278(1708), 982–984. DOI: 10.1098/rspb.2010.2667.

Pleasants, J.M., Hellmich, R.L., Dively, G.P., Sears, M.K., Stanley-Horn, D.E. *et al.* (2001) Corn pollen deposition on milkweeds in and near cornfields. *Proceedings of the National Academy of Sciences of the United States of America* 98(21), 11919–11924. DOI: 10.1073/pnas.211287498.

Prentice, K., Christiaens, O., Pertry, I., Bailey, A., Niblett, C. *et al.* (2017) RNAi-based gene silencing through dsRNA injection or ingestion against the African sweet potato weevil *Cylas puncticollis* (Coleoptera: Brentidae). *Pest Management Science* 73(1), 44–52. DOI: 10.1002/ps.4337.

Qiu, S., Adema, C.M. and Lane, T. (2005) A computational study of off-target effects of RNA interference. *Nucleic Acids Research* 33(6), 1834–1847. DOI: 10.1093/nar/gki324.

Romanowski, G., Lorenz, M.G. and Wackernagel, W. (1991) Adsorption of plasmid DNA to mineral surfaces and protection against DNase I. *Applied and Environmental Microbiology* 57(4), 1057–1061. DOI: 10.1128/AEM.57.4.1057-1061.1991.

Rosi-Marshall, E.J., Tank, J.L., Royer, T.V., Whiles, M.R., Evans-White, M. *et al.* (2007) Toxins in transgenic crop byproducts may affect headwater stream ecosystems. *Proceedings of the National Academy of Sciences of the United States of America* 104(41), 16204–16208. DOI: 10.1073/pnas.0707177104.

Santos, D., Mingels, L., Vogel, E., Wang, L., Christiaens, O. *et al.* (2019) Generation of virus- and dsRNA-derived siRNAs with species-dependent length in insects. *Viruses* 11(8), 738. DOI: 10.3390/v11080738.

Shukla, J.N., Kalsi, M., Sethi, A., Narva, K.E., Fishilevich, E. *et al.* (2016) Reduced stability and intracellular transport of dsRNA contribute to poor RNAi response in lepidopteran insects. *RNA Biology* 13(7), 656–669. DOI: 10.1080/15476286.2016.1191728.

Sijen, T., Fleenor, J., Simmer, F., Thijssen, K.L., Parrish, S. *et al.* (2001) On the role of RNA amplification in dsRNA-triggered gene silencing. *Cell* 107(4), 465–476. DOI: 10.1016/S0092-8674(01)00576-1.

Song, H., Zhang, J., Li, D., Cooper, A.M.W., Silver, K. *et al.* (2017) A double-stranded RNA degrading enzyme reduces the efficiency of oral RNA interference in migratory locust. *Insect Biochemistry and Molecular Biology* 86, 68–80. DOI: 10.1016/j.ibmb.2017.05.008.

Sun, Y., Sparks, C., Jones, H., Riley, M., Francis, F. *et al.* (2019) Silencing an essential gene involved in infestation and digestion in grain aphid through plant-mediated RNA interference generates aphid-resistant wheat plants. *Plant Biotechnology Journal* 17(5), 852–854. DOI: 10.1111/pbi.13067.

Taning, C.N.T., Christiaens, O., Berkvens, N., Casteels, H., Maes, M. *et al.* (2016) Oral RNAi to control *Drosophila suzukii*: laboratory testing against larval and adult stages. *Journal of Pest Science* 89(3), 803–814. DOI: 10.1007/s10340-016-0736-9.

Taning, C.N.T., Gui, S., De Schutter, K., Jahani, M., Castellanos, N.L. *et al.* (2020) A sequence complementarity-based approach for evaluating off-target transcript knockdown in *Bombus terrestris*, following ingestion of pest-specific dsRNA. *Journal of Pest Science* 13. DOI: 10.1007/s10340-020-01273-z.

Tank, J.L., Rosi-Marshall, E.J., Royer, T.V., Whiles, M.R., Griffiths, N.A. *et al.* (2010) Occurrence of maize detritus and a transgenic insecticidal protein (Cry1Ab) within the stream network of an agricultural landscape. *Proceedings of the National Academy of Sciences of the United States of America* 107(41), 17645–17650. DOI: 10.1073/pnas.1006925107.

Tian, L., Zeng, Y., Xie, W., Wu, Q., Wang, S. *et al.* (2019) Genome-wide identification and analysis of genes associated with RNA interference in *Bemisia tabaci*. *Pest Management Science* 75(11), 3005–3014. DOI: 10.1002/ps.5415.

Timmons, L. (2006) Delivery methods for RNA interference in *C. elegans*. In: Strange, K. (ed.) *C. elegans. Methods in Molecular Biology*. 351. Humana Press, Clifton, New Jersey, pp. 119–125.

US Environmental Protection Agency (2015) *Environmental Risk Assessment for a FIFRA Section 3 Limited Seed Increase Registration of DvSnf7 Double Stranded RNA (dsRNA) and Cry3Bb1 Bacillus Thuringiensis Insecticidal Protein as Expressed in MON 87411 Maize*. Washington, DC.

Whyard, S. (2018) Biological control of insects. *US Patent Application* 10(039), 287.

Whyard, S., Singh, A.D. and Wong, S. (2009) Ingested double-stranded RNAs can act as species-specific insecticides. *Insect Biochemistry and Molecular Biology* 39(11), 824–832. DOI: 10.1016/j.ibmb.2009.09.007.

Wynant, N., Santos, D., Verdonck, R., Spit, J., Van Wielendaele, P. *et al.* (2014) Identification, functional characterization and phylogenetic analysis of double stranded RNA degrading enzymes present in the gut of the desert locust, *Schistocerca gregaria*. *Insect Biochemistry and Molecular Biology* 46, 1–8. DOI: 10.1016/j.ibmb.2013.12.008.

Yoon, J.-S., Gurusamy, D. and Palli, S.R. (2017) Accumulation of dsRNA in endosomes contributes to inefficient RNA interference in the fall armyworm, *Spodoptera frugiperda*. *Insect Biochemistry and Molecular Biology* 90, 53–60. DOI: 10.1016/j.ibmb.2017.09.011.

Zhang, X., Zhang, J. and Zhu, K.Y. (2010) Chitosan/double-stranded RNA nanoparticle-mediated RNA interference to silence chitin synthase genes through larval feeding in the African malaria mosquito (*Anopheles gambiae*). *Insect Molecular Biology* 19(5), 683–693. DOI: 10.1111/j.1365-2583.2010.01029.x.

Zhang, K., Wei, J., Huff Hartz, K.E., Lydy, M.J., Moon, T.S. *et al.* (2020) Analysis of RNA interference (RNAi) biopesticides: double-stranded RNA (dsRNA) extraction from agricultural soils and quantification by RT-qPCR. *Environmental Science & Technology* 54(8), 4893–4902. DOI: 10.1021/acs.est.9b07781.

Zhou, X., Xu, F., Mao, H., Ji, J., Yin, M. *et al.* (2014) Nuclear RNAi contributes to the silencing of off-target genes and repetitive sequences in *Caenorhabditis elegans*. *Genetics* 197(1), 121–132. DOI: 10.1534/genetics.113.159780.

Zotti, M., Dos Santos, E.A., Cagliari, D., Christiaens, O., Taning, C.N.T. *et al.* (2018) RNA interference technology in crop protection against arthropod pests, pathogens and nematodes. *Pest Management Science* 74(6), 1239–1250. DOI: 10.1002/ps.4813.

Zwahlen, C., Hilbeck, A., Gugerli, P. and Nentwig, W. (2003) Degradation of the Cry1Ab protein within transgenic *Bacillus thuringiensis* corn tissue in the field. *Molecular Ecology* 12(3), 765–775. DOI: 10.1046/j.1365-294X.2003.01767.x.

13 Food and Feed Safety Assessment of RNAi Plants and Products

Hanspeter Naegeli[1]*, Gijs Kleter[2] and Antje Dietz-Pfeilstetter[3]
[1]University of Zürich, Institute of Veterinary Pharmacology and Toxicology, Zürich, Switzerland; [2]RIKILT Wageningen University & Research, Wageningen, The Netherlands; [3]Julius Kühn-Institut, Institute for Biosafety in Plant Biotechnology, Braunschweig, Germany

13.1 Introduction: Steps in the Risk Assessment

The risk assessment of genetically modified (GM) plants for food and feed use is based on a comparative approach (EFSA GMO Panel, 2011) where the composition as well as phenotypic and agronomic characteristics of the GM plant are compared with those of a conventional counterpart with a close genetic background and to additional non-GM comparator lines, which are assumed to have a history of safe use. Comparative risk assessment identifies effects intended by the genetic modification as well as possible unintended effects arising from transgene insertion into functional genome regions or from inadvertent impacts of the transgene product(s) on plant metabolic pathways. If differences and/or lack of equivalence between the GM plant and its comparator(s) above natural variation are identified, possible adverse effects on human and animal health have to be considered.

This type of hazard identification is the first step in the risk assessment of a GM plant. Intended and unintended differences in contrast to comparator(s) are then evaluated with respect to adverse health effects. This involves in

the first place toxicological and allergenicity assessment of newly expressed proteins (NEPs). In the case of RNA interference (RNAi) plants not expressing any new protein, these assessments are inapplicable. Instead, as the introduction of a gene silencing construct may cause silencing of 'off-target' genes, bioinformatics searches for 'off-target' sequences in the plant genome should be part of hazard characterization. If plant metabolic genes are silenced, unintended interferences with endogenous metabolic pathways are also possible and may cause alterations in metabolites and precursors of suppressed metabolic routes, justifying – on a case-by-case basis – analysis of specific RNAs or metabolites (EC, 2013).

An important aspect of risk assessment is the determination of exposure to the food and feed derived from GM plants, which involves identification of the population groups and animal species exposed as well as the extent of exposure. A starting point for the extent of exposure is the expression product of the introduced genetic modification, which is double-stranded RNA (dsRNA) in the case of RNAi plants. Its level, as well as levels of constituents altered as a result of the genetic modification, should be

*Corresponding author: hanspeter.naegeli@vetpharm.uzh.ch

© CAB International 2021. *RNAi for Plant Improvement and Protection*
(eds B. Mezzetti *et al.*)
DOI: 10.1079/9781789248890.0013

determined in plant parts used for food or feed. For estimating exposure from dietary intake or feed consumption, the stability of the dsRNA and derived small interfering RNAs (siRNAs) during storage and processing of plant material as well as during oral consumption need to be taken into account.

The final risk characterization of food and feed derived from GM RNAi plants is based on the results from the evaluation of potential adverse effects on human and animal health and from exposure assessment.

13.2 Potential Hazards of Food and Feed Derived from RNAi Plants

13.2.1 Adverse changes of plant metabolism

The principle of RNAi is used to modulate agricultural, phenotypic or compositional characteristics of plants by promoting gene silencing (Fire et al., 1998; Dykxhoorn et al., 2003; Frizzi and Huang, 2010). This strategy does not pose an inherent hazard to consumers or the environment, because it exploits gene regulation mechanisms that occur naturally in plants and animals. There are already manifold examples of spontaneously occurring RNAi-mediated genetic traits that were selected by conventional plant breeding. Such 'natural' gene silencing traits involve, for example, changes in the coat color of soybean seed (Tuteja et al., 2004) and maize stalk (Della Vedova et al., 2005), or mediate a low glutelin level in rice (Kusaba et al., 2003). So far, the RNAi strategy has been adopted in biotechnology-derived food crops to generate virus-resistant varieties (Sherman et al., 2015), optimize their agronomic performance (Ogita et al., 2003), provide pest and pathogen protection (Baum et al., 2007; Gordon and Waterhouse, 2007; Mao et al., 2007; Koch and Kogel, 2014), facilitate industrial processes like starch production (EFSA GMO Panel, 2006), improve the nutritional profile (Andersson et al., 2006; Regina et al., 2006) and reduce allergen levels (Le et al., 2006). Some prominently discussed RNAi-mediated products that have achieved market approval include the Flavr Savr™ tomato with reduced

polygalacturonase expression for delayed fruit softening (Redenbaugh et al., 1992), Plenish™ soybean with reduced omega-6 desaturase for high oleic acid content (EFSA GMO Panel, 2013) and Arctic™ apple with reduced polyphenol oxidase expression for delayed browning (Sherman et al., 2015; Waltz, 2015).

The intended decrease in the expression of a target gene may require safety considerations on a case-by-case basis. For example, the purpose of soybean with reduced omega-6 desaturase activity (also known as soybean 305423) is to obtain oil for frying and bakery with an increased content of heat-stable oleic acid (C18:1) at the expense of heat-labile polyunsaturated fatty acids (PUFAs) (C18:2 and C18:3). The consequences of this intended change in plant metabolism and composition need to be assessed to ascertain that the altered fatty acid profile does not impact on human and animal health in an exposure scenario where conventional vegetable oils are replaced with oil from soybean 305423 (EFSA GMO Panel, 2013). This assessment is focused on soybean oil and does not extend, for example, to soy milk and tofu for human consumption or defatted toasted meal for animal consumption, as such products are not expected to differ in composition between conventional soybean and soybean 305423, except for their altered fatty acid profile. However, the low contribution of fatty acids from these other soybean products to overall exposure is not anticipated to modify their nutritional impact. There is a detailed discussion of the risk assessment of an RNAi crop with altered metabolic composition in section 13.4.1, below.

Unintended effects caused by silencing genes in plant metabolic pathways

The engineering of plants with RNAi-mediated traits is achieved using the same transgenic techniques employed in the production of other GM crops grown and consumed widely today. In particular, RNAi plants are generated by inserting DNA sequences that lead to the expression of dsRNA or short hairpin RNA (shRNA), which are processed into siRNAs and microRNA (miRNA), respectively. These processed RNA molecules of 20–30 nt in length, collectively termed small RNA (sRNA), suppress gene expression at the transcriptional or post-transcriptional level, but are themselves designed to lack translation initiation signals and open reading frames necessary for protein biosynthesis (reviewed by Casacuberta *et al.*, 2015; Petrick *et al.*, 2013). Because the sRNA effectors are not translated to heterologous proteins, the risk assessment is focused on the direct and indirect consequences of the gene silencing machinery. The standard comparative analysis is well suited to detect possible unintended effects of RNAi-mediated silencing that may occur in addition to the intended gene expression changes.

An example of a potential indirect effect of RNAi-mediated silencing became apparent with the compositional analysis of the aforementioned high-oleic acid soybean 305423. In fact, the comparison between soybean 305423 and its non-GM (conventional) counterpart 'Jack' confirmed the expected change in fatty acid composition (increased levels of oleic acid at the expense of PUFAs), but also revealed an unexpected increase in the level of odd-chain fatty acids heptadecanoic acid (C17:0), heptadecenoic acid (C17:1) and nonadecenoic acid (C19:1). It is not known whether this effect results from off-target gene silencing (see below), from the manipulation of fatty acid synthesis pathways, from another unidentified response to the genetic modification, or as a consequence of either the simultaneous expression of a transgenic acetolactate synthase (ALS) enzyme (conferring herbicide tolerance) or the genetic background of the recipient soybean variety. In any case, a nutritional assessment came to the conclusion that the slight changes observed in the concentration of odd-chain fatty acids would not constitute a health hazard for humans and animals (EFSA GMO Panel, 2013).

Unintended effects caused by off-target gene suppression

In addition to the intended effects induced by expression of the non-coding RNA, unintentional changes may occur in the plant by suppression of genes that were not foreseen as RNAi targets. RNAi-mediated silencing is hybridization-dependent and therefore takes place in a sequence-specific manner. Nevertheless, suppression of genes with less than perfect sequence complementarity is possible (Senthil-Kumar

and Mysore, 2011). In some cases, indications for such off-target effects may come from the screen for agronomic performance and phenotypic characteristics or from the compositional analysis (see section 13.2.5, below), as changes of gene expression may impact on one or more of these routinely measured parameters and analytes. Genome-wide bioinformatics studies retrieving transcripts that match the newly expressed sRNA sequences would potentially indicate possible off-target effects (see section 13.2.3, below, for appropriate bioinformatics tools). However, the genomes of typical crops are at best only partially sequenced and known reference genomes do not take into account the sequence variability occurring between varieties (Ramon *et al.*, 2014; Casacuberta *et al.*, 2015). Despite these limitations of bioinformatics-based predictions, whole-genome homology searches may nonetheless reveal unintended silencing targets.

The EFSA GMO Panel acknowledged the limitations of bioinformatics searches for possible off-targets of sRNA produced by GM plants. A predictive strategy is nevertheless possible, due to the fact that plant miRNAs are usually perfectly or nearly perfectly complementary to their target transcripts (Pačes *et al.*, 2017). Thus, a set of parameters may allow for the prediction of RNAi off-targets in plants, whereas in humans and animals the extent of complementarity between the sRNA molecules and their targets is more flexible, thus preventing sufficiently reliable predictions (Pinzón *et al.*, 2017). Besides the abundance of each sRNA produced, the degree and position of base pairing between the sRNA and the target mRNA are the primary factors determining the efficiency of silencing (Rhoades *et al.*, 2002; Allen *et al.*, 2005; Pasquinelli, 2012; Liu *et al.*, 2014). Based on the current knowledge gained from the target specificity of natural miRNAs, the EFSA GMO Panel described in Annex II of the minutes of its 118th Plenary meeting (EFSA GMO Panel, 2017) a practical approach to identify sequences with potential off-target silencing. This procedure considers all 21 nt sRNA sequences that derive from a given dsRNA precursor and comply with the following rules:

- No more than 4 base mismatches with no gap or 3 mismatches and one gap in the alignment between the 21-mer sRNA sequence and a potential target mRNA transcript, whereby each G:U base mispair counts as half a mismatch.

- Only one gap can be present in the sequence alignment between the 21-mer sRNA sequence and a potential target transcript, and this single gap cannot be longer than one nucleotide.
- The sequence alignment should not reveal any mismatches or gap at position 10/11 of the sRNA sequence.
- The sequence alignment should also not reveal more than two mismatches (or no more than one mismatch and one gap) in the first 12 nucleotides from the 5′ end of the sRNA sequence.
- The minimum free energy of the imperfect duplex of the sRNA sequence with a potential target, divided by the minimum free energy of the perfect complement, should be > 0.75.

The ensuing risk assessment of potential off-target silencing in the plant should consider the abundance and the number of different sRNAs showing relevant similarity to the same transcript, as the potential for gene repression increases with multiple sRNA sequences being able to bind to the same mRNA molecule (Hannus *et al.*, 2014). Depending on the nature and function of the potential off-targets, the safety assessment may require extra studies in addition to the standardized agronomic/phenotypic characterization and compositional analysis.

13.2.2 Mechanisms and potential for non-target gene silencing in humans and livestock, including gut microbiome

Mammals have an RNA silencing machinery, which is distinct from that of plants and other animal orders. While in plants there is a complex RNAi system with different types of siRNAs and Dicer proteins and a distinct miRNA pathway, mammals have a single set of Dicer and Argonaute (AGO) proteins for both miRNA and siRNA pathways (Pačes *et al.*, 2017). This implies that in mammals siRNAs can function

in the same way as miRNA, i.e. bind to mRNAs depending on homologies to the 'seed region' which comprises nucleotides 2–8 from the 5′-end of the miRNA (Brennecke *et al.*, 2005) and therefore have less strict target specificity than siRNAs in plants. In fact, in mammals sRNAs that are perfectly complementary to a target mRNA sequence are loaded into an AGO2 RNA-induced silencing complex (RISC) guiding target RNA cleavage, while siRNAs and miRNAs with minimum seed region homology are loaded on all four mammalian AGO proteins, resulting in translational inhibition (Meister *et al.*, 2004; Gebert and MacRae, 2019). Lower requirements for sequence complementarities between miRNAs and mammalian mRNA make predictions of putative target sequences more difficult. As there is no distinct siRNA pathway in mammals, efficient induction of RNAi by long dsRNA, which has to be processed first into active siRNAs, is limited by poor Dicer activity in most mammalian cells (Nejepinska *et al.*, 2012; Flemr *et al.*, 2013; Pačes *et al.*, 2017).

Another specific feature of plant siRNAs and miRNAs which distinguishes them from siRNAs and miRNAs in mammals is their 3′-terminal methylation at the 2′-hydroxyl group (Li *et al.*, 2005; Yu *et al.*, 2005). 3′-terminal methylation probably protects small RNAs from degradation (Li *et al.*, 2005; Ren *et al.*, 2014) and may promote recognition by plant Argonaute proteins in RISC (Yu *et al.*, 2005). In contrast, mammalian AGO proteins preferably bind to non-methylated miRNAs (Tian *et al.*, 2011). On the other hand, it was shown by Ma *et al.* (2004) that 2′-OH methylation only moderately decreased the binding affinity of siRNA for the PAZ domain of a human AGO protein, while binding was heavily reduced by most other 2′-OH modifications at the 3′-terminal nucleotide. In line with this, Chau and Lee (2007) found no obvious effect of 2′-OH methylation on the efficiency of silencing in mammalian cells. However, these authors also showed that siRNAs derived from a plant hairpin transgene and extracted from transgenic plants were not effective for gene silencing in mammalian cells. Among other things, they attributed this lack of cross-species function to a putative plant-specific siRNA modification.

These molecular mechanisms indicate that there is no evidence that plant-produced dsRNA and siRNAs are functional in mammalian cells. Another limiting factor is the high number of miRNAs required to exert an effect on gene expression (Brown *et al.*, 2007). The expected unfavorable stoichiometry between exogenous small RNAs and mammalian mRNA targets will therefore further restrict gene silencing effects of dietary siRNAs in humans and livestock. In this context it also has to be mentioned that, in contrast to plants, fungi and nematodes, mammalian genomes do not possess an RNA-dependent polymerase (RdRP) homologue (Stein *et al.*, 2003; Maida and Masutomi, 2011), implying that there is no amplification of ingested siRNAs and that each exogenous siRNA effector molecule would have to be delivered with the diet. Nevertheless, there is still some controversy about the bioavailability of relevant amounts of functional exogenous, plant-derived miRNAs in mammalian plasma and tissues and their possible effects on endogenous gene expression (Zhang *et al.*, 2012b; Dickinson *et al.*, 2013; Witwer *et al.*, 2013; Pačes *et al.*, 2017) (see sections 13.3.2 and 13.3.3, below).

If siRNAs from ingested food or feed stay intact after entering the digestive tract of humans and livestock, they may have effects on gut microbiota. Although RNA taken up by microorganisms is generally degraded and used for bacterial nutrition, there is some evidence that faecal miRNAs derived from mammalian gut epithelial cells penetrate gut bacteria and co-localize with bacterial nucleic acids (Liu *et al.*, 2016). These authors showed that some of these miRNAs can regulate bacterial gene expression and thereby affect growth of certain bacterial species. Effects on gene expression in microorganisms encompassed decreases, as well as enhancements of transcripts, and were obviously distinct from RNA interference in eukaryotic organisms. Prokaryotes do not have an intrinsic RNAi machinery, but they produce non-coding sRNAs of around 100 nt that can up- or down-regulate mRNA stability and translation by base pairing to target mRNAs (Mayoral *et al.*, 2014; Wagner and Romby, 2015). Although stem-loop structures similar to eukaryotic precursor miRNAs have been detected for sRNAs from *Wolbachia*-infected insect cells and although sR-NAs from these bacteria were shown to regulate expression of *Wolbachia* genes as well as expression of host genes (Mayoral *et al.*, 2014), so far

there is no evidence that plant-derived dietary miRNAs have an effect on the gut microbiome.

13.2.3 Bioinformatics tools for prediction of off-target sequences of interfering RNA

Bioinformatics tools are available that may help identify potential 'off-target' binding sites in the transcriptome of the recipient plant. However, these have not been specifically developed for the purpose of safety assessment of GM crops. Algorithms searching for off-target effects as part of the optimization of design of siRNA/miRNA are offered as a single tool or, frequently, as part of a package. These include both accessible online web applications and open source, stand-alone software. Several of these tools predict which genes' mRNA transcripts will be targeted by sRNAs. The sequences of the latter can be entered by the user as such or as part of a larger cDNA sequence, often in FASTA format or with reference to a database accession. Such predictions assist in the selection and design of artificial siRNA or miRNA molecules that effectively bind a target with low off-target effects binding to mRNA transcripts of other genes (Lukasik and Zielenkiewicz, 2019). This also applies to the retrospective identification of targets of small RNAs that have been added to cells in massive functional screening experiments, i.e. 'miRNA screening', and have shown an effect (Lemons *et al.*, 2013). The predicted targets can then be compared and confirmed with parallel data on downregulated genes from, for example, transcriptomics. Target-identifying tools can also be used for annotation, namely for genes encoding sRNA precursors in genomics data, or for sRNAs that have been identified in transcriptomics studies. Another common purpose includes, amongst others, investigation of isoforms (e.g. single-nucleotide polymorphisms (SNPs)) in naturally occurring sRNAs (Lukasik and Zielenkiewicz, 2019). A large number of such applications have been brought together in portals such as Tools4Mirs (https://tools4mirs. org/, accessed 30 March 2020), which harbours 170 tools, including 59 software items and ten websites that can be used for target prediction. The user could, for example, use multiple tools for target prediction to reach a 'consensus' outcome on the likeliest targets.

A basic approach to identify possible 'off-target' genes for the siRNA/miRNA would be to search for sequence homologies between the cDNA sequence of interest and its counterparts from RNA transcriptome databases. Using BLASTn, for instance, the query sequence could be aligned with sequences from NCBI's RefSeq collections of mRNA transcripts from various organisms. The latter could contain data for the recipient plant species that has been genetically modified, or for humans, animals and other species representing environmental non-target organisms. The alignments in the BLAST outputs should then be judged for compliance with certain criteria that are known to affect the *in vivo* alignment and binding of mi/siRNA to the target mRNA within the RISC complex. For example, in animals a perfect match of the seed sequence of 6–8 bp at the miRNA molecule's 5′ end is required. Some 'wobbly' mismatches are tolerated in this seed sequence but they decrease efficacy. Mismatches are also tolerated to a limited extent in the guide part of the miRNA molecule. These reportedly prevent degradation by Slicer but still block translation of the bound mRNA. Other factors include conserved residues and guanine-cytasine (GC) contents, amongst others, based on experience gained with miRNAs for certain species. Such factors affecting the efficacy of target binding and inhibition of gene expression are commonly automated as part of the specialized algorithms. These other factors are also relevant for the purpose of risk assessment, given that the seed sequence alone is relatively small (starting at 6 nt), which would easily render hundreds of genomic sequences that could be recognized but still remain without any major impact on gene expression.

Mainstream target prediction tool websites that specifically also focus on RNA targeting in plants include, for example, psRNATarget (Dai and Zhao, 2011) and TAPIR (Bonnet *et al.*, 2010). On the psRNATarget website (http://plantgrn.noble.org/psRNATarget/analysis#, accessed 30 March 2020) users can enter the sequences of either the sRNA or target RNA and select various variables, such as the seed region (i.e. by default nt 2–13 being recognized as critical in plants), the penalty for mismatches and opening gaps in the seed and other regions, and whether or not bulges or caps should be allowed

in the structure of the miRNA–mRNA complex. The outputs thus list the various sRNAs or target genes, show the alignments with the matching parts of the sRNA and mRNA molecules compliant with the criteria and indicate whether the complex probably will be cleaved or inhibit translation. The algorithm underlying psRNATarget not only takes into account Crick–Watson base pairing using scoring matrices for matches, mismatches and gaps, but also features optionally an energy calculation for the unwinding of the adjacent RNA parts upon binding, which correlates with accessibility (Dai and Zhao, 2011).

Similarly, the TAPIR website offers a comparable search feature, yet both the miRNA and target sequences have to be entered at the same time. Variables that can be modified by the user include score and free energy, i.e. the binding energy of the mismatched sequences as compared with that of a perfectly matching pair.

Another tool that more specifically focuses on potential off-targets of small RNA in a wide array of organisms is pssRNAit (http://plantgrn.noble.org/pssRNAit/, accessed 30 March 2020). WMD3, a program for designing artificial miRNAs, also has a feature to BLAST a query sequence against DNA data sets from a large collection of plants (http://wmd3.weigelworld.org/cgi-bin/webapp.cgi?page=Blast;project=stdwmd, accessed 30 March 2020).

13.2.4 Possible non-specific effects of dsRNA and siRNA in mammals

Molecules of dsRNA as well as the derived siRNAs, which constitute the molecular effectors of gene silencing in RNAi plants, occur naturally in food or feed and, therefore, constitute a ubiquitous component of the diet for both humans and animals. Systemic exposure following consumption of plants containing dsRNA or siRNA is limited in higher organisms by extensive denaturation and degradation of ingested RNA and by biological barriers preventing their cellular uptake. Inflammatory responses have been observed following systemic administration of siRNA in animal models (Judge and MacLachlan, 2008; Robbins *et al.*, 2009). Such responses of the innate immune system are mediated by receptors that recognize nucleic acids such as Toll-like receptors (TLRs) or the RNA-binding protein kinase PKR. However, the observed inflammatory response might also be due to the delivery system or to chemical modifications introduced into the nucleic acid backbone to increase stability, rather than being elicited by the presence of native siRNA molecules (Heidel *et al.*, 2004; Ma *et al.*, 2005; Petrick *et al.*, 2013). In any case, inflammatory reactions upon oral exposure to siRNA or other nucleic acids are not expected.

13.2.5 Comparison of data requirements for safety assessment of food and feed from RNAi plants and from plants expressing recombinant proteins

A universally accepted strategy is in place for the evaluation of the safety of GM crops and their products used as food or feed. This general approach, described in relevant documents issued by international organizations (Codex, 2003; ILSI, 2004; EFSA GMO Panel, 2011), analyses both the safety of intended effects introduced by the genetic modification and possible unintended effects resulting inadvertently from the new trait or from the genetic transformation process. A common cornerstone of all GM plant safety evaluation guidelines is the comparative assessment. This entails an extensive analysis comparing the GM crop with a genetically close, conventional counterpart with a history of safe use. Common, recurrent features of this analysis include the following.

- A **molecular characterization** of the inserted DNA including sequences introduced, their copy number, orientation, possible rearrangements, etc., as well as their expression (e.g. mRNA or NEPs) in different plant tissues and in different developmental stages, and stability of inheritance. For RNAi-modified crops, particularly relevant is the expression of the RNA encoded by the inserted genes, as well as the mRNA of the genes targeted for silencing, whilst no new proteins are expected to be expressed. Horizontal gene transfer, which also has to be assessed, may only be relevant if genes

are introduced that convey a selective advantage to the recipient.

- An extensive **comparative compositional analysis** of the GM crop versus a conventional counterpart, grown in various locations representative of the conditions under which the crop is intended to be produced commercially. This quantitative analysis entails a wide range of macronutrients, micronutrients (e.g. vitamins, minerals), antinutrients and toxins, which are characteristic and relevant for the crop species and which are listed in consensus documents developed under the international frame of the Organization for Economic Co-operation and Development (for example, for maize composition see OECD, 2002). Despite considerable natural variability in nutrient/antinutrient content, the compositional analysis provides an indicator of possible unintended effects resulting from the genetic modification and, therefore, is also applicable to the risk assessment of RNAi plants. The outcome of this analysis should reflect the intended changes if the introduced trait is intended to affect plant composition, for example its fatty acid content (see section 13.4.1, below, for a specific case study).

- **Phenotypic and agronomic characteristics** of the crop are analysed in a similar way and this may reveal possible unintended changes caused by the genetic modification, as well as providing important data for the environmental risk assessment. The determination of agronomic/phenotypic characteristics may also help to confirm specific traits that are intended to improve plant growth, grain yield or protection from biotic and abiotic stresses (see section 13.4.2, below). Thus, this part of the comparative evaluation is also pertinent to the risk assessment of RNAi plants.

Based on these comparative tests, it can be decided if there is sufficient information to conclude the risk assessment or to proceed with assessment of additional data. Usually the items that are also addressed during risk assessment of GM crops include: (1) potential toxicity and allergenicity; and (ii) nutritional impact (Codex, 2003; EFSA GMO Panel, 2011; EC, 2013).

For **potential toxicity and allergenicity** of NEPs and other compounds introduced or whose levels have been altered by the genetic modification, common items of the 'weight of evidence' approach include a bioinformatics-based comparison of the amino acid sequence of NEPs with those of known toxic and allergenic proteins. This is because all known food allergens are proteins, which raises the question as to whether any NEP could indeed become an allergen (or a toxin). The query sequences also include those that are hypothetically formed from open reading frames (ORFs) present in the insert and crossing its borders with the host's genomic DNA. This latter comparison would also still be applicable for hypothetical peptides encoded by the ORFs within the inserted construct encoding silencing RNA.

Other commonly assessed factors that are specific to proteins and not to non-coding RNA include:

- Information on the gene donor: is there a known history of toxicity or allergenicity, i.e. the propensity to cause toxic or allergic reactions? Is it known if these properties are linked with the product of the gene used?
- Resistance of NEPs to *in vitro* degradation by the digestive proteolytic enzyme pepsin, which indicates a greater likelihood of *in vivo* passage of the protein through the gastrointestinal tract, and possibility to cause toxicity or interact with the immune system in the consumer or animal.
- *In vivo* toxicity trials in laboratory animals with the NEPs or any other compound altered or introduced by the genetic modification and administered to the animals in purified form.

Toxicity testing with whole GM food and feed products in experimental animals should only be performed as a last resort given the inherent insensitivities and other practical and ethical limitations, and with a clear hypothesis of potential adverse effects. Nevertheless, a unique feature of the European legislation is the mandatory requirement for 90-day feeding studies, which need to be provided even in the absence of any hazard or risk hypothesis (Devos *et al.*, 2016). For food allergenicity testing, there are no validated animal models yet.

The **nutritional impact** of any intended and unintended changes in the nutrient profile of the host crop caused by the genetic modification may be particularly relevant if there are substantive changes in nutrient levels (e.g. beyond background variability) and if the particular crop is known to be a relevant source of the specific nutrient. In such cases, it should be estimated to what extent these changes will affect the intake of the particular nutrient by consumers and domestic animals. To estimate the intake, the quantitative data on the altered nutrient levels from the compositional analyses need to be combined with data on the intake of the crop and derived product, such as from the EFSA Food Consumption Database. In some rare cases, it may be necessary to extend the data with new studies in representative animal models.

In summary, the paradigm of comparative assessment is well suited for the safety evaluation of RNAi plants. Unlike the vast majority of GM crops currently on the market, which have been designed to express heterologous proteins (so-called NEPs) that confer a desired phenotype like herbicide tolerance, pest protection or increased yield, RNAi-mediated traits involve the expression of non-coding RNA without NEP biosynthesis. Therefore, NEP-related aspects of the safety assessment process are not applicable. This includes the search for homology of NEPs with known protein toxins and allergens, their digestibility (as allergenic or toxic proteins may be refractory to degradation by digestive enzymes) and, in the absence of a proven history of safe use, rodent studies to test the potential oral toxicity of NEPs. However, in order to predict unintended effects from off-target gene silencing, EU Regulation No. 503/2013 requests that for the authorization of RNAi plants in the EU an *in silico* bioinformatics analysis is carried out to identify potential off-target genes in the plant genome (EC, 2013).

Any additional studies considering intended or unintended effects of sRNA should be considered as needed on a case-by-case basis. For example, the use of RNAi as an insecticidal tool raises the question of whether sRNAs that are lethal to insects could also harm humans and farm or companion animals. There is a detailed discussion of the risk assessment of an RNAi crop conferring insecticidal properties in section 13.4.2, below.

13.3 Exposure Assessment

13.3.1 Expression level of dsRNA and siRNAs in plants

The first determinant for exposure of humans and farm animals to siRNAs from consumption of plant-derived food or feed is the level of dsRNA expression in the respective GM RNAi crop which provides the basis for a maximal estimate for exposure, assuming a worst-case scenario with no barriers to bioavailability. The expression level of transgenes in different plant tissues is dependent on the regulatory sequences, but may also be affected by environmental changes and the age of the plant (Meyer *et al.*, 1992; van der Hoeven *et al.*, 1994). Transgenes including dsRNA constructs introduced into GM plants are often under control of a strong constitutive promoter like the cauliflower mosaic virus 35S promoter (P35S), accounting for a high constitutive expression in all plant tissues. Nuclear expressed dsRNA, however, is to a large part processed in plant cells into siRNAs by Dicer-like (DCL) proteins from the plant RNAi machinery (Chau and Lee, 2007; Frizzi and Huang, 2010; Zhang *et al.*, 2015). Thus, when quantifying specific RNA levels relevant for dietary intake, both dsRNA and siRNAs have to be taken into account. Chau and Lee (2007) found that in transgenic tobacco plants expressing a hairpin construct under P35S control, the hairpin-specific siRNA level was about 50 ng/g leaf tissue. Petrick *et al.* (2013) calculated a daily dietary exposure to transgene-derived siRNA from a putative RNAi soybean product of 45 µg/kg for adults, assuming a transgene-derived siRNA rate of 1.5% of the total RNA and a maximum amount of total RNA in grain tissue of 986.6 µg RNA/g. However, this transgene-specific siRNA percentage and the resulting exposure estimate seems to be too high. Ivashuta *et al.* (2009) reported endogenous small RNAs (21–24 nt) as a whole to be present at levels of maximally 1.61 µg/g soybean grain, which corresponds to clearly less than 1.5% of total plant RNA, with similar levels found in conventional maize and rice grain. Using a validated quantification assay based on the QuantiGene Plex 2.0 (Affymetrix)

technology (Armstrong *et al.*, 2013), Bachman *et al.* (2016) detected DvSnf7 dsRNA in transgenic insecticidal maize MON 87411 at levels up to 0.17 ng/g dry weight in grain, while the mean level in leaf was 14.4 ng/g fresh weight. There was no information, though, on the proportions of long dsRNA originating from the transcript versus small siRNAs resulting from DCL processing.

For RNAi plants conferring resistance against certain insects via host-induced gene silencing (HIGS), for example MON 87411 expressing dsRNA targeting an essential insect transcript, it has to be considered that dsRNAs require a certain minimal length in order to be taken up efficiently and become biologically active in insects (Bolognesi *et al.*, 2012). This implies that processing of dsRNA into siRNAs, which has been shown to occur readily for nuclear expressed dsRNAs, needs to be kept at a minimum. This is especially important for certain insect groups like Lepidoptera, which are less susceptible to RNAi (Terenius *et al.*, 2011) and therefore require delivery of a very large amount of dsRNA to be efficiently targeted. One way to prevent the rapid turnover of dsRNAs in plants is the construction of transplastomic plants where dsRNA accumulates in the chloroplasts, thereby being protected from Dicer (Zhang *et al.*, 2017). In contrast to nuclear transformants, no detectable levels of siRNAs were found in transplastomic *Nicotiana benthamiana* plants, but only large amounts of the unspliced hairpin RNA (Bally *et al.*, 2016). Similar results were obtained by Zhang *et al.* (2015) for tobacco and potato lines expressing insect gene-specific dsRNAs from the plastid genome. Moreover, differences between plant tissues were reported. While in transplastomic potatoes specific dsRNA transcripts were below the detection limit in tubers, levels of insect gene-specific dsRNAs up to 0.4 % of the total cellular RNA accumulated in leaves (Zhang *et al.*, 2015). Assuming a total RNA amount of around 500 µg/g leaf tissue, this corresponds to 2 µg of dsRNA, which is about 150-fold higher compared with the amount of DvSnf7 dsRNA reported by Bachman *et al.* (2016). Transplastomic crop plants are thus distinct from nuclear transformants due to the restriction of substantial dsRNA production to chloroplast-containing photosynthetic tissues and due to deviant amounts of dsRNA and siRNAs.

13.3.2   Oral exposure from dietary intake

All foods and feeds contain naturally occurring coding and non-coding RNA, but animal tissues generally have a higher RNA concentration than plants (Jonas *et al.*, 2001). The overall content of RNA in plant-derived food and feed is in the order of 1 mg/g tissue, but up to 95% by weight of this total amount consists of highly abundant transfer RNA (tRNA), ribosomal RNA (rRNA) and mRNA. As outlined above, siRNA and miRNA make up less than 5% of the RNA content of plant tissues. Although present at such minor levels, siRNA and miRNA sequences found in plant tissues (for example, cereal or soybean seeds) display a high similarity or even identity to genomic regions of humans and livestock animals (Lassek and Montag, 1990; Heisel *et al.*, 2008; Ivashuta *et al.*, 2009). This observation opens the possibility that dietary sRNA, of natural occurrence or inserted into RNAi plants, may elicit biological responses in humans or animals. It should be noted, however, that even with the usually intended overexpression of the transgene-derived sRNA in RNAi plants, this additional sRNA represents only a very minute fraction of the total dietary RNA occurring in food and feed.

Many lines of evidence demonstrate that the mammalian digestive tract provides an extremely effective barrier to the local or systemic uptake of exogenous RNA molecules, which are inherently unstable in their natural form. First, the degradation of ingested RNA begins in saliva, which is a rich source of ribonucleases (Bardoń and Shugar, 1980; Park *et al.*, 2006). Secondly, the harsh milieu in the stomach with low pH promotes further RNA degradation as well as depurination (Loretz *et al.*, 2006; O'Neill *et al.*, 2011). The dominant gastric enzyme pepsin, which was thought to be protein-specific, has been shown to digest effectively nucleic acids including RNA (Liu *et al.*, 2015). Thirdly, pancreatic nucleases, phosphodiesterases and nucleoside phosphorylases, secreted into the intestinal lumen, degrade ingested RNA into

oligonucleotides, nucleotides and free bases (Jain, 2008; O'Neill *et al.*, 2011). Also, the intestinal epithelium, like any other cellular membrane, presents a physical barrier to hydrophilic compounds like nucleic acids (Khatsenko *et al.*, 2000). Any RNA that may bypass cellular membranes by transport into intestinal cells through endocytosis will be targeted to endosomal vesicles, sequestered into lysosomes and thereby degraded by lysosomal nucleases (Gilmore *et al.*, 2004). Thus, the systemic absorption of orally ingested RNA is negligible. For a 20mer DNA oligonucleotide, constructed with stabilizing phosphorothionate linkages to prevent degradation in the gastrointestinal tract, oral bioavailability in rats was at best 0.3% after gavage administration (Nicklin *et al.*, 1998). Kendal D. Hirschi and colleagues reported on an unusual sRNA from herbs, flowers and vegetables with a comparably high stability in gastrointestinal fluids (Yang *et al.*, 2016, 2018). This atypical sRNA of 20 nt, denoted as MIR2911 although it is a product of 26S rRNA breakdown, arises abundantly from ribosome degradation in macerated plant tissues. When tested in cabbage extracts, around 0.1% of input MIR2911 survived a 60 min *in vitro* incubation in gastrointestinal fluid, whereas for comparison MIR168 (see below) was digested under identical conditions around 1000 times more efficiently. MIR2911 was reported to occur at femtomolar concentrations in the plasma of rodents fed vegetable-enriched diets and was found in human plasma (Yang *et al.*, 2015a, Yang *et al.*, 2015b), where it appears to be stabilized against degradation and elimination by some host factors (Yang *et al.*, 2016, 2017, 2018). However, a follow-up report considered that MIR2911 was probably misidentified in human plasma as a plant-derived sRNA, but instead matches human genome sequences (Witwer, 2018).

Other authors found only trace levels at best of exogenous dietary miRNA molecules in the plasma of mice or humans (Chen *et al.*, 2013; Snow *et al.*, 2013; Dickinson *et al.*, 2013; Witwer *et al.*, 2013; Huang *et al.*, 2018), confirming the generally ineffective uptake and transfer of miRNA from food or feed to recipient organisms. A major problem is the use of exceptionally sensitive methods allowing for the detection of few nucleic acid molecules, which gives rise to false positive results due to non-specific amplification or sample contamination (Zhang *et al.*, 2012a; Tosar *et al.*, 2014). A survey of publicly available sequencing data sets found foreign miRNA in human body fluids and tissues, although at low abundance. Intriguingly, there is no enrichment of foreign RNA in human tissues that are most directly exposed to dietary intake, like liver, and there is no depletion of foreign sequences in compartments that are comparably well separated from the bloodstream, like for example the brain. The majority of foreign miRNA detected in this survey originates from rodents, which are common laboratory animals but do not contribute to human nutrition. It was, therefore, concluded that the apparent detection of foreign sRNA sequences in mammalian/human body fluids or tissue results from technical artifacts or misidentification (Kang *et al.*, 2017; Witwer, 2018).

The detection of plant-derived or other dietary sRNA entering the bloodstream of animals or humans (see for example Zhang *et al.*, 2012b; Yuan *et al.*, 2016) is prone to artifacts and fails independent experimental reproduction, and the general consensus is that only a very small fraction, if any, of ingested sRNA will be absorbed into the circulation. Additionally, minor traces of sRNA that might be absorbed into the blood are not spared from degradation and excretion. For siRNA molecules, a rapid breakdown in human plasma has been described with nearly 75% degradation within 2 min of incubation (Layzer *et al.*, 2004). Samples of siRNA injected intravenously into mice exhibited a short half-life in the range of only a few minutes and were subject to rapid hepatic and renal clearance (Vaishnaw *et al.*, 2010; Christensen *et al.*, 2013). It can be concluded from the above findings that plant sRNA molecules never reach sufficiently high concentrations and stability to exert biologically relevant effects in mammals and humans. In the unlikely event that traces of ingested RNA molecules are absorbed from the gastrointestinal tract, not degraded within the cardiovascular system and not readily eliminated through the liver or kidneys, these remaining RNA molecules, in the absence of any delivery vehicle and amplification mechanism, would not be able to cross lipid membranes, escape lysosomal degradation and, hence, reach the cytoplasm of cells (Sioud, 2005; Manjunath and Dykxhoorn, 2010). Taken together, the instability in biological fluids and matrices, in combination with biological barriers, reduces the likelihood that ingested sRNA will display local or systemic biological activities in mammals and there is currently no reason to believe that this conclusion may not also apply to other vertebrates, including birds and fish (EFSA GMO Panel, 2018). There is also no basis for the hypothesis that engineered sRNA in GM feed and foods may develop different nutritional properties than the background of natural sRNAs already present in all GM and conventional crop.

13.3.3 Likelihood of transfer of dsRNA or siRNA from plant to mammalian cells

Mammalian cells do not efficiently take up dsRNA or sRNA. The genetic basis for RNA uptake mechanisms has been investigated in detail in the nematode *Caenorhabditis elegans*. RNAi-mediated gene silencing is induced by soaking worms in siRNA-containing solutions (Tabara *et al.*, 1998; Maeda *et al.*, 2001) or by feeding them with bacteria that express dsRNA (Timmons *et al.*, 2001; Newmark *et al.*, 2003). A straightforward RNA uptake in *C. elegans* is mediated by transmembrane protein channels like SID-1, SID-2 and SID-5 (SID, systemic RNA interference-deficient) that promote RNA endocytosis, transfer of RNA into the cytoplasm of cells and RNA spread from cell to cell (Winston *et al.*, 2002, 2007; Feinberg and Hunter, 2003; Jose and Hunter, 2007; McEwan *et al.*, 2012). Two mammalian proteins, referred to as SIDT1 and SIDT2 (SID transmembrane family member 1 and 2), have been annotated as homologues of the RNA transporter SID-1. However, these mammalian proteins have more similarity with the *C. elegans* cholesterol uptake protein CHUP-1 than with SID-1. Also, SIDT1 and SIDT2 contain putative cholesterol-binding motifs known as cholesterol recognition/interaction amino acid consensus (CRAC) domains. Accordingly, the expression of SIDT1 and SIDT2 in human cells in culture demonstrated that they function as transmembrane cholesterol transporters, but are unable to mediate the intracellular uptake of dsRNA or miRNA (Méndez-Acevedo *et al.*, 2017). Another aspect of RNAi observed in *C. elegans* is the amplification of sRNA, a mechanism that is restricted to plants, fungi, some nematodes and some other invertebrates.

Amplification takes place when sRNA molecules hybridize to target RNA sequences and prime these targets to be copied by RNA-dependent RNA polymerase (RdRP) (Hunter *et al.*, 2006; Mittelbrunn and Sánchez-Madrid, 2012). Such an amplification of sRNA is absent in insects and vertebrates, including mammals and humans (Tomari and Zamore, 2005; Miller *et al.*, 2012).

Nonetheless, reports from the Nanjing University in China suggested that natural plant miRNA, once ingested by animals or humans, may exert local activities in the intestinal mucosa and even systemic effects. The authors switched the regular feed of mice to a diet consisting entirely of unprocessed rice and, already 3–6 h later, detected several rice miRNAs at femtomolar concentrations in the bloodstream and liver (Zhang *et al.*, 2012b). One of these detected miRNAs (miR168a) displays sequence identity with a region of the mouse gene coding for LDLRAP1 (low-density lipoprotein receptor adapter protein 1). Although the respective mRNA was not affected following the consumption of rice, the authors reported that the level of LDLRAP1 protein was lower in the liver of mice fed rice than in controls receiving a standard rodent diet. The authors also reported elevated LDL-associated cholesterol levels in plasma and attributed this effect to the suppressed LDLRAP1 expression resulting from miR168a uptake. This finding is contradicted by a 90-day feeding study as well as by a three-generation study, both with a 70% inclusion of rice into the diet of rats. In these studies, no LDL changes were detected relative to control groups (Zhou *et al.*, 2011, 2012). Another study, published by Dickinson *et al.* (2013), attempted to replicate the findings of Zhang and colleagues. After feeding mice with rice-containing diets at an inclusion rate of up to 75%, these authors detected only trace levels, if any, of plant-derived miRNA molecules in plasma. Zhang *et al.* (2012a) also described the presence of miR168a in the serum of Chinese persons with high dietary intake of rice. However, traces of plant-derived siRNA or miRNA taken up into the cells of the gastrointestinal epithelium, or systemically absorbed, would not be sufficient in terms of their concentration to trigger the gene silencing machinery or exert any other biologically relevant effect. In summary, humans and farm or companion animals do not have the mechanisms to take up and amplify dietary RNA in a way that their genes would be subjected to foreign sRNA-mediated regulation. Given the history of safe consumption of nucleic acids including RNA, oral toxicity studies of dsRNA and derived sRNA are currently not warranted (EFSA GMO Panel, 2018).

13.4   RNAi-specific Risk Assessment

In view of the above considerations, there is no reason to expect that GM plants tested in depth by the usual comparative and, if necessary, nutritional assessments are any less safe than conventional comparators just because a particular trait is generated by the RNAi pathway. This conclusion is supported by two specific examples outlined below.

13.4.1   Case study of an RNAi crop with altered metabolite composition

RNAi-based GM crops that have previously been filed for regulatory approval, particularly in North America but also elsewhere, include papaya, potato and cucurbits with transgenes aimed at silencing genes of invading viruses, and maize with resistance against an infesting insect (corn rootworm). In addition to these crops, GM soybean and potato that had been modified with silencing constructs targeting endogenous genes involved in biosynthesis of fatty acids and starch as well as apple and tomato for the suppression of plant genes involved in enzymatic browning (polyphenol oxidase) and ripening (polygalacturonidase), respectively, have been submitted for approval. A case in point for such compositionally altered crops is high-oleic soybean, which will be explored in further detail in this section. Whilst the risk assessment of this crop included the regular, recurrent items summarized in section 13.2.5 above, we will discuss here some features that were specific to this type of compositionally altered RNAi crop.

The modification in soybean 305423 targeted the biosynthesis of PUFAs. PUFA biosynthesis includes several subsequent steps of enzymatic dehydrogenation, which introduces double bonds in the carbon chain of the fatty acid, hence the 'unsaturation'. Oleic acid

(C18:1), for example, is the mono-unsaturated form of stearic acid (C18:0) with a backbone chain of 18 carbon atoms length, and a double bond between the 9th and 10th carbon atoms (C9-C10). It is a substrate for the fatty acid dehydrogenase enzyme FAD2 (oleoyl phosphatidylcholine dehydrogenase). Its FAD2-2 and FAD2-3 isomers are constitutively expressed throughout the plant, whilst FAD2-1 is strongly expressed during embryogenesis in seeds. FAD2 converts oleic acid to the PUFA linoleic acid (C18:2), by introducing a second double bond between the 12th and 13th carbon atoms. Linoleic acid, in turn, can be further enzymatically transformed to the PUFA linolenic acid (C18:3) with three double bonds.

For industrial purposes, such as for frying and bakery, it is desirable to increase the oxidative stability of PUFA-rich vegetable oils. This can be done by decreasing the content of PUFAs, which are particularly prone to oxidation. This way, the oils will tend to become rancid less quickly. Whilst catalytic hydrogenation was historically used for this purpose, there are potential consumer health issues with the *trans*-fatty acids that may be formed during this process. As an alternative, the use of high-oleic acid mutants of oilseed crops with decreased levels of PUFAs would help to avoid these issues.

Around the globe, for example, there is widespread cultivation of high-oleic sunflower varieties, which originate from a mutant created through chemical mutagenesis, in which the *FAD2-1* gene has been partially duplicated, causing gene silencing (Schuppert *et al.*, 2006). Also, transgenesis exploiting RNA interference has been applied to introduce constructs silencing expression of *FAD2-1* in experimental and commercial oilseed crop lines. Examples include soybean, cotton, Indian mustard, carinata, flax and camelina (Kinney and Knowlton, 1998; Chapman *et al.*, 2001; Sivaraman *et al.*, 2004; Du *et al.*, 2018). A more recent, pre-commercial example is super high-oleic safflower, which has been modified with transgenes encoding hpRNAs that target the *FAD2-2* and *FATB* genes (Wood *et al.*, 2018). Genome editing using TALENs or CRISPR-Cas9 has also been applied to achieve similar results, introducing mutations in *FAD2* genes to create high-oleic variants of oilseed rape, rice and soybean, for example (Haun *et al.*, 2014; Abe *et al.*, 2018; Okuzaki

et al., 2018; Al Amin *et al.*, 2019). A genome-edited soybean line with deletions created with TALENs in the FAD2-1 gene was recently introduced into the US market, and its oil is offered to the food industry under the trade name 'Calyno®' for frying and dressing, and use as a sauce (FDA, 2019).

As an example of a risk assessment of high-oleic soybean, transgenic GM soybean line 305423 (tradename 'Plenish®') assessed by the EFSA GMO Panel contained two types of modifications, namely: (i) a gene-silencing construct targeting the *FAD2-1* gene imparting the high-oleic phenotype; and (ii) the *ALS* gene encoding a mutant acetolactate synthase (ALS) gene mediating resistance to ALS-inhibiting herbicide active substances.

With regard to safety, soybean 305423 had to be assessed as a GM crop, in line with the internationally harmonized principles of comparative safety assessment. Soybean 305423 had been transformed with a DNA construct containing a partial sequence of the soybean *fad2-1* gene aimed at silencing the expression of the host's endogenous counterpart. This gene was under the control of the promoter and terminator sequences from the Kunitz trypsin inhibitor gene 3 (*Kti3*) coding for the antinutrient and allergenic trypsin inhibitor protein, thus allowing for seed-specific expression of the inserted genes. The inserted DNA also carried the *gm-hra* gene encoding an ALS enzyme conferring herbicide resistance (EFSA GMO Panel, 2013).

Molecular characterization of soybean 305423 showed that DNA had been inserted at four distinct sites, with a relatively complex pattern comprising, for example, seven copies of the *fad2-1* fragment and five of the *KTi3* terminator. Expression data (Northern blotting) indeed showed inhibited expression of *fad2-1* in seeds of soybean 305423 as well as of *KTi3* as a corollary effect. The latter can be attributed to silencing caused by the presence of *KTi3*-related sequences (promoter, terminator) in the introduced DNA. Bioinformatics-supported comparisons of the amino acid sequences hypothetically formed from the ORFs of the inserts and flanks with known toxic and allergenic proteins did not reveal any relevant similarities. An experiment testing the stability of inheritance indicated recombination between the *KTi3* promoter elements at one of the integration sites in a single progeny plant, accounting for loss of the *gm-hra* gene cassette, but this was not considered as relevant for safety assessment (EFSA GMO Panel, 2013).

Compositional analysis of seeds from experimental field sites showed that, whilst non-transgenic soybeans contained 19% oleic acid, 55% linoleic acid and 8% linolenic acid, these figures had markedly changed in 305423 soybeans to 73% (+54%), 4% (−5%), and 4% (−4%), respectively. This confirmed that the fatty acid composition had indeed changed from a preponderance of the PUFAs linoleic and linolenic acids to that of the MUFA oleic acid, as intended (EFSA GMO Panel, 2013).

The potential nutritional impact of these changes was assessed, taking into account that consumers need to attain adequate intakes of linoleic and linolenic acids. It was assumed that the new oil from soybean 305423 would totally replace conventional vegetable oils in targeted foods. Based on consumption data, the intakes of the various types of fatty acid by different consumer groups, ranging from toddlers to elderly people, were estimated. The reductions in PUFA intakes thus obtained under these conservative scenarios did not raise health concerns, though post-market monitoring for verification of these consumption data was also recommended (EFSA GMO Panel, 2013).

In conclusion, the risk assessment of soybean 305423 as an example of a crop with silenced endogenous genes shows that various generic factors come in play. This could also be translated to other traits achieved through RNAi, such as disease resistance based on suppression of the plant host's intrinsic vulnerability factors. For example, Northern blotting or other means of RNA expression analysis will be important to confirm the targeted suppression of the endogenous gene of interest. Moreover, it should be verified if endogenous genes bearing similarity with inserted elements (such as observed for *KTi3* on soybean 305423) are also affected, though not the aim of the modification, as a corollary effect. Bioinformatics-supported comparisons of the amino acid sequences of peptides that could be hypothetically formed from ORFs with sequences of allergenic and toxic proteins will help to identify potential safety issues should the inserted DNA sequences unexpectedly be translated into peptides. Extensive comparative analysis of compositional, phenotypic and agronomic characteristics will help to identify any potential intended and unintended effects of the modification. For the changes identified, the impact on toxicity, allergenicity and nutritional value of the modified RNAi crop should be assessed.

13.4.2 The case of insecticidal RNAi maize

RNAi-mediated silencing, first observed in nematodes and plants (Dougherty *et al.*, 1994; Fire *et al.*, 1998; Brodersen and Voinnet, 2006; Jones-Rhoades *et al.*, 2006; Vazquez, 2006), was later demonstrated also in some insects (Ghildiyal *et al.*, 2008). Several reports demonstrated the proof-of-principle that it is possible to induce an RNAi-mediated suppression of essential genes by feeding parasitic nematodes (Huang *et al.*, 2006; Yadav *et al.*, 2006; Fairbairn *et al.*, 2007) or the larvae of insect pests with GM plants engineered to express specific dsRNA precursors (Baum *et al.*, 2007; Mao *et al.*, 2007). In maize MON 87411, this principle is employed for the management of western corn rootworm (WCR), *Diabrotica virgifera virgifera*, the most important maize pest in the US 'corn belt'. A single insert has been introduced to express a modified version of *Bacillus thuringiensis* Cry3Bb1 protein, a glyphosate-tolerant 5-enolpyruvylshikimate-3-phosphate synthase (EPSPS) enzyme and an expression cassette containing two sequences of the *D. virgifera (Dv)Snf7* gene coding for an essential vacuolar sorting protein. The reader is referred to the relevant scientific opinion of the EFSA GMO Panel for the risk assessment of the Cry3Bb1 and EPSPS proteins (EFSA GMO Panel, 2018). The Snf7 dsRNA expression cassette consists of two fragments of the coding sequence of the *DvSnf7* gene in an inverted repeat configuration flanked by the e35S promoter from cauliflower mosaic virus, the heat shock protein 70 intron from *Zea mays* and the 3′ untranslated sequence of the E9 gene from *Pisum sativum*. The *DvSnf7* inverted repeat sequence generates a 240 bp precursor with hairpin structure that is processed to generate siRNA molecules. When the WCR larvae feed on MON 87411 maize, silencing of the *DvSnf7* gene leads to insect lethality, thus protecting the plant from root damage.

The mechanism of RNA uptake and systemic spread in the WCR is poorly understood but may involve SID-like (SIL) proteins and also clathrin-mediated endocytosis. An RdRP-mediated amplification is absent in the WCR (Huvenne and Smagghe, 2010; Fishilevich *et al.*, 2016).

As outlined in section 13.2.1 above, in the EU a bioinformatics analysis is required according to Regulation (EU) No. 503/2013 to identify potential off-target genes that may be influenced in their expression by the siRNA approach. Following the recommendations of the EFSA GMO Panel for an RNAi off-target search in plants, it was found that none of the maize transcripts in the available databases showed perfect match to any of the siRNAs possibly produced. A few maize transcripts have sequences matching the siRNAs with one to four mismatches. Some of these sequences presented matches for more than one (up to five) possible siRNA. However, a scrutiny of the anticipated function of the proteins encoded by these mRNAs matching the siRNA sequences indicated that off-target effects, if they took place, would not raise safety concerns, because the possible depletion of these potential targets is not expected to affect agronomic, phenotypic, compositional and nutritional characteristics of the GM maize. This conclusion is confirmed by the comparative analysis of maize MON 87411 and non-GM comparators. A field trial for the assessment of agronomic and phenotypic characteristics did not reveal any statistically significant differences between maize MON 87411 and this conventional counterpart. Also, no changes in the composition of grains were detected, i.e. concentrations of none of the 78 tested maize constituents were significantly different in maize MON 87411 compared with its conventional counterpart and also present at levels outside the equivalence range defined by non-GM reference varieties grown in the same field trial. Of course, an off-target gene silencing may also theoretically occur in organisms exposed to the RNAi plant, for example upon food and feed consumption. As described in sections 13.3.2 and 13.3.3 above, dietary dsRNA and sRNA are, however, rapidly denatured, depurinated and degraded after ingestion, due to the particular milieu of the gastrointestinal tract and the presence of multiple digestive enzymes in humans, mammals and other vertebrates. Further biological barriers like cellular membranes or lysosomes limit the uptake of dsRNA and sRNA. Therefore, it is not expected that sRNAs with *DvSnf7* sequences are able to exert any biological effects once maize MON 87411 is ingested by humans, or by farm or companion animals (EFSA GMO Panel, 2018). A 28-day oral repeated-dose study in mice with DvSnf7 dsRNA, conducted in accordance with the principles laid down in the OECD Test Guideline 407, lends further support to the above conclusion. In this study, the DvSnf7 dsRNA was administered by daily oral gavage at doses of 1, 10 and 100 mg/kg body weight. No treatment-related effects were observed in the animal body weights, food intake, clinical parameters, clinical chemistry values, haematology, gross pathology and histopathology (Petrick *et al.*, 2016). Considering the possible Snf7 dsRNA and sRNA content of maize MON 87411, which is difficult to assess quantitatively, the authors of this toxicity study calculated that a human would need to eat 60 million kilograms of maize MON 87411 per day to reach the dose of 100 mg/kg body weight that in the mouse study remained without any effects. The lack of biological activity of ingested dsRNA or sRNA is also documented by a previous 28-day toxicity study in mice using dsRNA of 218 bp, or a pool of four 21-mer siRNA molecules, targeting a mouse vacuolar ATPase transcript. The daily dose administered by gavage was 64 mg/kg for the dsRNA and 48 mg/kg for the siRNA. This 28-day toxicity study revealed no adverse effects and, importantly, no changes of vacuolar ATPase expression in any tissue, including the gastric mucosa (Petrick *et al.*, 2015), thus supporting the notion that no consequences are expected from the dietary uptake of dsRNA or siRNA present in food or feed. Taking into account all of the above, maize MON 87411 is considered equivalent, with respect to its food and feed safety and its nutritional profile, to non-GM maize counterparts.

13.5 Conclusion

The concept of gene silencing in GM plants based on the principles of RNAi has been exploited from the early days of commercial crop biotechnology. Applications straddle traits of both agronomic importance, such as disease and pest resistance, and of consumer and producer benefit, such as oilseeds with altered fatty acid composition. The internationally harmonized risk assessment approach for the food safety of GM crops can also be applied well to the subcategory with RNAi-based gene-silencing traits, notwithstanding some special features, such as the lack of newly expressed proteins. Moreover, the issues of off-target effects of the silencing RNAs within the plant, as well as the hypothetical uptake by consumers after ingestion of foods derived from RNAi-based GM crops, has been at the focus of scientific discourse. The current state of knowledge indicates that cross-kingdom interactions of consumed plant sRNA with the intrinsic RNAi machinery of humans and farm animals is a highly remote possibility at best, with unlikely impacts of any potential health concern. The featured case studies both underscore the applicability of current guidelines of the EFSA GMO Panel, enshrined in Implementing Regulation No. 503/2013, and more generally, those of the international Codex Alimentarius of the FAO/WHO Food Standards Programme.

Acknowledgement

Financial support from the Dutch Ministry of Agriculture, Nature, and Food Quality under the Statutory and Supportive Tasks program (WOT-2) for GK's contribution is gratefully acknowledged.

References

Abe, K., Araki, E., Suzuki, Y., Toki, S. and Saika, H. (2018) Production of high oleic/low linoleic rice by genome editing. *Plant Physiology and Biochemistry* 131, 58–62. DOI: 10.1016/j.plaphy.2018.04.033.

Al Amin, N., Ahmad, N., Wu, N., Pu, X., Ma, T. *et al.* (2019) Crispr-Cas9 mediated targeted disruption of FAD2-2 microsomal omega-6 desaturase in soybean (*glycine max.*L). *BMC Biotechnology* 19(1), 9. DOI: 10.1186/s12896-019-0501-2.

Allen, E., Xie, Z., Gustafson, A.M. and Carrington, J.C. (2005) MicroRNA-directed phasing during trans-acting siRNA biogenesis in plants. *Cell* 121(2), 207–221. DOI: 10.1016/j.cell.2005.04.004.

Andersson, M., Melander, M., Pojmark, P., Larsson, H., Bülow, L. *et al.* (2006) Targeted gene suppression by RNA interference: an efficient method for production of high-amylose potato lines. *Journal of Biotechnology* 123(2), 137–148. DOI: 10.1016/j.jbiotec.2005.11.001.

Armstrong, T.A., Chen, H., Ziegler, T.E., Iyadurai, K.R., Gao, A.-G. *et al.* (2013) Quantification of transgene-derived double-stranded RNA in plants using the QuantiGene nucleic acid detection platform. *Journal of Agricultural and Food Chemistry* 61(51), 12557–12564. DOI: 10.1021/jf4031458.

Bachman, P.M., Huizinga, K.M., Jensen, P.D., Mueller, G., Tan, J. *et al.* (2016) Ecological risk assessment for DvSnf7 RNA: a plant-incorporated protectant with targeted activity against western corn rootworm. *Regulatory Toxicology and Pharmacology* 81, 77–88. DOI: 10.1016/j.yrtph.2016.08.001.

Bally, J., McIntyre, G.J., Doran, R.L., Lee, K., Perez, A. *et al.* (2016) In-plant protection against *Helicoverpa armigera* by production of long hp RNA in chloroplasts. *Frontiers in Plant Science* 7(e47534), 1453. DOI: 10.3389/fpls.2016.01453.

Bardoń, A. and Shugar, D. (1980) Properties of purified salivary ribonuclease, and salivary ribonuclease levels in children with cystic fibrosis and in heterozygous carriers. *Clinica Chimica Acta* 101(1), 17–24. DOI: 10.1016/0009-8981(80)90051-0.

Baum, J.A., Bogaert, T., Clinton, W., Heck, G.R., Feldmann, P. *et al.* (2007) Control of coleopteran insect pests through RNA interference. *Nature Biotechnology* 25(11), 1322–1326. DOI: 10.1038/nbt1359.

Bolognesi, R., Ramaseshadri, P., Anderson, J., Bachman, P., Clinton, W. *et al.* (2012) Characterizing the mechanism of action of double-stranded RNA activity against Western corn rootworm (*Diabrotica virgifera virgifera* LeConte). *PLoS ONE* 7(10), e47534. DOI: 10.1371/journal.pone.0047534.

Bonnet, E., He, Y., Billiau, K. and Van de Peer, Y. (2010) TAPIR, a web server for the prediction of plant microRNA targets, including target mimics. *Bioinformatics* 26(12), 1566–1568. DOI: 10.1093/bioinformatics/btq233.

Brennecke, J., Stark, A., Russell, R.B. and Cohen, S.M. (2005) Principles of microRNA-target recognition. *PLoS Biology* 3, e85–418. DOI: 10.1371/journal.pbio.0030085.

Brodersen, P. and Voinnet, O. (2006) The diversity of RNA silencing pathways in plants. *Trends in Genetics* 22(5), 268–280. DOI: 10.1016/j.tig.2006.03.003.

Brown, B.D., Gentner, B., Cantore, A., Colleoni, S., Amendola, M. *et al.* (2007) Endogenous microRNA can be broadly exploited to regulate transgene expression according to tissue, lineage and differentiation state. *Nature Biotechnology* 25(12), 1457–1467. DOI: 10.1038/nbt1372.

Casacuberta, J.M., Devos, Y., du Jardin, P., Ramon, M., Vaucheret, H. *et al.* (2015) Biotechnological uses of RNAi in plants: risk assessment considerations. *Trends in Biotechnology* 33(3), 145–147. DOI: 10.1016/j.tibtech.2014.12.003.

Chapman, K.D., Austin-Brown, S., Sparace, S.A., Kinney, A.J., Ripp, K.G. *et al.* (2001) Transgenic cotton plants with increased seed oleic acid content. *Journal of the American Oil Chemists' Society* 78(9), 941–947. DOI: 10.1007/s11746-001-0368-y.

Chau, B.L. and Lee, K.A.W. (2007) Function and anatomy of plant siRNA pools derived from hairpin transgenes. *Plant Methods* 3(1), 13. DOI: 10.1186/1746-4811-3-13.

Chen, X., Zen, K. and Zhang, C.-Y. (2013) Lack of detectable oral bioavailability of plant microRNA after feeding in mice. *Nature Biotechnology* 31(11), 967–969. DOI: 10.1038/nbt.2741.

Christensen, J., Litherland, K., Faller, T., van de Kerkhof, E., Natt, F. *et al.* (2013) Metabolism studies of unformulated internally 3H-labeled short interfering RNAs in mice. *Drug Metabolism and Disposition* 41(6), 1211–1219. DOI: 10.1124/dmd.112.050666.

Codex (2003) *Guideline for the conduct of food safety assessment of foods derived from recombinant-DNA plants (CAC/GL 45-2003).* Codex Alimentarius Commission, Joint FAO/WHO Food Standards Program. Food and Agriculture Organization, Rome, Italy.

Dai, X. and Zhao, P.X. (2011) psRNATarget: a plant small RNA target analysis server. *Nucleic Acids Research* 39(suppl. 2), W155–W159. DOI: 10.1093/nar/gkr319.

Della Vedova, C.B., Lorbiecke, R., Kirsch, H., Schulte, M.B., Scheets, K. *et al.* (2005) The dominant inhibitory chalcone synthase allele C2-Idf (inhibitor diffuse) from *Zea mays* (L.) acts via an endogenous RNA silencing mechanism. *Genetics* 170(4), 1989–2002. DOI: 10.1534/genetics.105.043406.

Devos, Y., Naegeli, H., Perry, J.N. and Waigmann, E. (2016) 90-day rodent feeding studies on whole GM food/feed: is the mandatory EU requirement for 90-day rodent feeding studies on whole GM food/feed fit for purpose and consistent with animal welfare ethics? *EMBO Reports* 17, 942–945.

Dickinson, B., Zhang, Y., Petrick, J.S., Heck, G., Ivashuta, S. *et al.* (2013) Lack of detectable oral bioavailability of plant microRNA after feeding in mice. *Nature Biotechnology* 31(11), 965–967. DOI: 10.1038/nbt.2737.

Dougherty, W.G., Lindbo, J.A., Smith, H.A., Parks, T.D., Swaney, S. (1994) RNA-mediated virus resistance in transgenic plants: exploitation of a cellular pathway possibly involved in RNA degradation. *Molecular Plant-Microbe Interactions* 7(5), 554–552. DOI: 10.1094/MPMI-7-0554.

Du, C., Chen, Y., Wang, K., Yang, Z., Zhao, C. *et al.* (2018) Strong co-suppression impedes an increase in polyunsaturated fatty acids in seeds overexpressing FAD2. *Journal of Experimental Botany* 70(3), 985–994. DOI: 10.1093/jxb/ery378.

Dykxhoorn, D.M., Novina, C.D. and Sharp, P.A. (2003) Killing the messenger: short RNAs that silence gene expression. *Nature Reviews Molecular Cell Biology* 4(6), 457–467. DOI: 10.1038/nrm1129.

EC (2013) Commission implementing regulation (EU) NO 503/2013 of 3 April 2013 on applications for authorisation of genetically modified food and feed in accordance with regulation (EC) No. 1829/2003 of the European Parliament and of the Council and amending Commission regulations (EC) NO 641/2004 and (EC) NO 1981/2006. *Official Journal of the European Union* L157, 1–48.

EFSA GMO Panel (2006) Opinion of the scientific panel on genetically modified organisms related to the notification (reference C/SE/96/3501) for the placing on the market of genetically modified potato EH92-527-1 with altered starch composition, for cultivation and production of starch, under part C of Directive 2001/18/EC from BASF plant science. *EFSA Journal* 323, 1–20.

EFSA GMO Panel (2011) Guidance for risk assessment of food and feed from genetically modified plants. *EFSA Journal* 9(5), 2150. DOI: 10.2903/j.efsa.2011.2150.

EFSA GMO Panel (2013) Scientific opinion on application EFSA-GMO-NL2007-45 for the placing on the market of herbicide-tolerant, high-oleic acid, genetically modified soybean 305423 for food and fees uses, import and processing under regulation (EC) NO 1829/2003 from pioneer. *EFSA Journal* 11, 3499.

EFSA GMO Panel (2017) Annex II of the minutes of the 118th GMO plenary meeting: internal note on the strategy and technical aspects for small RNA plant off-target bioinformatics studies. Available at: https://www.efsa.europa.eu/sites/default/files/event/171025-m.pdf (accessed 30 March 2020).

EFSA GMO Panel (2018) Scientific opinion on the assessment of genetically modified maize MON 87411 for food and feed uses, import and processing, under regulation (EC) NO 1829/2003 (application EFSA-GMO-NL-2015-124). *EFSA Journal* 16, 5310.

Fairbairn, D.J., Cavallaro, A.S., Bernard, M., Mahalinga-Iyer, J., Graham, M.W. *et al.* (2007) Host-delivered RNAi: an effective strategy to silence genes in plant parasitic nematodes. *Planta* 226(6), 1525–1533. DOI: 10.1007/s00425-007-0588-x.

FDA (2019) Biotechnology notification file No. 000164: FAD2KO, high oleic acid soybean. Food and Drug Administration, Center for Veterinary Medicine, Rockville, Maryland. Available at: https://www.fda.gov/media/120660/download (accessed 30 March 2020).

Feinberg, E.H. and Hunter, C.P. (2003) Transport of dsRNA into cells by the transmembrane protein SID-1. *Science* 301(5639), 1545–1547. DOI: 10.1126/science.1087117.

Fire, A., Xu, S., Montgomery, M.K., Kostas, S.A., Driver, S.E. *et al.* (1998) Potent and specific genetic interference by double-stranded RNA in *Caenorhabditis elegans*. *Nature* 391(6669), 806–811. DOI: 10.1038/35888.

Fishilevich, E., Vélez, A.M., Storer, N.P., Li, H., Bowling, A.J. *et al.* (2016) RNAi as a management tool for the western corn rootworm, *Diabrotica virgifera virgifera*. *Pest Management Science* 72(9), 1652–1663. DOI: 10.1002/ps.4324.

Flemr, M., Malik, R., Franke, V., Nejepinska, J., Sedlacek, R. *et al.* (2013) A retrotransposon-driven dicer isoform directs endogenous small interfering RNA production in mouse oocytes. *Cell* 155(4), 807–816. DOI: 10.1016/j.cell.2013.10.001.

Frizzi, A. and Huang, S. (2010) Tapping RNA silencing pathways for plant biotechnology. *Plant Biotechnology Journal* 8(6), 655–677. DOI: 10.1111/j.1467-7652.2010.00505.x.

Gebert, L.F.R. and MacRae, I.J. (2019) Regulation of microRNA function in animals. *Nature Reviews Molecular Cell Biology* 20(1), 21–37. DOI: 10.1038/s41580-018-0045-7.

Ghildiyal, M., Seitz, H., Horwich, M.D., Li, C., Du, T. *et al.* (2008) Endogenous siRNAs derived from transposons and mRNAs in *Drosophila* somatic cells. *Science* 320(5879), 1077–1081. DOI: 10.1126/science.1157396.

Gilmore, I.R., Fox, S.P., Hollins, A.J., Sohail, M. and Akhtar, S. (2004) The design and exogenous delivery of siRNA for post-transcriptional gene silencing. *Journal of Drug Targeting* 12(6), 315–340. DOI: 10.1080/10611860400006257.

Gordon, K.H.J. and Waterhouse, P.M. (2007) RNAi for insect-proof plants. *Nature Biotechnology* 25(11), 1231–1232. DOI: 10.1038/nbt1107-1231.

Hannus, M., Beitzinger, M., Engelmann, J.C., Weickert, M.-T., Spang, R. *et al.* (2014) siPools: highly complex but accurately defined siRNA pools eliminate off-target effects. *Nucleic Acids Research* 42(12), 8049–8061. DOI: 10.1093/nar/gku480.

Haun, W., Coffman, A., Clasen, B.M., Demorest, Z.L., Lowy, A. *et al.* (2014) Improved soybean oil quality by targeted mutagenesis of the fatty acid desaturase 2 gene family. *Plant Biotechnology Journal* 12(7), 934–940. DOI: 10.1111/pbi.12201.

Heidel, J.D., Hu, S., Liu, X.F., Triche, T.J. and Davis, M.F. (2004) Lack of interferon response in animals to naked siRNAs. *Nature Biotechnology* 22(12), 1579–1582. DOI: 10.1038/nbt1038.

Heisel, S.E., Zhang, Y., Allen, E., Guo, L., Reynolds, T.L. *et al.* (2008) Characterization of unique small RNA populations from rice grain. *PLoS ONE* 3(8), e2871. DOI: 10.1371/journal.pone.0002871.

Huang, G., Allen, R., Davis, E.L., Baum, T.J. and Hussey, R.S. (2006) Engineering broad root-knot resistance in transgenic plants by RNAi silencing of a conserved and essential root-knot nematode parasitism gene. *Proceedings of the National Academy of Sciences* 103(39), 14302–14306. DOI: 10.1073/pnas.0604698103.

Huang, H., Davis, C. and Wang, T. (2018) Extensive degradation and low bioavailability of orally consumed corn miRNAs in mice. *Nutrients* 10(2), 215. DOI: 10.3390/nu10020215.

Hunter, C.P., Winston, W.M., Molodowitch, C., Feinberg, E.H., Shih, J. *et al.* (2006) Systemic RNAi in *Caenorhabditis elegans*. *Cold Spring Harbor Symposia on Quantitative Biology* 71, 95–100. DOI: 10.1101/sqb.2006.71.060.

Huvenne, H. and Smagghe, G. (2010) Mechanisms of dsRNA uptake in insects and potential of RNAi for pest control: a review. *Journal of Insect Physiology* 56(3), 227–235. DOI: 10.1016/j.jinsphys.2009.10.004.

ILSI (2004) Nutritional and safety assessments of foods and feeds nutritionally improved through bio-technology: an executive summary. *Comprehensive Reviews in Food Science and Food Safety* 3, 35–104.

Ivashuta, S.I., Petrick, J.S., Heisel, S.E., Zhang, Y., Guo, L. *et al.* (2009) Endogenous small RNAs in grain: Semi-quantification and sequence homology to human and animal genes. *Food and Chemical Toxicology* 47(2), 353–360. DOI: 10.1016/j.fct.2008.11.025.

Jain, K. (2008) Stability and delivery of RNA via the gastrointestinal tract. *Current Drug Delivery* 5(1), 27–31. DOI: 10.2174/156720108783331023.

Jonas, D.A., Elmadfa, I., Engel, K.-H., Heller, K.J., Kozianowski, G. *et al.* (2001) Safety considerations of DNA in food. *Annals of Nutrition and Metabolism* 45(6), 235–254. DOI: 10.1159/000046734.

Jones-Rhoades, M.W., Bartel, D.P. and Bartel, B. (2006) MicroRNAs and their regulatory roles in plants. *Annual Review of Plant Biology* 57(1), 19–53. DOI: 10.1146/annurev.arplant.57.032905.105218.

Jose, A.M. and Hunter, C.P. (2007) Transport of sequence-specific RNA interference information be-tween cells. *Annual Review of Genetics* 41(1), 305–330. DOI: 10.1146/annurev.genet.41.110306.130216.

Judge, A. and MacLachlan, I. (2008) Overcoming the innate immune response to small interfering RNA. *Human Gene Therapy* 19(2), 111–124. DOI: 10.1089/hum.2007.179.

Kang, W., Bang-Berthelsen, C.H., Holm, A., Houben, A.J.S., Müller, A.H. *et al.* (2017) Survey of 800+ data sets from human tissue and body fluid reveals xenomRs are likely artifacts. *RNA* 23(4), 433–445. DOI: 10.1261/rna.059725.116.

Khatsenko, O., Morgan, R., Truong, L., York-Defalco, C., Sasmor, H. *et al.* (2000) Absorption of antisense oligonucleotides in rat intestine: effect of chemistry and length. *Antisense and Nucleic Acid Drug Development* 10(1), 35–44. DOI: 10.1089/oli.1.2000.10.35.

Kinney, A.J. and Knowlton, S. (1998) Designer oils: the high oleic acid soybean. In: Roller, S. and Harlander, S. (eds) *Genetic Modification in the Food Industry: A Strategy for Food Quality Improvement.* Boston, Massachusetts, Springer US, pp. 193–213.

Koch, A. and Kogel, K.-H. (2014) New wind in the sails: improving the agronomic value of crop plants through RNAi-mediated gene silencing. *Plant Biotechnology Journal* 12(7), 821–831. DOI: 10.1111/pbi.12226.

Kusaba, M., Miyahara, K., Iida, S., Fukuoka, H., Takano, T. *et al.* (2003) Low glutelin content1: A dominant mutation that suppresses the glutelin multigene family via RNA silencing in rice. *The Plant Cell* 15(6), 1455–1467. DOI: 10.1105/tpc.011452.

Lassek, E. and Montag, A. (1990) Nucleic acids components in carbohydrate-rich food. *Zeitschrift für Lebensmittel-Untersuchung und -Forschung* 190, 17–21.

Layzer, J.M., McCaffrey, A.P., Tanner, A.K., Huang, Z., Kay, M.A. (2004) *In vivo* activity of nuclease-resistant siRNAs. *RNA* 10(5), 766–771. DOI: 10.1261/rna.5239604.

Le, L., Mahler, V., Lorenz, Y., Scheurer, S., Biemelt, S. *et al.* (2006) Reduced allergenicity of tomato fruits harvested from Lyc e 1-silenced transgenic tomato plants. *Journal of Allergy and Clinical Immunology* 118(5), 1176–1183. DOI: 10.1016/j.jaci.2006.06.031.

Lemons, D., Maurya, M.R., Subramaniam, S. and Mercola, M. (2013) Developing microRNA screening as a functional genomics tool for disease research. *Frontiers in Physiology* 4, 223. DOI: 10.3389/fphys.2013.00223.

Li, J., Yang, Z., Yu, B., Liu, J. and Chen, X. (2005) Methylation protects miRNAs and siRNAs from a $3'$-end uridylation activity in *Arabidopsis*. *Current Biology* 15(16), 1501–1507. DOI: 10.1016/j.cub.2005.07.029.

Liu, Q., Wang, F. and Axtell, M.J. (2014) Analysis of complementary requirements for plant microRNA targeting using a *Nicotiana benthamiana* quantitative transient assay. *The Plant Cell* 26(2), 741–753. DOI: 10.1105/tpc.113.120972.

Liu, Y., Zhang, Y., Dong, P., An, R., Xue, C. *et al.* (2015) Digestion of nucleic acids starts in the stomach. *Scientific Reports* 5(1), 11936. DOI: 10.1038/srep11936.

Liu, S., da Cunha, A. P., Rezende, R. M., Cialic, R., Wei, Z. *et al.* (2016) The host shapes the gut microbiota via fecal microRNA. *Cell Host & Microbe* 19(1), 32–43. DOI: 10.1016/j.chom.2015.12.005.

Loretz, B., Föger, F., Werle, M. and Bernkop-Schnürch, A. (2006) Oral gene delivery: strategies to im-prove stability of pDNA towards intestinal digestion. *Journal of Drug Targeting* 14(5), 311–319. DOI: 10.1080/10611860600823766.

Lukasik, A. and Zielenkiewicz, P. (2019) An overview of miRNA and miRNA target analysis tools. In: de Folter, S. (ed.) *Plant MicroRNAs: Methods and Protocols.* Springer, New York, New York, pp. 65–87.

Ma, J.-B., Ye, K. and Patel, D.J. (2004) Structural basis for overhang-specific small interfering RNA recognition by the PAZ domain. *Nature* 429(6989), 318–322. DOI: 10.1038/nature02519.

Ma, Z., Li, J., He, F., Wilson, A., Pitt, B. *et al.* (2005) Cationic lipids enhance siRNA-mediated interferon response in mice. *Biochemical and Biophysical Research Communications* 330(3), 755–759. DOI: 10.1016/j.bbrc.2005.03.041.

Maeda, I., Kohara, Y., Yamamoto, M. and Sugimoto, A. (2001) Large-scale analysis of gene function in *Caenorhabditis elegans* by high-throughput RNAi. *Current Biology* 11(3), 171–176. DOI: 10.1016/S0960-9822(01)00052-5.

Maida, Y. and Masutomi, K. (2011) RNA-dependent RNA polymerases in RNA silencing. *Biological Chemistry* 392(4), 299–304. DOI: 10.1515/bc.2011.035.

Manjunath, N. and Dykxhoorn, D.M. (2010) Advances in synthetic siRNA delivery. *Discovery Medicine* 9, 418–430.

Mao, Y.-B., Cai, W.-J., Wang, J.-W., Hong, G.-J., Tao, X.-Y. *et al.* (2007) Silencing a cotton bollworm P450 monooxygenase gene by plant-mediated RNAi impairs larval tolerance of gossypol. *Nature Biotechnology* 25(11), 1307–1313. DOI: 10.1038/nbt1352.

Mayoral, J.G., Hussain, M., Joubert, D.A., Iturbe-Ormaetxe, I., O'Neill, S.L. and Asgari, S. *et al.* (2014) *Wolbachia* small noncoding RNAs and their role in cross-kingdom communications. *Proceedings of the National Academy of Sciences* 111(52), 18721–18726. DOI: 10.1073/pnas.1420131112.

McEwan, D. L., Weisman, A. S. and Hunter, C. P. (2012) Uptake of extracellular double-stranded RNA by SID-2. *Molecular Cell* 47(5), 746–754. DOI: 10.1016/j.molcel.2012.07.014.

Meister, G., Landthaler, M., Patkaniowska, A., Dorsett, Y., Teng, G. *et al.* (2004) Human Argonaute2 mediates RNA cleavage targeted by miRNAs and siRNAs. *Molecular Cell* 15(2), 185–197. DOI: 10.1016/j.molcel.2004.07.007.

Meyer, P., Linn, F., Heidmann, I., Meyer, H., Niedenhof, I. *et al.* (1992) Endogenous and environmental factors influence 35S promoter methylation of a maize A1 gene construct in transgenic Petunia and its colour phenotype. *Molecular and General Genetics MGG* 231(3), 345–352. DOI: 10.1007/BF00292701.

Miller, S.C., Miyata, K., Brown, S.J. and Tomoyasu, Y. (2012) Dissecting systemic RNA interference in the red flour beetle *Tribolium castaneum*: parameters affecting the efficiency of RNAi. *PLoS ONE* 7(10), e47431. DOI: 10.1371/journal.pone.0047431.

Mittelbrunn, M. and Sánchez-Madrid, F. (2012) Intercellular communication: diverse structures for exchange of genetic information. *Nature Reviews Molecular Cell Biology* 13(5), 328–335. DOI: 10.1038/nrm3335.

Méndez-Acevedo, K.M., Valdes, V.J., Asanov, A. and Vaca, L. (2017) A novel family of mammalian transmembrane proteins involved in cholesterol transport. *Scientific Reports* 7(1), 7450. DOI: 10.1038/s41598-017-07077-z.

Nejepinska, J., Malik, R., Filkowski, J., Flemr, M., Filipowicz, W. *et al.* (2012) dsRNA expression in the mouse elicits RNAi in oocytes and low adenosine deamination in somatic cells. *Nucleic Acids Research* 40(1), 399–413. DOI: 10.1093/nar/gkr702.

Newmark, P.A., Reddien, P.W., Cebria, F. and Sanchez Alvarado, A. (2003) Ingestion of bacterially expressed double-stranded RNA inhibits gene expression in planarians. *Proceedings of the National Academy of Sciences* 100(Supplement 1), 11861–11865. DOI: 10.1073/pnas.1834205100.

Nicklin, P.L., Bayley, D., Giddings, J., Craig, S.J., Cummins, L.L. *et al.* (1998) Pulmonary bioavailability of a phosphorothioate oligonucleotide (CGP 64128A): comparison with other delivery routes. *Pharmaceutical Research* 15(4), 583–591. DOI: 10.1023/A:1011934011690.

OECD (2002) *Consensus document on compositional considerations for new varieties of maize* (Zea mays)*: key food and feed nutrients, anti-nutrients and secondary plant metabolites*. Organisation for Economic Co-operation and Development, Paris. Available at: http://www.oecd.org/chemicalsafety/biotrack/46815196.pdf (accessed 30 March 2020).

Ogita, S., Uefuji, H., Yamaguchi, Y., Koizumi, N. and Sano, H. (2003) RNA interference: producing decaffeinated coffee plants. *Nature* 423, 823.

Okuzaki, A., Ogawa, T., Koizuka, C., Kaneko, K., Inaba, M. *et al.* (2018) CRISPR/Cas9-mediated genome editing of the fatty acid desaturase 2 gene in *Brassica napus*. *Plant Physiology and Biochemistry* 131, 63–69. DOI: 10.1016/j.plaphy.2018.04.025.

O'Neill, M.J., Bourre, L., Melgar, S. and O'Driscoll, C.M. (2011) Intestinal delivery of non-viral gene therapeutics: physiological barriers and preclinical models. *Drug Discovery Today* 16(5-6), 203–218. DOI: 10.1016/j.drudis.2011.01.003.

Pačes, J., Nič, M., Novotný, T. and Svoboda, P. (2017) *Literature review of baseline information to support the risk assessment of RNAi-based GM plants*. EFSA Supporting Publication 14(6), EN-1246. European Food Safety Authority, Parma, Italy, p. 314.

Park, N.J., Li, Y., Yu, T., Brinkman, B.M.N. and Wong, D.T. (2006) Characterization of RNA in saliva. *Clinical Chemistry* 52(6), 988–994. DOI: 10.1373/clinchem.2005.063206.

Pasquinelli, A.E. (2012) MicroRNAs and their targets: recognition, regulation and an emerging reciprocal relationship. *Nature Reviews Genetics* 13(4), 271–282. DOI: 10.1038/nrg3162.

Petrick, J.S., Brower-Toland, B., Jackson, A.L. and Kier, L.D. (2013) Safety assessment of food and feed from biotechnology-derived crops employing RNA-mediated gene regulation to achieve desired traits: a scientific review. *Regulatory Toxicology and Pharmacology* 66(2), 167–176. DOI: 10.1016/j. yrtph.2013.03.008.

Petrick, J.S., Moore, W.M., Heydens, W.F., Koch, M.S., Sherman, J.H. *et al.* (2015) A 28-day oral toxicity evaluation of small interfering RNAs and a long double-stranded RNA targeting vacuolar ATPase in mice. *Regulatory Toxicology and Pharmacology* 71(1), 8–23. DOI: 10.1016/j.yrtph.2014.10.016.

Petrick, J.S., Frierdich, G.E., Carleton, S.M., Kessenich, C.R., Silvanovich, A. *et al.* (2016) Corn rootworm-active DvSnf7: repeat dose oral toxicology assessment in support of human and mammalian safety. *Regulatory Toxicology and Pharmacology* 81, 57–68. DOI: 10.1016/j. yrtph.2016.07.009.

Pinzón, N., Li, B., Martinez, L., Sergeeva, A., Presumey, J. *et al.* (2017) MicroRNA target prediction programs predict many false positives. *Genome Research* 27(2), 234–245. DOI: 10.1101/gr.205146.116.

Ramon, M., Devos, Y., Lanzoni, A., Liu, Y., Gomes, A. *et al.* (2014) RNAi-based GM plants: food for thought for risk assessors. *Plant Biotechnology Journal* 12(9), 1271–1273. DOI: 10.1111/pbi.12305.

Redenbaugh, K., Hiatt, B., Martineau, B., Kramer, M. and Sheehy, R. (1992) *Safety Assessment of Genetically Engineered Fruits and Vegetables: a Case Study of the Flavr Savr Tomato*. CRC Press, Boca Raton, Florida, p. 288.

Regina, A., Bird, A., Topping, D., Bowden, S., Freeman, J. *et al.* (2006) High-amylose wheat generated by RNA interference improves indices of large-bowel health in rats. *Proceedings of the National Academy of Sciences* 103(10), 3546–3551. DOI: 10.1073/pnas.0510737103.

Ren, G., Xie, M., Zhang, S., Vinovskis, C., Chen, X. *et al.* (2014) Methylation protects microRNAs from an AGO1-associated activity that uridylates 5′ RNA fragments generated by AGO1 cleavage. *Proceedings of the National Academy of Sciences* 111(17), 6365–6370. DOI: 10.1073/pnas.1405083111.

Rhoades, M.W., Reinhart, B.J., Lim, L.P., Burge, C.B., Bartel, B. *et al.* (2002) Prediction of plant micro RNA targets. *Cell* 110(4), 513–520. DOI: 10.1016/S0092-8674(02)00863-2.

Robbins, M., Judge, A. and MacLachlan, I. (2009) siRNA and innate immunity. *Oligonucleotides* 19(2), 89–102. DOI: 10.1089/oli.2009.0180.

Schuppert, G.F., Tang, S., Slabaugh, M.B. and Knapp, S.J. (2006) The sunflower high-oleic mutant ol carries variable tandem repeats of FAD2-1, a seed-specific oleoyl-phosphatidyl choline desaturase. *Molecular Breeding* 17(3), 241–256. DOI: 10.1007/s11032-005-5680-y.

Senthil-Kumar, M. and Mysore, K.S. (2011) Caveat of RNAi in plants: the off-target effect. *Methods in Molecular Biology* 744, 13–25.

Sherman, J.H., Munyikwa, T., Chan, S.Y., Petrick, J.S., Witwer, K.W. *et al.* (2015) RNAi technologies in agricultural biotechnology: the toxicology forum 40th annual summer meeting. *Regulatory Toxicology and Pharmacology* 73(2), 671–680. DOI: 10.1016/j.yrtph.2015.09.001.

Sioud, M. (2005) On the delivery of small interfering RNAs into mammalian cells. *Expert Opinion on Drug Delivery* 2(4), 639–651. DOI: 10.1517/17425247.2.4.639.

Sivaraman, I., Arumugam, N., Sodhi, Y.S., Gupta, V., Mukhopadhyay, A. *et al.* (2004) Development of high oleic and low linoleic acid transgenics in a zero erucic acid *Brassica juncea* L. (Indian mustard) line by antisense suppression of the fad2 gene. *Molecular Breeding* 13(4), 365–375. DOI: 10.1023/B:MO LB.0000034092.47934.d6.

Snow, J.W., Hale, A.E., Isaacs, S.K., Baggish, A.L. and Chan, S.Y. (2013) Ineffective delivery of diet-derived microRNAs to recipient animal organisms. *RNA Biology* 10(7), 1107–1116. DOI: 10.4161/rna.24909.

Stein, P., Svoboda, P., Anger, M. and Schultz, R.M. (2003) RNAi: mammalian oocytes do it without RNA-dependent RNA polymerase. *RNA* 9(2), 187–192. DOI: 10.1261/rna.2860603.

Tabara, H., Grishok, A. and Mello, C.C. (1998) RNAi in *C. elegans*: soaking in the genome sequence. *Science* 282(5388), 430–431. DOI: 10.1126/science.282.5388.430.

Terenius, O., Papanicolaou, A., Garbutt, J.S., Eleftherianos, I., Huvenne, H. *et al.* (2011) RNA interference in Lepidoptera: an overview of successful and unsuccessful studies and implications for experimental design. *Journal of Insect Physiology* 57(2), 231–245. DOI: 10.1016/j.jinsphys.2010.11.006.

Tian, Y., Simanshu, D.K., Ma, J.-B. and Patel, D.J. (2011) Structural basis for piRNA 2′-O-methylated 3′-end recognition by Piwi PAZ (Piwi/Argonaute/Zwille) domains. *Proceedings of the National Academy of Sciences* 108(3), 903–910. DOI: 10.1073/pnas.1017762108.

Timmons, L., Court, D.L. and Fire, A. (2001) Ingestion of bacterially expressed dsRNAs can produce specific and potent genetic interference in *Caenorhabditis elegans*. *Gene* 263(1-2), 103–112. DOI: 10.1016/S0378-1119(00)00579-5.

Tomari, Y. and Zamore, P.D. (2005) Perspective: machines for RNAi. *Genes & Development* 19(5), 517–529. DOI: 10.1101/gad.1284105.

Tosar, J.P., Rovira, C., Naya, H. and Cayota, A. (2014) Mining of public sequencing databases supports a non-dietary origin for putative foreign miRNAs: underestimated effects of contamination in NGS. *RNA* 20(6), 754–757. DOI: 10.1261/rna.044263.114.

Tuteja, J.H., Clough, S.J., Chan, W.-C. and Vodkin, L.O. (2004) Tissue-specific gene silencing mediated by a naturally occurring chalcone synthase gene cluster in *Glycine max*. *The Plant Cell* 16(4), 819–835. DOI: 10.1105/tpc.021352.

Vaishnaw, A.K., Gollob, J., Gamba-Vitalo, C., Hutabarat, R., Sah, D. *et al.* (2010) A status report on RNAi therapeutics. *Silence* 1, 14. DOI: 10.1186/1758-907X-1-14.

van der Hoeven, C., Dietz, A. and Landsmann, J. (1994) Variability of organ-specific gene expression in transgenic tobacco plants. *Transgenic Research* 3, 159–166. DOI: 10.1007/BF01973983.

Vazquez, F. (2006) Arabidopsis endogenous small RNAs: highways and byways. *Trends in Plant Science* 11(9), 460–468. DOI: 10.1016/j.tplants.2006.07.006.

Wagner, E.G.H. and Romby, P. (2015) Small RNAs in bacteria and archaea: who they are, what they do, and how they do it. *Advances in Genetics* 90, 133–208. DOI: 10.1016/bs.adgen.2015.05.001.

Waltz, E. (2015) Nonbrowning GM apple cleared for market. *Nature biotechnology* 33(4), 326–327. DOI: 10.1038/nbt0415-326c.

Winston, W.M., Molodowitch, C. and Hunter, C.P. (2002) Systemic RNAi in *C. elegans* requires the putative transmembrane protein SID-1. *Science* 295(5564), 2456–2459. DOI: 10.1126/science.1068836.

Winston, W.M., Sutherlin, M., Wright, A.J., Feinberg, E.H. and Hunter, C.P. (2007) *Caenorhabditis elegans* SID-2 is required for environmental RNA interference. *Proceedings of the National Academy of Sciences* 104(25), 10565–10570. DOI: 10.1073/pnas.0611282104.

Witwer, K.W. (2018) Alternative miRNAs? Human sequences misidentified as plant miRNAs in plant studies and in human plasma. *F1000Research* 7, 244. DOI: 10.12688/f1000research.14060.1.

Witwer, K.W., McAlexander, M.A., Queen, S.E. and Adams, R.J. (2013) Real-Time quantitative PCR and droplet digital PCR for plant miRNAs in mammalian blood provide little evidence for general uptake of dietary miRNAs: limited evidence for general uptake of dietary plant xenomiRs. *RNA Biology* 10(7), 1080–1086. DOI: 10.4161/rna.25246.

Wood, C.C., Okada, S., Taylor, M.C., Menon, A., Mathew, A. *et al.* (2018) Seed-specific RNAi in safflower generates a superhigh oleic oil with extended oxidative stability. *Plant Biotechnology Journal* 16(10), 1788–1796. DOI: 10.1111/pbi.12915.

Yadav, B.C., Veluthambi, K. and Subramaniam, K. (2006) Host-generated double stranded RNA induces RNAi in plant-parasitic nematodes and protects the host from infection. *Molecular and Biochemical Parasitology* 148(2), 219–222. DOI: 10.1016/j.molbiopara.2006.03.013.

Yang, J., Farmer, L.M., Agyekum, A.A.A. and Hirschi, K.D. (2015a) Detection of dietary plant-based small RNAs in animals. *Cell Research* 25(4), 517–520. DOI: 10.1038/cr.2015.26.

Yang, J., Farmer, L.M., Agyekum, A.A.A., Elbaz-Younes, I. and Hirschi, K.D. (2015b) Detection of an abundant plant-based small RNA in healthy consumers. *PLoS ONE* 10(9), e0137516. DOI: 10.1371/journal.pone.0137516.

Yang, J., Hotz, T., Broadnax, L., Yarmarkovich, M., Elbaz-Younes, I. *et al.* (2016) Anomalous uptake and circulatory characteristics of the plant-based small RNA MIR2911. *Scientific Reports* 6, 26834. DOI: 10.1038/srep26834.

Yang, J., Kongchan, N., Primo Planta, C., Neilson, J.R. and Hirschi, K.D. (2017) The atypical genesis and bioavailability of the plant-based small RNA MIR2911: bulking up while breaking down. *Molecular Nutrition & Food Research* 61(9), 1600974. DOI: 10.1002/mnfr.201600974.

Yang, J., Elbaz-Younes, I., Primo, C., Murungi, D. and Hirschi, K.D. (2018) Intestinal permeability, digestive stability and oral bioavailability of dietary small RNAs. *Scientific Reports* 8(1), 10253. DOI: 10.1038/s41598-018-28207-1.

Yu, B., Yang, Z., Li, J., Minakhina, S., Yang, M. *et al.* (2005) Methylation as a crucial step in plant micro RNA biogenesis. *Science* 307(5711), 932–935. DOI: 10.1126/science.1107130.

Yuan, T., Huang, X., Woodcock, M., Du, M., Dittmar, R. *et al.* (2016) Plasma extracellular RNA profiles in healthy and cancer patients. *Scientific Reports* 6, 19413. DOI: 10.1038/srep19413.

Zhang, Y., Wiggins, B.E., Lawrence, C., Petrick, J., Ivashuta, S. *et al.* (2012a) Analysis of plant-derived miRNAs in animal small RNA datasets. *BMC Genomics* 13, 381. DOI: 10.1186/1471-2164-13-381.

Zhang, L., Hou, D., Chen, X., Li, D., Zhu, L. *et al.* (2012b) Exogenous plant MIR 168a specifically targets mammalian LDLRAP1: evidence of cross-kingdom regulation by microRNA. *Cell Research* 22(1), 107–126. DOI: 10.1038/cr.2011.158.

Zhang, J., Khan, S.A., Hasse, C., Ruf, S., Heckel, D.G. *et al.* (2015) Full crop protection from an insect pest by expression of long double-stranded RNAs in plastids. *Science* 347(6225), 991–994. DOI: 10.1126/science.1261680.

Zhang, J., Khan, S.A., Heckel, D.G. and Bock, R. (2017) Next-generation insect-resistant plants: RNAi-mediated crop protection. *Trends in Biotechnology* 35(9), 871–882. DOI: 10.1016/j.tibtech.2017.04.009.

Zhou, X.H., Dong, Y., Xiao, X., Wang, Y., Xu, Y. *et al.* (2011) A 90-day toxicology study of high-amylose transgenic rice grain in Sprague-Dawley rats. *Food and Chemical Toxicology* 49(12), 3112–3118. DOI: 10.1016/j.fct.2011.09.024.

Zhou, X.H., Dong, Y., Wang, Y., Xiao, X., Xu, Y. *et al.* (2012) A three generation study with high-lysine transgenic rice grain in Sprague-Dawley rats. *Food and Chemical Toxicology* 50(6), 1902–1910. DOI: 10.1016/j.fct.2012.04.001.

14 Regulatory Aspects of RNAi in Plant Production

Werner Schenkel* and Achim Gathmann

Federal Office of Consumer Protection and Food Safety, Braunschweig, Germany

14.1 Introduction

Technologies based on RNA interference (RNAi) may be used in plant production in different contexts. With respect to applicable regulations, a major distinction is to be made between plants producing small RNA molecules due to modifications of the genome and topically applied plant protection products (PPPs) based on double-stranded RNA (dsRNA).

The first group may be further divided into those using RNAi technology to achieve changes in the plant's metabolism and those where plant-produced RNA molecules are intended to impact other organisms that interact with the plant.

For PPPs, relevant aspects are whether the product contains living organisms or only purified molecules. The intended use of the product is another relevant aspect with respect to regulation. It is expected that PPPs will be among the first products utilizing the RNAi mechanism in the European Union (EU).

Based on these considerations, it is clear that the main relevant regulatory frameworks are in the areas of genetically modified organisms and plant protection products.

14.2 Regulation of Modified RNAi Plants

A meaningful utilization of RNAi effects in plants is generally only possible by modifying the plant's genome in a way that does not occur naturally by mating and/or natural recombination. Based on this premise, these plants fall within the scope of Directive (EC) 2001/18 in the EU and are therefore regulated as genetically modified organisms (GMOs) (EC, 2001). Any person intending to place such products on the market or to carry out a deliberate release into the environment of a GMO for any other purposes than placing on the market within the Community requires authorization to do so. It is important to note that Directive (EC) 2001/18 covers the authorization of the deliberate release of living organisms but does not cover any product produced from a GMO if it no longer contains living organisms. Most agricultural crops are not authorized under directive (EC) 2001/18, since food and feed products are covered by Regulation (EC) 1829/2003 (EC, 2003). For this reason, there are few genetically modified plants conceivable where developers would seek authorization under Directive (EC) 2001/18 only.

*Corresponding author: werner.schenkel@bvl.bund.de

© CAB International 2021. *RNAi for Plant Improvement and Protection*
(eds B. Mezzetti *et al.*)
DOI: 10.1079/9781789248890.0014

Crops for fibre or energy production might be such cases if they cannot be used for food or feed too, but a more relevant group of products is ornamentals. There has been an application for carnations with altered flower colour due to silencing of a gene in the anthocyan pathway. The carnation with unique identifier IFD-25958-3 was authorized for use as a cut flower in 2015 (EC, 2015). Although there are a few instances where RNAi plants have been assessed directly under Directive (EC) 2001/18, Annex II of the Directive is of high importance, as it lays down the principles of the environmental risk assessment that are also followed for the assessment of applications under Regulation (EC) 1829/2003, especially if cultivation is within the scope of the application.

Under Regulation (EC) 1829/2003, authorization may be granted for: (a) genetically modified plants for food or feed uses; (b) food or feed containing or consisting of genetically modified plants; and (c) food produced from or containing ingredients produced from genetically modified plants or feed produced from such plants (EC, 2003). It should be noted that, in contrast to Directive (EC) 2001/18, products that no longer contain a living GMO are covered by Regulation (EC) 1829/2003.

Detailed rules for the implementation of Regulation (EC) 1829/2003 are laid down in Commission Regulation (EC) No. 641/2004 and Commission Implementing Regulation (EU) No 503/2013 (EC, 2004, 2013c). These regulations provide rules concerning applications for authorizations. Commission Implementing Regulation (EU) No. 503/2013 especially details procedures on the preparation and presentation of data for applications and is therefore of high relevance for applicants and risk assessors. In this implementing regulation the only direct reference to RNAi can be found within the European legislation on genetically engineered organisms. Under the section on toxicology, RNAi is covered indirectly by the mention of gene silencing as a genetic modification with potential toxicological impact. Annex I describes specifically all the information that an application shall contain. Within the section on molecular characterization, information on the expression of the insert is requested. Under point 1.2.2.3.(e), data requirements for gene silencing and RNAi approaches are specified as follows:

When justified by the nature of the insert (such as silencing approaches or where biochemical pathways have been intentionally modified), specific RNA(s) or metabolite(s) shall be analysed.

For silencing approaches by RNAi expression, potential 'off target' genes should be searched by in silico analysis to assess if the genetic modification could affect the expression of other genes which raise safety concerns ...

Under Regulation (EC) 1829/2003, two genetically modified soybeans (MON87705 and DP305423) and one maize (MON87411) producing small RNA molecules due to modifications of the genome have been authorized for placing on the market. In these cases, all uses with the exception of cultivation have been approved. In both soybean events the composition of fatty acids and oils has been changed. The change in composition has been achieved by a silencing approach targeting genes of the fatty acid metabolism of the modified plant itself. In contrast to this internal silencing effect, the construct in maize MON87411 results in the expression of dsRNA that targets an essential gene in a different species, namely *Diabrotica virgifera*, the corn rootworm, thus conferring resistance to this coleopteran pest.

As delineated above, products containing dsRNA that are not to be used in the food or feed sector and do not contain living organisms are not covered by EU legislation on GMOs. Such products, however, may be subject to other regulations, depending on the intended activity and use. Within the area of plant production, the most relevant examples of such products will be PPPs.

14.3 Regulation of PPPs Utilizing RNAi Mechanisms

Double-stranded RNA might be a new class of active substances in externally applied PPPs. From the scientific literature, such products may be used to control a range of different pathogens and pests.

In general, each active substance and any product placed on the market to protect plants needs an authorization. In the EU, the legal basis for this is provided by Regulation (EC) No.

1107/2009 (EC, 2009). The regulation foresees a two-step approach. In the first step the active substance must be assessed in an EU-wide process led by the European Food Safety Authority (EFSA) and approved by the EU Commission. In the second step the PPP containing the active substance is assessed by the Member States (MS). With Regulation (EC) No. 1107/2009 a zonal approach was introduced to streamline the authorization process (EC, 2009). The EU has been divided into three zones. The northern zone comprises the Scandinavian and Baltic countries, the central zone the countries of central and Eastern Europe and the southern zone the countries contiguous to the Mediterranean Sea plus Bulgaria. In a zone, one Member State (zonal rapporteur Member State) (zRMS) assesses the risk of a PPP for its whole zone. MS of this zone are obliged to follow the conclusion of the assessment of the zRMS. However, MS can claim national specificities and decide on specific risk management options for their country.

Data requirements for assessing active substances are laid down in Regulation (EC) No. 283/2013 (EC, 2013a) and for PPP in Regulation (EC) No. 284/2013 (EC, 2013b), respectively. Furthermore, the implementing Regulation (EU) No. 546/2011 (EC, 2011a) defines uniform principles for evaluation and authorization of PPPs. The aim is to ensure a high level of protection of human and animal health and the environment in all Member States. Additionally, guidance documents produced by the Organisation for Economic Co-operation and Development (OECD), the European and Mediterranean Plant Protection Organization (EPPO) and EFSA give detailed guidance on the methodological requirements for the risk assessment active substances and PPPs.

The above-mentioned documents define the data requirements and the decision criteria in detail. In consequence, a required data set is more or less fixed. It can be expected that RNA-based PPPs will have different properties than the chemicals mostly used as active substances in PPPs until now. Therefore, adaptations of the data requirement for the risk assessment might be reasonable for different reasons. For example, non-target arthropods are exposed to an active substance by contact, but for dsRNA-based PPPs usually oral exposure is needed to cause effects. A further aspect might be that new kinds

of risks have to be addressed, such as off-target effects. Additionally, the models used for the prediction of environmental fate are designed to assess specific chemicals, but are probably not applicable for assessing dsRNA-based products. Therefore, new tools might be introduced into the risk assessment of PPPs. One of these new methods might be bioinformatics supporting risk assessment.

Although the initial focus of the regulations was set on chemicals as active substances, the legislature has had other substances already in mind to protect plants. Therefore, article 77 of the Regulation (EC) No. 1107/2009 (EC, 2009) emphasizes the possibility that

> ... the Commission may ... adopt or amend technical and other guidance documents such as explanatory notes or guidance documents on the content of the application concerning micro-organisms, pheromones and biological products, for the implementation of this regulation. The Commission may ask the Authority to prepare or to contribute to such guidance documents.

The reason might be that other data requirements for these product classes are needed, compared with chemicals. In fact, for microorganisms, pheromones and botanicals, specific data requirements were developed and implemented (EC, 2011b, 2014a, b, 2016). However, discussions on adaptation of the data requirements are at an early stage (OECD, 2020) so that no specific guidance documents for dsRNA as active substance or dsRNA-based PPPs are available in the EU.

However, the effectiveness of dsRNA-based PPPs will, inter alia, depend on the stability of the dsRNA in the environment and transport into the cells of target organisms. Producers are likely to develop formulations containing synergists or co-formulants that stabilize dsRNA in the environment or to enhance transport into the target cells. Such targeted formulations of PPPs will also require consideration in the risk assessment.

The properties of dsRNA-based PPPs raise several additional questions related to procedural issues, which have to be clarified. For example: how much difference in the nucleotide sequences is acceptable to be considered as one active substance? Furthermore, would two dsRNAs differing in one nucleotide be considered as two active substances?

For economic reasons dsRNA would probably be produced in microbial production systems using genetically modified microorganisms. Questions regarding the variability in the product and the purity regarding the nucleotide sequence are as yet unanswered. Additionally, it must be guaranteed that the genetically modified microorganisms are completely inactivated. Otherwise an additional authorization for placing GMOs on the market is needed. It is important to note that products which do not contain a living organism and are not to be used as food or feed, like topically applied PPPs, are not regulated under GMO regulations within the EU.

Answers to these questions are urgently required in order to develop clear criteria to characterize the active substances and PPPs based on dsRNA.

The authorities responsible for placing PPPs on the market are confronted with a new class of products. The thoughts we highlighted in this chapter are a preliminary list of open questions and it is time to discuss these regulatory and biosafety issues intensively in order to define an adequate framework and design appropriate risk assessment procedures. Active ingredients and PPPs containing dsRNA are in the pipeline and the first applications can be expected within a few years.

References

EC (2001) Directive 2001/18/EC of the European Parliament and of the Council of 12 March 2001 on the deliberate release into the environment of genetically modified organisms and repealing Council Directive 90/220/EEC – Commission Declaration. *Official Journal of the European Union L* 106, 1–39.

EC (2003) Regulation (EC) No. 1829/2003 of the European Parliament and of the Council of 22 September 2003 on genetically modified food and feed. *Official Journal of the European Union L* 268, 1–23.

EC (2004) Commission Regulation (EC) No. 641/2004 of 6 April 2004 on detailed rules for the implementation of Regulation (EC) No. 1829/2003 of the European Parliament and of the Council as regards the application for the authorisation of new genetically modified food and feed, the notification of existing products and adventitious or technically unavoidable presence of genetically modified material which has benefited from a favourable risk evaluation. *Official Journal of the European Union L* 102, 14–25.

EC (2009) Regulation (EC) no. 1107/2009 of the European parliament and of the Council of 21 October 2009 concerning the placing of plant protection products on the market and repealing Council Directives 79/117/EEC and 91/414/EEC. *Official Journal of the European Union L* 309, 1–50.

EC (2011a) Commission Regulation (EU) No. 546/2011 of 10 June 2011 implementing Regulation (EC) No. 1107/2009 of the European Parliament and of the Council as regards uniform principles for evaluation and authorisation of plant protection products. *Official Journal of the European Union L* 155, 127–175.

EC (2011b) Guidance Document on the assessment of new substances falling into the group of Straight Chain Lepidopteran Pheromones (SCLPs) included in Annex I of Council Directive 91/414/EEC (SANCO/5272/2009 rev. 3, 28 October 2011).

EC (2013a) Commission Regulation (EU) No. 283/2013 of 1 March 2013 setting out the data requirements for active substances, in accordance with Regulation (EC) No. 1107/2009 of the European Parliament and of the Council concerning the placing of plant protection products on the market. *Official Journal of the European Union L* 93, 1–84.

EC (2013b) Commission Regulation (EU) No. 284/2013 of 1 March 2013 setting out the data requirements for plant protection products, in accordance with Regulation (EC) No. 1107/2009 of the European Parliament and of the Council concerning the placing of plant protection products on the market. *Official Journal of the European Union L* 93, 85–152.

EC (2013c) Commission Implementing Regulation (EU) No. 503/2013 of 3 April 2013 on applications for authorisation of genetically modified food and feed in accordance with Regulation (EC) No. 1829/2003 of the European Parliament and of the Council and amending Commission Regulations (EC) No. 641/2004 and (EC) No. 1981/2006. *Official Journal L* 157, 1–48.

EC (2014a) Guidance Document for the assessment of the equivalence of technical grade active ingredients for identical microbial strains or isolates approved under Regulation (EC) no. 1107/2009 (SANCO/12823/2012 – rev. 4, 12 December 2014).

EC (2014b) Guidance document on botanical active substances used in plant protection products (SANCO/11470/2012 – rev. 8, 20 March 2014).

EC (2015) Commission implementing decision (EU) 2015/692 of 24 April 2015 concerning the placing on the market, in accordance with Directive 2001/18/EC of the European Parliament and of the Council, of a carnation (*Dianthus caryophyllus* L., line 25958) genetically modified for flower colour. *Official Journal L* 112, 44–47.

EC (2016) Guidance Document on semiochemical active substances and plant protection products (SANTE/12815/2014 rev. 5, 2 May 2016).

OECD (2020) OECD Conference on RNAi-based Pesticides. Available at: https://www.oecd.org/chemical-safety/pesticides-biocides/conference-on-rnai-based-pesticides.htm (accessed 25 February 2020).

15 The Economics of RNAi-based Innovation: from the Innovation Landscape to Consumer Acceptance

Vera Ventura[1]* and Dario G. Frisio[2]

[1]Department of Civil, Environmental, Architectural Engineering and Mathematics, Università degli Studi di Brescia, Italy; [2]Department of Environmental Science and Policy, Università degli Studi di Milano, Italy

15.1 Introduction

RNA interference (RNAi) is an innovative technology of gene silencing which offers great opportunities for the development of sustainable solutions for crop protection (Palmgren *et al.*, 2015; Borel, 2017; Limera *et al.*, 2017; Zotti *et al.*, 2018). The most original aspect related to the economics of RNAi is the opening of a completely new innovation scenario consisting of new formulations of RNAi-based products for topical use, which are considered to be able to meet the need to find safer and more effective strategies for pest control and combat agricultural losses (Mitter *et al.*, 2017; Wang and Jin, 2017; Niu *et al.*, 2018). The possibility of substituting agrochemicals with more natural molecules is seen as the major advantage of these new technologies, which provide contributions towards a more sustainable agriculture (Collinge, 2018). In this context, academic interest in the economic aspects of this new technology is growing rapidly, suggesting that this innovative set of technologies is going to reshape the state of the art of the agricultural biotechnology (agbiotech) sector under multiple aspects, including the market structure (Bonny, 2017) and, most probably, public acceptance.

15.2 Market Potential of RNAi Innovation

After decades of debate on genetically modified organisms (GMOs), one of the most controversial 'science and society' issues able to divide scientific community and public opinion, a new wave of techniques has replaced the previous transgenic approach to plant breeding, introducing the possibility of imitating natural genetic recombination and thus avoiding the introduction of foreign genetic material. Among them, the economic landscape of RNAi-based innovation has been analysed. Frisio and Ventura (2019) investigated the structure of the global patent landscape of RNAi agricultural applications, identifying significant differences in the role of private and public research and evidencing the specialization of some universities and the rising power of Chinese research. Results revealed that China's pattern of innovation is able to stay at the forefront in most modern

*Corresponding author: vera.ventura@unibs.it

© CAB International 2021. *RNAi for Plant Improvement and Protection*
(eds B. Mezzetti *et al.*)
DOI: 10.1079/9781789248890.0015

agricultural biotechnologies, in stark contrast to the European scenario, where the regulatory landscape continues to impede the exploitation of agbiotech inventions. Mat Jalaluddin *et al.* (2019) provided an analysis of the global trend of RNAi-based product commercialization, using both bibliometric and patent data. They outlined that resistance against viruses, fungi and insect pests are the priorities for research activity and that the global market is rapidly moving toward huge investments in this field, with potential positive impacts on the development of RNAi technologies. These technologies could have very promising opportunities for being developed and applied in a broad range of agrifood products as well as in the formulation of innovative methods for biocontrol.

15.3   The Frontiers of Innovation: RNAi for Biocontrol

A new wave of RNA-based commercial products is ready to reach the market, with the first plant protectant product (to control rootworm) approved in the USA (EPA, 2017). Thus, the identification of the global scenario for RNAi technology innovation applied to biotic control, using patent data as indicators of innovation output, can provide some useful insights about this specific innovation scenario and its future applications (Chi-Ham *et al.*, 2010; Frisio *et al.*, 2010; Lundin, 2011; Egelie *et al.*, 2016)

The analysis has been carried out by mining the Questel-Orbit patent database through specific keywords for the identification of those inventions regarding the use of RNAi technology for plant biotic resistance. For this purpose, a set of keywords related to the term 'RNAi' have been searched in the 'title and abstract' field. The search has been limited to a number of International Patent Classification (IPC) and Cooperative Patent Classification (CPC) classes associated with biopesticides (IPC code A01N and CPC code Y02A-040). Time coverage of data is limited to the past 10 years (2010–2019). The original data set contains information about worldwide innovation in agricultural RNAi-based inventions, amounting to a total of 641 patent families. Then, with the aim of extracting from the data set only those inventions

specifically developed for plant protection, a text-mining analysis has been performed through double check in the patent title, abstract, claims and technical concepts, to identify those inventions referring to biotic control for agricultural application. The final data set is composed of 223 patent families, corresponding to 1224 single patents. In some cases, data elaboration has been performed making the distinction between inventions and patents. The term invention relates to the first filing of a patent application, anywhere in the world (usually in the applicant's domestic patent office). The statistics are based on the count of single inventions that provide information on the origin of the invention itself. Conversely, the term patent also refers to the set of patents filed in several countries that are related to the same invention, thus representing the so-called patent family. This variable is more indicative of the spread of innovation and its market, as the size of patent family is considered a proxy for the value of the invention.

Time trends outlined that RNAi technology applied to plant resistance is a field of innovation that has witnessed a good development globally in recent years, with an annual average number of new inventions equivalent to 22, corresponding to 122 patent applications. Nevertheless, Fig. 15.1 shows a peak in the numbers of patent filings in 2014 and a subsequent decline starting from 2015. Since patent applications are normally published after 18 months, data can be considered complete until 2017. The data set is composed of 223 inventions, whose legal status is 'alive' for 96% of cases, while the only nine inventions classified as 'dead' have been at some stage revoked, or lapsed. The analysis of the evolution over time of patent trends based on the nationality of the assignee indicates that, on a global level, the three main countries involved in this innovation sector are China (41.7%), the USA (26%) and the European Union (EU) (20%). The European data are quite surprising, since previous studies focusing on the analysis of the more global patent landscape of RNAi technology for plant improvement (Frisio and Ventura, 2019) revealed the marginal role of European players in producing innovations in this sector. This probably means that, amongst the different applications of RNAi technology, European research and development (R&D) activity shows greater competitiveness in the implementation

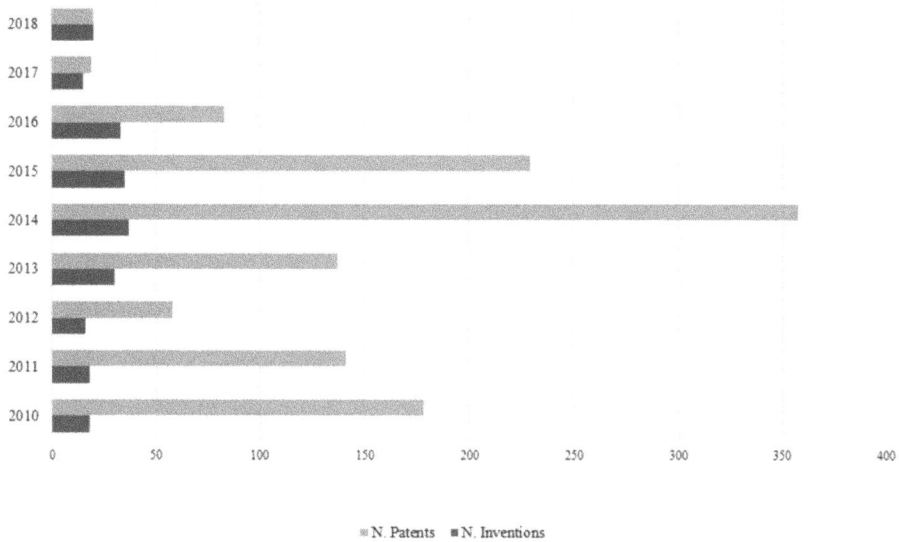

Fig. 15.1. Time trend for plant-RNAi inventions. (Source: own elaboration on Questel-Orbit data.)

of RNAi-based solutions for biotic resistance. The major contribution to the European innovation capacity derives from Germany, accounting for 26% of EU patents, principally applied by the agbiotech firms Bayer and BASF and, for public research, by the Max Planck and the Fraunhofer research institutes. The relevant role of Chinese applicants is most probably due to the massive investments in public research made by a government that considered agbiotech innovation a national priority. Nevertheless, the importance of Chinese applications dramatically decreases when considering the diffusion of inventions, represented by the number of patents filed in foreign patent systems, for which China accounts for only 8% of the total patenting activity.

The analysis of the type of assignee (Table 15.1) reveals that almost 47% of inventions are produced by public research, a value ten points greater than the private sector (35%). Moreover, nearly 18% of inventions derive from collaboration between public and private assignees. However, statistics related to the share of patents show that private players are more capable of exploiting inventions through their protection in different patent systems, as the value of the private sector's contribution moves from 35% of inventions to 55% of patents. It can be deduced that public sector R&D is competitive in producing innovative ideas and products for the application of RNAi technology in agriculture, but misses the opportunity to implement innovations in the form of more

Table 15.1 Analysis of the type of assignee. (Source: own elaboration on Questel-Orbit data.)

		% Share of inventions	% Share of patents
Single Assignee			
	Public	46.6	13.3
	Private	35.4	55.1
Multiple Assignee			
	Public–Private	14.3	29.2
	Public–Public	3.6	2.3

market-oriented solutions. A more detailed classification indicates that the public sector is principally composed of academic institutions, while the private sector is composed of the 'Big Four' agbiotech companies for 35% of the total data set, with an additional 25% represented by other biotech companies.

The top player is Dow Agrosciences (merged with Du Pont in 2017), the seed company most interested in investing in the development of this technology. Notably, this firm shares several patents with three public research institutions, showing a great level of public–private collaboration activity. Apart from the former 'Big Six' agbiotech companies, top assignees (Table 15.2) are small–medium firms specializing in very specific innovation sectors. For example, FuturaGene Ltd focuses on sustainable wood production, Forrest Innovation Ltd aims at providing eco-friendly solutions for mosquito vector control, RNAgri was born as a start-up specifically focused on RNAi-based products for modern agriculture. Considering the content of inventions, the innovative nature of this specific

use of RNAi technology emerges from the fact that 65% of patents do not have a single plant as target (30% plant not specified, 25% multiple applications and 10% multiple major crops). The remaining patents have maize as the major target plant (14%), followed by wheat and rice.

As for the analysis of the type of plant resistance, Fig. 15.2 shows that the main trait is insect resistance (79% of inventions), which is an impressive share indicating that this technology is considered to be more effective or even more easily applicable for insect control. Fungal control is included in 6% of patent application and relates to resistance to *Magnaporthe grisea*, *Botrytis*, *Verticillium* and *Zymoseptoria* species. Considering the minor categories, virus resistance accounts for 5% of patents, while nematode resistance (principally to the Heteroderidae family) represents 4% of the applications.

Finally, with regards to the subset of insect resistance, the analysis of the target species (Fig. 15.3) reveals that 32% of inventions relate to Hemiptera. The great majority of these patents derive from China and are intended to confer

Table 15.2 Top players. (Source: own elaboration on Questel-Orbit data.)

Applicant		No. of inventions	No. of patents
Dow Agrosciences llc		24	274
	with Fraunhofer Institute	18	175
	with University of Nebraska	9	147
	with University of Sidney	1	19
Syngenta - DevGen		13	140
BASF		6	50
Bayer CropScience Ag – Monsanto Co		8	62
	with Universitaet Hohenheim	1	9
FuturaGene Ltd		4	32
Forrest Innovations Ltd		1	26
AB Seeds		2	22
University of Queensland		2	20
United States Department of Agriculture		7	17
RNAgri		2	15
Nemgenix Pty Ltd		2	13
Caas (Institute of Crop Sciences)		10	11

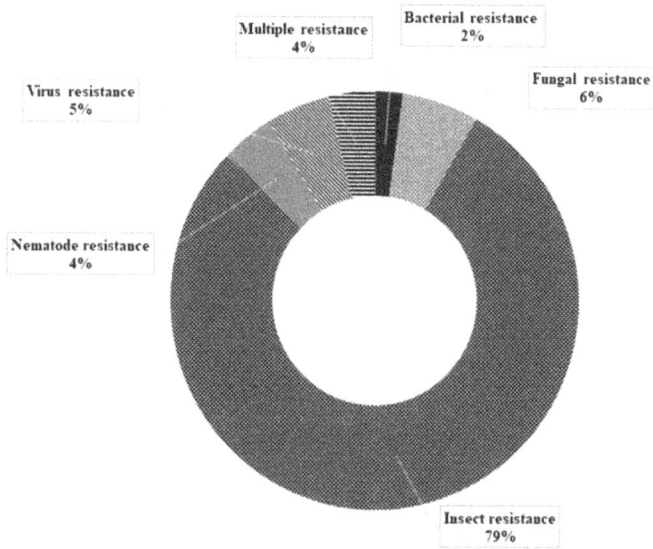

Fig. 15.2. Trait analysis. (Source: own elaboration on Questel-Orbit data.)

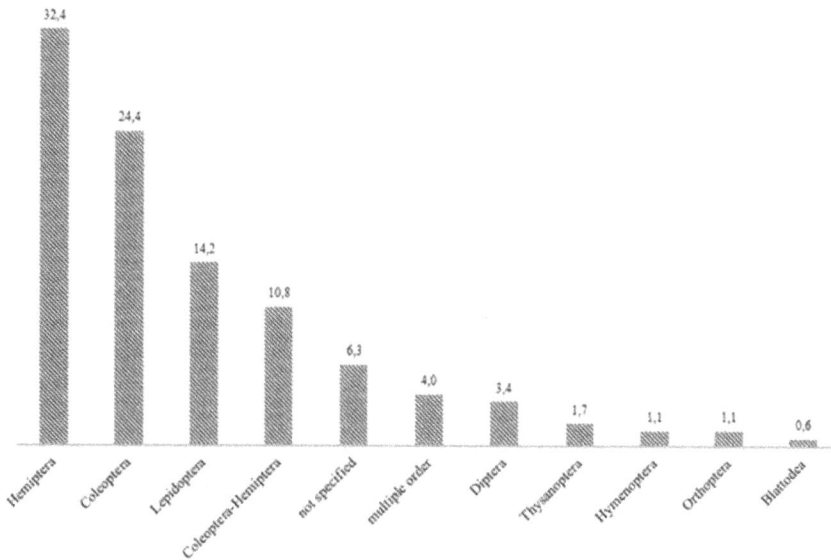

Fig. 15.3. Main targets for insect resistance.

resistance to the Aphididae family. The second type of insect resistance targets Coleoptera, almost entirely represented by the resistance to *Diabrotica* in maize. An additional 10% of patents aim at conferring resistance to both the Hemiptera and Coleoptera, while 14.2% are aimed at resistance to Lepidoptera. With regard to the type of application, the analysis showed that 24% of inventions contain a specific mention of the spray/topical application of the RNAi-based product.

15.4 RNAi: Stakeholder and Consumer Perceptions

Despite the fact that technological innovation plays a crucial role in enhancing the global sustainability of food chains and meeting changing consumers' needs and choices, growing evidence suggests that consumers tend to appreciate technology applications in many fields of their everyday life but tend to reject innovation when applied to the food domain. For this reason, academic research is focusing on the identification of the drivers of consumers' acceptance of innovative products, in order to find the most appropriate tools to mitigate consumer scepticism and resistance to these new technologies. In relation to new breeding techniques for crop improvement, public opinion has always shown one of the highest levels of rejection, principally based on the perceived unnaturalness of crop genetic modification (Mielby *et al.*, 2013; Kronberger *et al.*, 2014). Nevertheless, the literature suggests that not all the biotechnology solutions are perceived as being the same by consumers. Shew *et al.* (2017) showed that respondents valued CRISPR and GM food similarly and substantially less than conventional food, which could be detrimental for meeting future food demand. They also concluded that RNAi may be a better market alternative to more traditional biotechnologies such as GM crops expressing *Bacillus thuringiensis* (Bt) insect resistance. Topical application on plants avoids the need for genetic modification of plants, which could decrease consumer scepticism. Britton and Tonsor (2019) investigated the acceptance of a hypothetical RNAi beef product, concluding that consumers require a discount for buying the

innovative product compared with conventional ones. Nevertheless, they also stated that the way RNAi technology is framed in food labels could have an influence on its acceptance. Results could support policy makers in understanding the current determinants of consumer attitudes toward RNAi technologies, in particular the role played by communication. If the information gap represents one of the main barriers to consumer acceptance, policies including information campaigns or educational programmes could be recommended to make consumers more aware and informed during food choice. This aspect has been confirmed by the outcome of a meeting with stakeholders (seed companies, farmer associations, producers) organized by iPlanta in October 2018 in Brussels. The meeting offered the opportunity to exchange knowledge on RNAi technology, biosafety and socio-economic impacts. All stakeholders attending the meeting showed a high interest towards this innovative technology, especially as a potential solution for farmers' needs, but also expressed concerns mostly related to consumer acceptance of RNAi-based products. The meeting outlined the importance of defining common ground to discuss solutions with scientists and stakeholders and for engaging with consumers to reduce the knowledge gaps.

15.5 Conclusions

RNAi for plant biotic resistance is a field of innovation that has been receiving increasing interest in recent years, showing promising future applications and developments. Innovation is being produced by both public and private players. As for the latter category, some emerging small–medium firms are gaining market share by developing tailored solutions for specific problems. In this initial stage of development, insect management is the trait that is receiving the greatest attention in relation to RNAi technology, but new solutions for pest control reveal broad opportunities for the creation of new products for the agbiotech industry. A more comprehensive analysis of the economic costs and benefits for their production in the European Union will have to take into account certain aspects of the innovation supply chain.

Specifically, one of the major issues is how these new highly specific molecules will be classified in the existing EU regulation system (chemicals, bioregulators, biostimulants or biopesticides) (Taning *et al.*, 2019). If properly communicated to consumers, and inserted in the correct legal framework, the economic perspective of RNAi technology in the EU will lead to a growing market, rich in opportunities for all the actors of the agri-food chain.

References

Bonny, S. (2017) Corporate concentration and technological change in the global seed industry. *Sustainability* 9(9), 1632. DOI: 10.3390/su9091632.

Borel, B. (2017) When the pesticides run out. *Nature* 543, 302–304.

Britton, L.L. and Tonsor, G.T. (2019) Consumers' willingness to pay for beef products derived from RNA interference technology. *Food Quality and Preference* 75, 187–197. DOI: 10.1016/j.foodqual.2019.02.008.

Chi-Ham, C.L., Clark, K.L. and Bennett, A.B. (2010) The intellectual property landscape for gene suppression technologies in plants. *Nature Biotechnology* 28(1), 32–36. DOI: 10.1038/nbt0110-32.

Collinge, D.B. (2018) Transgenic crops and beyond: how can biotechnology contribute to the sustainable control of plant diseases? *European Journal of Plant Pathology* 152(4), 977–986. DOI: 10.1007/s10658-018-1439-2.

Egelie, K.J., Graff, G.D., Strand, S.P. and Johansen, B. (2016) The emerging patent landscape of CRISPR-Cas gene editing technology. *Nature Biotechnology* 34(10), 1025–1031. DOI: 10.1038/nbt.3692.

EPA (2017) EPA registers innovative tool to control corn rootworm. US environmental protection agency. Available at: www.archive.epa.gov/newsreleases/epa-registers-innovative-tool-control-corn-rootworm (accessed 18 November 2020).

Frisio, D.G. and Ventura, V. (2019) Exploring the patent landscape of RNAi-based innovation for plant breeding. *Recent Patents on Biotechnology* 13(3), 207–216. DOI: 10.2174/1872208313666190204121109.

Frisio, D., Ferrazzi, G., Ventura, V. and Vigani, M. (2010) Public vs. private agbiotech research in the United States and European Union, 2002–2009. *AgBioForum* 13(4), 333–342.

Kronberger, N., Wagner, W. and Nagata, M. (2014) How natural is 'more natural'? The role of method, type of transfer, and familiarity for public perceptions of cisgenic and transgenic modification. *Science Communication* 36(1), 106–130. DOI: 10.1177/1075547013500773.

Limera, C., Sabbadini, S., Sweet, J.B. and Mezzetti, B. (2017) New biotechnological tools for the genetic improvement of major woody fruit species. *Frontiers in Plant Science* 8, 1418. DOI: 10.3389/fpls.2017.01418.

Lundin, P. (2011) Is silence still golden? Mapping the RNAi patent landscape. *Nature Biotechnology* 29(6), 493–497. DOI: 10.1038/nbt.1885.

Mat Jalaluddin, N.S., Othman, R.Y. and Harikrishna, J.A. (2019) Global trends in research and commercialization of exogenous and endogenous RNAi technologies for crops. *Critical Reviews in Biotechnology* 39(1), 67–78. DOI: 10.1080/07388551.2018.1496064.

Mielby, H., Sandøe, P. and Lassen, J. (2013) Multiple aspects of unnaturalness: are cisgenic crops perceived as being more natural and more acceptable than transgenic crops? *Agriculture and Human Values* 30(3), 471–480. DOI: 10.1007/s10460-013-9430-1.

Mitter, N., Worrall, E.A., Robinson, K.E., Li, P., Jain, R.G. *et al.* (2017) Clay nanosheets for topical delivery of RNAi for sustained protection against plant viruses. *Nature Plants* 3(2), 1–10. DOI: 10.1038/nplants.2016.207.

Niu, J., Taning, C.N.T., Christiaens, O., Smagghe, G. and Wang, J.J. (2018) Rethink RNAi in insect pest control: challenges and perspectives. *Advances in Insect Physiology* 55, 1–17.

Palmgren, M.G., Edenbrandt, A.K., Vedel, S.E., Andersen, M.M., Landes, X. *et al.* (2015) Are we ready for back-to-nature crop breeding? *Trends in Plant Science* 20(3), 155–164. DOI: 10.1016/j.tplants.2014.11.003.

Shew, A.M., Danforth, D.M., Nalley, L.L., Nayga, R.M., Tsiboe, F. *et al.* (2017) New innovations in agricultural biotech: consumer acceptance of topical RNAi in rice production. *Food Control* 81, 189–195. DOI: 10.1016/j.foodcont.2017.05.047.

Taning, C.N., Arpaia, S., Christiaens, O., Dietz-Pfeilstetter, A., Jones, H. *et al.* (2019) RNA-based biocontrol compounds: current status and perspectives to reach the market. *Pest Management Science* 76(3), 841–845. DOI: 10.1002/ps.5686.

Wang, M. and Jin, H. (2017) Spray-induced gene silencing: a powerful innovative strategy for crop protection. *Trends in Microbiology* 25(1), 4–6. DOI: 10.1016/j.tim.2016.11.011.

Zotti, M., Dos Santos, E.A., Cagliari, D., Christiaens, O., Taning, C.N.T. *et al.* (2018) RNA interference technology in crop protection against arthropod pests, pathogens and nematodes. *Pest Management Science* 74(6), 1239–1250. DOI: 10.1002/ps.4813.

16 Future Plant Solutions by Interfering RNA and Key Messages for Communication and Dissemination

Hilde-Gunn Opsahl-Sorteberg*

BIOVIT, NMBU, N 1432 – Ås, Norway

Abstract

Communication is an increasing prerequisite to justify academic existence and value, and for project funding of all kinds to show relevance and value, including the future of European networks like COST Actions. Academia is slowly adapting to this expectation and learning the profession of communication. Language and vocabulary are key issues in communication, and particularly to reach the many important non-scientific audiences. Therefore, this chapter starts with a description of some new plant breeding technologies relevant for communicating, in general terms, the science behind plant improvement. This is followed by selected examples of the application of these techniques to improve current and future crop varieties. Finally, key messages gathered from the European iPLANTA project for policy makers, non-specialists and specially interested citizens are communicated. This is to show a wider audience how RNAi can contribute to sustainable food solutions and food security with minimal environmental impacts.

16.1 Plant Breeding Tools to Meet Future Sustainable Food Security

Plant breeding uses a range of techniques to develop new varieties of plants that cope with current and future environmental stresses, including climate challenges, pests and diseases, in order to produce optimal feed and food products. These plant varieties typically have improved yield and quality, combined with improved use of natural resources. They are therefore more sustainable and contribute towards feeding the growing global population. Plants need to meet increasing challenges from pests and diseases, changing climates and demands to reduce inputs of water and fertilizer. However, to meet all these challenges we need to substantially increase plant breeding efforts and use a range of new scientific tools. These new tools include new breeding technologies that allow greater levels of genetic modification of plants, to introduce desired characteristics not achieved before (Madre and Agostino, 2017). These plant breeding technologies are discussed below, including the introduction of RNA interference (RNAi) into plants, which allow genes to be turned up or down.

*hildop@nmbu.no

16.1.1 Conventional plant breeding

All plant varieties grown and eaten today have been genetically changed (modified) by plant breeding. Most current varieties have been changed by what are known as conventional methods like selection, crossing selected beneficial parents, laboratory techniques like doubling the chromosomes from gamete cells like pollen to get pure-bred offspring with two identical alleles of each gene from one parent (the pollen donor), or mutating cells to increase genetic variation and get new characteristics, like pink grapefruit.

16.1.2 Genetically modified plants

Genetically modified organisms (GMOs) can be defined biologically or legally, and not necessarily be classified the same in both 'worlds'. If the genetic change has been done by the extraction of DNA from an organism and adding it to another organism by laboratory techniques (termed transgenesis), they are defined as GMO under international United Nations (UN) protocols and hence by national and European Union (EU) legislation. When defined according to law there are different legal interpretations in different countries, sometimes embracing a range of non-transgenic technologies as well as the biological GMO. In the EU, GM regulations were interpreted by the EU Court of Justice in 2018 to include plants that have been mutated by certain gene editing techniques that change plant gene arrangements, but not those that have been mutated by other means such as irradiation (Heitz, 2020). Other countries, such as the USA, China, Canada and Australia, do not consider that gene edited mutants should be legally classified as GMOs. Discussions are ongoing to achieve international consensus on these definitions. The regulation and assessment of GMOs in the EU, particularly GM RNAi plants, are discussed in Chapters 13, 14 and 15.

16.1.3 Mutation breeding techniques

Mutations spontaneously occur in nature and many have been used to produce new plant types and varieties. Conspicuous examples are variegated ornamental plants and contorted or dwarf types. Mutation frequencies can be increased by using chemicals to disrupt cell division so that uneven numbers of genes occur in cells, or by irradiation, which damages genes. The mutated plants are examined for desirable types which are then selected, tested and propagated to produce new varieties. Examples include most dwarf and semi-dwarf wheat varieties currently grown, which put more resources into grain production than vegetative parts as in taller varieties. Genetic technologies have enabled the genetic components of most crops to be characterized so that genes can be precisely modified and edited to change their expression. For example, genes producing plant toxins and allergens can be inhibited or removed to improve plant quality using these gene editing techniques.

16.1.4 RNA interference

Since the discovery of RNAi (Baulcombe, 2019 and references therein), the mechanism has gained increasing recognition due to its important applications in medicine and feed/food production. RNAi is a central tool in functional genomics, since it allows basic studies of all genes, which is important to understand gene function and genomic interaction between genes/DNA and RNA sequences. The results and uses are from low to full downregulation of single genes or gene families, in order to change plant characteristics and improve plant varieties while protecting natural resources (Christiaens et al., 2018). RNAi can lead to improved plant protection against pests, diseases and environmental stresses. Food and feed quality can be improved to reduce losses along the food chain and provide better nutritional value for consumers. In the USA a corn (maize) variety has been commercialized with resistance to the root worm pest and a papaya has been bred with virus resistance. Oilseed crops such as soybean have had their oils modified to contain improved fatty acid profiles; and spoilage of fruits such as apple during storage has been reduced. Also, allergens are being removed from some crops, such as wheat.

16.2   Applications of RNAi for Gene Regulation and Public Acceptance of New Plant Foods

Some RNAi products have been approved for marketing globally, but in Europe most have only been approved for animal feed use. The major challenge in Europe is the extensive regulatory demands to get a GMO marketed in most countries. This means that only large companies can afford the extra costs of producing large amounts of information to demonstrate safety and stability of the products, reducing the opportunities for small and medium-sized actors. In addition, the EU has the problem that a number of countries are blocking the cultivation of approved GM crops and there is political and social opposition to the consumption of GM foods that have been assessed for safety.

Globally there are marked differences in the public and political acceptance of GMOs between continents, such as America and Europe. This is not based on scientific evidence, but on perceptions of the different consumers often driven by non-governmental organizations (NGOs) that oppose the use of GM technology in foods (though not in medicine). Many of the new products of GM plants have nutritional and quality benefits of direct value to consumers and yet public acceptance is problematic. Recent examples are the products of the companies Impossible Foods (IFs) and Beyond Meat, which illustrate how different perceptions, public acceptance and regulations affect consumer availability, and how beneficial products can affect and change consumer choices. Recently, we have seen consumers shifting towards more environmentally friendly food choices and an increasing awareness of the environmental impacts of livestock farming in particular.

The goal for IFs is to develop a green alternative to the 95% human population that prefers eating meat due to its flavour. The company found that haemoglobin causes the meatiness characteristic of beef, and that it could be mimicked by adding haeme from soybean roots. However, it would require harvesting and extraction from large amounts of roots. By contrast, transferring the haeme-producing gene to laboratory cell cultures allows efficient production of large amounts of haeme. This is added to the pure vegetable components of the impossible burger and, in tests, people can not tell the difference between a regular meat burger and an IFs burger. Consumers welcomed this new choice and it has become a number one seller at Burger King and the stores selling it. IFs has been upfront about why it depends on a GM product to add the meat flavor to its 100% veggie-burger, and this has been accepted by US consumers. In Europe the approval of this GM product might be blocked by political and activist groups unless public perceptions and approval are changed. Europe so far only has the Beyond Meat burger, since this is not using GM. However, it does not have the meaty flavor caused by haeme, so appeals mostly to vegetarians.

IFs is already hitting the global food market in the USA beyond all expectations (Gravier, 2019; Fontanazza, 2020). When looking ahead to completely new food production solutions that possibly increase sustainability beyond any previous food alternatives, Solar Foods is a powerful example. The company uses microbes to produce proteins directly using atmospheric conditions, water and solar energy. Such solutions depend on advanced understanding of functions and availability of genes and precise regulation of the selected genes. In addition, we see emerging companies like Solar Foods from Finland making food from air and water, possibly 20,000 times more sustainable than current food production (Southey, 2019; Solar Foods, 2020). Headlines like 'New food solutions will save the planet but kill farming' have been produced by journalist George Monbiot in The Guardian newspaper (Monbiot, 2020). IFs achieved 'Generally Regarded As Safe' (GRAS) approval for restaurant provision from mid-2018, retail sales in 2019 and started home-deliveries in 2020.

16.3   How RNAi Communication Can Hit the Goals of Relevance, Surprise, Solution, Challenge and Obstacle

Successful communication is achieved when relevance is combined with a selected message being received and understood by the target audience. This takes clear wording and messages. Additionally, communication must fit into attention spans and compete with many platforms in a rapidly changing media world.

From updated professional media courses given by the European Co-operation in Science and Technology (COST, created in 1971) academy, the messages are that, for social media, Facebook is still the platform reaching most people with 2.3 billion users, followed by YouTube with 1.9, WhatsApp 1.5 and FB Messenger with 1.3 billion users (ABS CBN News, 2019). YouTube's rapid rise is due to videos being more efficient at reaching target audiences than still photos and written texts. LinkedIn is of special value to professionals; despite having a lower number of users, it provides a platform for a well-educated and influential audience. It also publishes short stories and papers in addition to job market information. Twitter and Instagram are widely used but are mainly picture driven and less text oriented, with correlated limitations.

16.3.1 Relevance

There is growing global consumer demand for meeting climate change goals while feeding our growing population. This will require increased production of high-quality foods with reduced levels of inputs on existing land surfaces. Such sustainable production demands some plant production and food systems to replace animal sources. This additionally meets the increased market trend for plant-based food. Greta Thunberg is a strong advocate for this drive, being selected as one of the most influential leaders in 2019 by *Time* and *Nature* magazines (Alter *et al.*, 2019; Nature, 2019). Her message that her generation's future is lost unless meat consumption is dramatically reduced is helping to drive food shifts towards meat replacements. Applying RNAi technology to plant production will contribute to meeting these new food demands. RNAi products improving plants and protecting them from external damaging factors like pests and climate change will be very relevant for achieving these sustainability goals.

16.3.2 Surprise

Our brain is designed to save power and run on default unless surprised (Luna and Renninger, 2015). Therefore, to successfully gain the attention of an audience, surprise and telling the story while attention is held is an important factor. This can be done with videos providing a few key messages to increase impact, or by piggybacking on popular podcast hosts. Podcasts have longer airtime and professionals can adapt the messages and wording to deliver more complicated messages.

A good example of a successful plant food message success is how the IFs burger has achieved a market share beyond any expectations (Gravier, 2019; Fontanazza, 2020). In addition, messages from emerging companies like Solar Foods stating they can make food from air and water (Southey, 2019; Solar Foods, 2020) have prompted headlines like 'New food solutions will save the planet but kill farming' (Monbiot, 2020), making a large impact.

16.3.3 Solutions

While achieving attention, it is important to provide solutions to challenges arising in the minds of audiences. For example, nutritionally enhanced foods provide solutions to allergies, vitamin deficiencies and alternatives to meat. RNAi plants provide solutions to controlling pests and diseases without the need for pesticides. Thus, the new crop varieties provide clear advantages for production and quality. Covid-19 vaccines based on mRNA demanding -70 °C storage can be avoided if replacing the unstable mRNA with short RNAi, showing how RNAi can contribute with extremely high impact solutions.

16.3.4 Challenges

To meet future food requirements, we require improved productivity and quality of crops and protection against pests, diseases and environmental stresses using minimal inputs and on limited land areas. Plant breeding can meet this challenge if it is permitted to use the wide range of new technologies. RNAi is an important technology to meet this challenge, either activated through plant breeding or in developing new biological plant protection treatments (see Chapters 9 and 11).

16.3.5   Obstacles

The main obstacles to meeting the challenges and providing solutions should be presented clearly, together with strategies for overcoming them. For GM plants, and RNAi in particular, the main challenges are public perception and regulations. Regulatory frames should be soundly science based and determine whether a new variety is safe for consumption and its environmental impacts are the same as or less than those for similar varieties. However, a major additional obstacle in some countries is that there is political interference in the regulatory process and non-acceptance of the scientific findings. This is often driven by lobbying organizations who are opposed to many aspects of new scientific development and have a powerful influence on public perception, politicians and decision makers. In addition, regulations are often lagging behind scientific progress so that they are not fit for assessing new technologies. Good science communication is thus required to demonstrate clearly the present and future roles that new varieties can play in meeting sustainable food solutions. In addition, we need updated regulatory processes that permit improved plant solutions and we need to assess expected new plant breeding technologies and their products (Hartung and Schiemann, 2014; Zetterberg and Edvardsson Björnberg, 2017).

16.3.6   Consumer choices

Future food solutions will depend on available tools, including plant breeding technology, and an understanding of the choices paramount to keep working democratic principles for anchoring decisions for the common good. Consumer trends show rapid shifts demanding real changes to increase sustainable food production, while protecting the environment and meeting future climate changes. IFs, for example, is clear on the need for gene technology and gene transfer to make true 'meat-like' alternatives to satisfy current meat consumers. Other plant breeding solutions such as RNAi will be required to develop sustainable production of nutritious foods to provide choice to growing populations.

16.3.7   The wording best used in communicating RNAi

In order to improve communication with non-scientific audiences and the general public it is important that the language used is clear and succinct and does not use words that have double meanings or emotive effects. Therefore, in relation to RNAi, 'genetic interference' must be clearly explained to avoid negative reactions and should be described as on/off switches for genes to prevent the expression of undesirable characteristics. Scientists also typically say RNAi causes 'knockdown' of gene expression which results in increasing/reducing protein production. The analogy with on/off switches could be taken further to describe the system as being like a dimmer switch which can increase or decrease the expression of a gene. Gene silencing is also used to describe RNAi activity and this could be described as analogous to the volume switch on a radio where the sound level can be regulated. Also 'mutant' is considered a negative word that people may associate with, say, Zombie films. It should be explained that mutation is simply a change, which can be for the good and produce many desirable variants, such as those described above.

16.4   RNAi: Key Messages

The selected messages below are collected from the COST Action iPlanta working groups. Some of the messages overlap as the working groups themes do to secure full coverage of RNAi thematics, and the messages are formed and worded to reach a wide audience.

16.4.1   Exploiting RNAi to improve plant production

- RNAi is a biological phenomenon exploited by scientists to develop molecular tools for controlling pests and diseases, by changing the expression of desirable or undesirable genes, to secure future plant production. The plant's RNA can be guided to target genes in pests and diseases on the plant by inhibiting essential gene(s) in the pest or pathogen. This process is very selective and generally considered much safer than alternative methods of control, providing promising new solutions for crop protection.

- In addition, RNAi is being developed as externally applied treatments (e.g. sprays) which are highly selective to target pest species, have little or no effect on non-target and beneficial organisms and low environmental persistence. Hence biological pesticides provide a very efficient alternative technical solution for sustainable pest and disease control.

decreased, likely due to the cost and strict regulation of GMOs. This is blocking the potential of RNAi plants to aid adaptation of future crop and reforestation material to meet future requirements and climate change challenges.

- For external RNAi treatments of pests and diseases that does not involve GM plants, appropriate regulatory frameworks, including science-based risk assessment procedures, have not been developed and therefore are needed (see Chapters 9 and 11).

16.4.2 Applications

- The understanding of RNAi is now adequate to allow safe use for commercial applications as a tool to treat human and plant diseases and to develop improved crop varieties in agriculture and horticulture.
- In crops, some applications of RNAi are by genetically modified plants, others are through non-GM methods that are regulated differently.
- Robust pre-market regulatory procedures exist for the risk assessment and authorization of GMO RNAi plants (see below).
- Several crops have been modified by RNAi to change their quality characteristics and some have been commercialized in Europe. The amylogen potato with modified starch for industrial production has been approved for cultivation in the EU and imports of food/feed commodity crops such as soy and maize with changed nutritional characteristics have been approved.
- For pest control, the main commercial crop application is for control of root worm in maize in the USA. The main commercialized application in trees has been in papaya for control of ringspot virus. Plum with resistance to sharka virus has also been approved in North America but not elsewhere. Also, RNAi has been developed in trees such as poplar, prunus and citrus with great promise but not yet commercialized.
- Despite the promising applications of RNAi, in Europe the number of field trials has

16.4.3 Biosafety of RNAi

- RNA interference is widely present in plants and animals, meeting criteria for a history of safe use. In addition, dsRNA activity in higher animals is limited by existing biological barriers, meaning RNAi is safe in feed and food.
- RNAi can selectively target pest organisms, including viruses, and is a promising alternative to chemical control and increasing agriculture's sustainability.
- Biosafety assessments using bioinformatic tools allow design of specific dsRNA, avoiding adverse effects on known non-target organisms (NTOs). However, this depends on information about the NTO's genome sequence, so that until we have genome sequences for higher numbers of NTOs, initial bioinformatic analysis needs further confirmation with tests of NTO sensitivity.

16.4.4 Socio-economics of RNAi

- Consumer preferences are rapidly changing towards more sustainable production, affecting previous consumption models. Increasing demands for environmentally friendly food products like plant-based solutions can be delivered by RNAi technology.
- RNAi solutions can additionally have positive economic impacts for agriculture,

by both reducing costs and increasing productivity.

- Current European regulation and its operation is a major barrier for the introduction of RNAi-based products, hampering innovation capacity and competitiveness of EU firms and negatively affecting consumers and the environment.

Websites

- Beyond Meat:https://www.beyondmeat.com
- Impossible Foods (IFs): https://impossiblefoods.com
- Solar Foods:https//solarfoods.fi

References

ABS CBN News (2019) Filipinos still world's top social media users. Available at: https://news.abs-cbn.com/focus/01/31/19/filipinos-still-worlds-top-social-media-user-study (accessed 30 October 2020).

Alter, C., Haynes, S. and Worland, J. (2019) *Time* 2019 Person of the Year: Greta Thunberg. Available at: https://time.com/person-of-the-year-2019-greta-thunberg (accessed 30 October 2010).

Baulcombe, D.C. (2019) How virus resistance provided a mechanistic foundation for RNA silencing. *The Plant Cell* 31(7), 1395–1396. DOI: 10.1105/tpc.19.00348.

Christiaens, O., Dzhambazova, T., Kostov, K., Arpaia, S. and Reddy Joga, M. (2018) *Literature review of baseline information on RNAi to support the environmental risk assessment of RNAi-based GM plants.* Vol. 15, Issue 5. EFSA Supporting Publications, European Safety Authority, Parma, Italy, p. 173.

Fontanazza, M. (2020) How plant-based foods are changing the supply chain. Available at: https://food-safetytech.com/news_article/how-plant-based-foods-are-changing-the-supply-chain/ (accessed 30 October 2020).

Gravier, E. (2019) Impossible foods hired mobile phone execs to meet high demand. Available at: https://www.cnbc.com/2019/11/04/impossible-foods-hired-mobile-phone-execs-to-meet-high-demand.html (accessed 30 October 2020).

Hartung, F. and Schiemann, J. (2014) Precise plant breeding using new genome editing techniques: opportunities, safety and regulation in the EU. *The Plant Journal* 78(5), 742–752. DOI: 10.1111/tpj.12413.

Heitz, A. (2020) Viewpoint: French court ruling that already approved mutagenized crops should be heavily restricted as GMOs reaffirms need to revamp Europe's antiquated biotech regulations. Available at: https://geneticliteracyproject.org/2020/02/14/viewpoint-french-court-ruling-that-already-approved-mutagenized-crops-should-be-heavily-restricted-as-gmos-reaffirms-need-to-revamp-europes-antiquated-biotech-regulations/ (accessed 30 October 2020).

Luna, T. and Renninger, L. (2015) *Surprise: Embrace the Unpredictable and Engineer the Unexpected.* TarcherPerigree, New York, p. 112.

Madre, Y. and Agostino, V.D. (2017) New plant-breeding techniques: what are we talking about? farm Europe. Available at: https://www.farm-europe.eu/travaux/new-plant-breeding-techniques-what-are-we-talking-about/ (accessed 30 October 2020).

Monbiot, G. (2020) Lab-grown food will soon destroy farming – and save the planet. Available at: https://www.theguardian.com/commentisfree/2020/jan/08/lab-grown-food-destroy-farming-save-planet (accessed 30 October 2020).

Nature (2019) Nature's 10. Ten people who mattered in science in 2019. Greta Thunberg. Available at: www.nature.com/immersive/d41586-019-03749-0/index.html (accessed 30 October 2020).

Solar Foods (2020) Now you see it. Available at: https://solarfoods.fi/solein/#now-you-see-it (accessed 30 October 2020).

Southey, F. (2019) Solar foods makes protein out of thin air. Available at: https://www.foodnavigator.com/Article/2019/07/15/Solar-Foods-makes-protein-out-of-thin-air-This-is-the-most-environmentally-friendly-food-there-is (accessed 30 October 2020).

Zetterberg, C. and Edvardsson Björnberg, K. (2017) Time for a new EU regulatory framework for GM crops? *Journal of Agricultural and Environmental Ethics* 30(3), 325–347. DOI: 10.1007/s10806-017-9664-9.

Glossary

ABC: ATP-binding cassette

agbiotech: agriculture biotechnology

AGO, Ago: Argonaute

AM: assisted migration

amplification of RNAi: amplification might be required for efficient RNA-mediated silencing. In *Caenorhabditis elegans* and in plants, primary siRNAs can act as primers for the synthesis of additional dsRNA, using the target mRNA as a template, in a reaction catalysed by a RNA-dependent RNA polymerase (RdRP). The newly synthesized dsRNA is then cleaved by Dicer to generate secondary siRNAs, thereby amplifying RNA silencing.

Argonaute (AGO, Ago): a family of evolutionarily conserved genes. Their protein products are involved in various RNA interference processes because of being part of the RNA-induced silencing complex (RISC).

BLAST: Basic Local Alignment Search Tool

Bt: *Bacillus thuringiensis*

CAS: CRISPR-associated

cDNA: complementary DNA

CDS: coding sequence

COST: European Cooperation in Science and Technology programme

CP: coat protein

CPB: Colorado potato beetle

CQD: carbon quantum dot

CRAC: cholesterol recognition/interaction amino acid consensus

CRISPR: clustered regularly interspaced short palindromic repeats

CMPP: cell membrane penetrating peptide

DCL: Dicer-like proteins in plants

DdRP: DNA-dependent RNA polymerase

Dicer (Dcr): a ribonuclease III enzyme; a double-stranded RNA-specific endonuclease that processes dsRNAs to 20–25 nt siRNAs during RNA interference and excises miRNAs from precursor miRNA-hairpins.

diRNA: defective interfering RNA derived from RNA viruses

dsRNA: double-stranded RNA, i.e. RNA with two ribonucleic acid strands. dsRNAs longer than 30 nucleotides are the precursors of the siRNA that can trigger RNAi.

easiRNA: epigenetically activated 21 nt small interfering RNA, a type of siRNA

EFSA: European Food Safety Authority

EMBRAPA: Empresa Brasileira de Pesquisa Agropecuaria (Brazilian Agricultural Research Corporation)

endo-siRNA: endogenous siRNA, produced from endogenous dsRNA; involved in genome protection and gene regulation

EPPO: European and Mediterranean Plant Protection Organization

ERA: environmental risk assessment

ERF: ethylene responsive factor

EU: European Union

exo-siRNAs: exogenous siRNA derived from exogenous dsRNA; involved in antiviral defence

FAD: fatty acid dehydrogenase

FAO: The Food and Agriculture Organization of the United Nations

GA: gibberellic acid

GE: gene editing; genetically engineered

gene silencing: Any mechanism that silences a gene, such as various sequence homology-dependent silencing mechanisms (RNAi, PTGS, TGS, VIGS).

GFP: green fluorescent protein

GM: genetically modified

GMO: genetically modified organism. Organism in which *in vitro* prepared DNA is incorporated into its genome early in development. The newly inserted DNA is present in both somatic and germ cells, is expressed in one or more tissues and is inherited in a Mendelian fashion.

HC-Pro: helper-component proteinase

HDR: homology-directed repair

HIGS: host-induced gene silencing

hpRNA: hairpin RNA; a structure in which adjacent segments of RNA fold together and are stabilized by base pairing, creating a loop of single-strand RNA. Short hairpin RNAs can be engineered to suppress the expression of desired genes in cells. hpRNAs can be transcribed from RNA polymerase II promoters *in vivo*, thus permitting the construction of continuous cell lines.

HTS: high-throughput sequencing

ihpRNA: intron-spliced hairpin RNA

indels: insertions and deletions

IPC: International Patent Classification

iPlanta: COST action CA15223 to study RNAi genetic improvement methods, funded by the European Union

IPM: integrated pest management

LA: linoleic acid

LDH: layered double hydroxide

LDLRAP1: low-density lipoprotein receptor adapter protein 1

lncRNA: long non-coding RNA

MEEC: maximum expected environmental concentration

MIGS: miRNA-induced gene silencing

miPDC: miRNA precursor deposit complex

miPEPS: miRNA encoded peptides

miRNA: microRNA. Small RNAs that interact with components shared by the RNA-induced silencing complex (RISC). miRNAs play a central role in the regulation of gene expression in cells.

miRBase: miRNA database

miRISC: miRNA-induced silencing complex

miRLC: miRISC loading complex

miRNPs: microRNA ribonucleoproteins

MN: meganuclease

mRNA: messenger RNA, the RNA template for protein synthesis. mRNA is formed by transcription of the template DNA strand, followed by the excision of introns and the joining of exons to form mature mRNA. The mRNA is next translated to polypeptides making up proteins.

MS: Member States

NBS: nucleotide binding site

NBT: new breeding technique

ncRNA: non-coding RNA

NEP: newly expressed protein

NGS: next-generation sequencing

NHEJ: non-homologous end joining

nt: nucleotide

NTO: non-target organism

ODM: oligonucleotide-directed mutagenesis

OECD: Organisation for Economic Co-operation and Development

Off-target gene silencing effects: Suppression of genes other than the target gene. Off-target effects have been correlated with the concentration of siRNAs, as well as similarities between the off-target transcripts and the 5' ends of siRNAs.

ORF: open reading frame

PAMP: pathogen-associated molecular pattern

PAZ: Piwi/Argonaute/Zwille

PDR: pathogen-derived resistance

PF: problem formulation

phasiRNA: phased siRNA

PIP: prolactin-induced protein

piRNA: PIWI-interacting RNA

PIWI: P-element induced wimpy testis, a domain in particular AGO proteins

PPO: polyphenol oxidase

PPP: plant protection product

pre-miRNA: miRNA precursor; a small hairpin precursor before Dicer cleavage

pre-siRNA: siRNA precursor

pri-miRNA: primary miRNA; a primary miRNA transcript before processing to pre-miRNA

PRINT: particle replication in non-wetting templates

PTD–DRBD: peptide transduction domain–dsRNA-binding domain

PTGS: post-transcriptional gene silencing. The transcription of the gene is unaffected; however, gene expression is suppressed because mRNA is degraded and/or not translated. PTGS is involved in regulation of gene expression and provides nucleotide sequence-specific protection against a variety of foreign genetic elements, including viruses.

PUFA: polyunsaturated fatty acid

qRT-PCR: quantitative reverse transcriptase polymerase chain reaction

R&D: research and development

RC: RNA control

RdDM: RNA-directed DNA methylation

rDNA: ribosomal DNA – a part of the genome encoding ribosomal RNA

RdR6: RNA-dependent RNA polymerase VI, involved in the amplification of RNAs in some organisms such as *C. elegans* and plants.

RenSeq: resistance gene enrichment sequencing

RISC: RNA-induced silencing complex. An enzyme complex containing aspecific Argonaute protein that uses the sequence encoded by one of the siRNA strands or the miRNA guide strand to target mRNA of complementary sequence for degradation and/or translational repression.

RLC: RISC loading complex

RNAi: RNA interference. The RNAi pathway is based on two steps, each involving ribonuclease enzymes. In the first step, the trigger RNA (either dsRNA or pre-miRNA) is processed into short RNAs (sRNAs) by the RNase III enzyme Dicer. In the second step, siRNAs are loaded into the effector complex RISC. The siRNA is unwound during RISC assembly and the single-stranded RNA hybridizes with mRNA target. Gene silencing is a result of nucleolytic degradation or translational repression of the targeted mRNA by the slicing enzymatic activity of a specific Argonaute protein (Slicer).

rRNA: ribosomal RNA

rsd: RNAi spreading defective

SAGO: siRNA-specific Argonaute

SCR: southern corn rootworm

shRNA: short hairpin RNA

sid: spreading RNA interference defective

SID: systemic interference deficiency

SIGS: spray-induced gene silencing

siRNA: small (or short) interfering RNA. Short pieces of dsRNA, approximately 21–25 base pairs long, involved in RNAi. In the majority of cases, siRNA duplexes composed of equal length sense and antisense strands, are paired in a manner to have a 2 nt 3′ overhang.

Slicer: enzyme that cleaves mRNA in the RNAi RISC complex; action of an Argonaute protein involved in cleaving mRNA

SNP: single nucleotide polymorphism

SPc: star polycation

sRNA: small RNA, encompassing both siRNA and miRNA

ssRNA: single-stranded RNA

TALEN: transcription activator-like effector nuclease

tasiRNA: trans-acting siRNA

TGS: transcriptional gene silencing. Gene expression is reduced by a block at the transcriptional level. Transcriptional repression is caused by epigenetic modifications like DNA methylation and chromatin modification. The major role of TGS is the suppression of transposon activity. In transgenic plants TGS is mediated by dsRNA with homology to promoter sequences.

TGN: Trans-Golgi network

TILLING: targeting induced local lesion in genomes

TLR: Toll-like receptor

TO: target organism

tRNA: transfer RNA

transgene: a gene transferred into another organism using genetic engineering (genetic modification) techniques.

transgenic organism: an organism in which a DNA from another organism (a transgene) is stably incorporated into its genome. The transgene is present in both somatic and germ cells, is expressed in one or more tissues, and is inherited in a Mendelian fashion.

USDA-APHIS: US Department of Agriculture Animal and Plant Health Inspection Service

UTR: untranslated region

VIGS: virus-induced gene silencing; silencing that is induced by the presence of viral genomic RNA. Only replication-competent viruses cause silencing, indicating that dsRNA molecules might be the inducing agents. Restriction of virus growth in plants is mediated by PTGS, which can be initiated by production of dsRNA replicative intermediates.

viRNA: viral RNA

VLP: virus-like particle, virion-like particle

VNP: virus-based nanoparticle

vsiRNA: virus-derived small interfering RNA

VSR: viral suppressor of RNAi

WAGO: worm-specific Argonaute
WCR: western corn rootworm
ZFN: zinc-finger nuclease
ZRMS: zonal rapporteur MS

Abbreviations for virus names

ALSV: apple latent spherical virus
AMV: alfalfa mosaic virus
BBTV: banana bunchy top virus
BGMV: bean golden mosaic virus
BSMV: barley stripe mosaic virus
CaMV: cauliflower mosaic virus
CMV: cucumber mosaic virus
CPV: cypovirus (cytoplasmic polyhedrosis virus)
CTV: citrus tristeza virus
CymMV: cymbidium mosaic virus
PMMoV: pepper mild mottled virus
PNRSV: prunus necrotic ringspot virus
PPV: plum pox virus
PRSV: papaya ringspot virus
PSBMV: pea seedborne mosaic virus
PVY: potato virus Y
SCMV: sugarcane mosaic virus
TEV: tobacco etch virus
TMLCV: tomato yellow leaf curl virus
TMV: tobacco mosaic virus
TSWV: tomato spotted wilt virus
TuMV: turnip mosaic virus
TYLCV: tomato yellow leaf curl
TYMV: turnip yellow mosaic virus
WMV: watermelon mosaic virus
ZYMV: zucchini yellow mosaic virus

Index
